THE TRUTH OF

THE ORIGIN OF THE UNIVERSE

A NOVEL PERCEPTION OF THE BIG BANG AND THE UFO'S

VOLUME-3

COLOUR EDITION

THE TRUTH OF

THE ORIGIN OF THE UNIVERSE

A NOVEL PERCEPTION OF THE BIG BANG AND THE UFO'S

Dr. Sabrie Soloman
Ph.D., Sc.D., MBA, PE

KHANNA BOOK PUBLISHING CO. (P) LTD.
Publisher of Engineering and Computer Books
4C/4344, Ansari Road, Darya Ganj, New Delhi-110002
Phone : 011-23244447-48 **Mobile:** +91-9910909320
E-mail : contact@khannabooks.com
Website : www.khannabooks.com

ISBN: 978-93-5538-746-2

THE TRUTH OF THE ORIGIN OF THE UNIVERSE (VOLUME-3)
by **Dr. Sabrie Soloman**

First Edition: Dec 2024

Published by:
Khanna Book Publishing Co. (P) Ltd.
CIN: U22110DL1998PTC095547

Visit us at: www.khannabooks.com
Write us at: contact@khannabooks.com

DEDICATIONS

William Thomson – The voice of his words of wisdom echoed through the core of the inner chambers of my heart, engraving channels of living knowledge, which generated floods of insight to illuminate my life`s journey.

William Thomson`s pure and innocent soul has reached the gates of heavens to bring me knowledge of Eternity. His living loving example compelled me to follow even his shadow to touch Eternity, while my feet are still on Earth!

His knowledge imparted on my mind inspiring me to write the "**The Origin of The Universe**" book – distinct from most other science books, which composed by worldwide prominent scientists highlighting unjustified and unproven hypotheses. Through **William Thomson** the factual truth of the Universe is recognized and logically revealed to my modest intellect to compose "**The Truth of The Origin of The Universe**," Thus, I humbly dedicate this book to him.

Joseph Rondinelli – While the size of the universe may not be comprehended, so also the goodness of the heart of my friend Joseph Rondinelli may not be grasped. Joseph permitted me to touch the limitless boundaries of the goodness of his loving nature. His friendship with me permitted me to observe a living loving example to follow.

Joseph Rondinelli is bountifully living loving those whom he cherished. He brought heavens highest values to my humble living earth. While I was temporarily incapacitated, he was my feet, hands, and even eyes sitting for hours by my side, debating our intellectual capacities to understand life under the domain of our **Creator, The Origin of Our Universe**.

When my mind acted as an interrupted volcano spouting novel thoughts of **The Origin of The Universe**, he came to my dwelling to rescue me out of the dense-smoke of my scattered knowledge --suffocating my mind. He instantly changed my environment to a more tranquil atmosphere enabling me once again recollect my thoughts, helping me to **Create Order out of Chaos**. For this and much more I am forever, indebted.

FOREWORD

Shortly before Frances R. Havergal, at age 36, wrote the hymn, "Lord, Speak to Me" she wrote in a letter, "... if I am to write any good, a great deal of living must go to a very little writing." Before putting pen to paper, a great deal of living has gone into what you are about to read.

When Dr. Soloman asked me to write a forward to his book on the origin of the universe, I was humbled and honored because this book is of unusual importance by an author with an exceptional life of service and greatly respected by all who know him and his work. I pray that those who read his 2,800 pages, 32 chapters will find its message clearly stated, giving clarity to a subject that has too long been delt with on a surface and non-scientific level.

Understanding the origin of the universe is key to understanding everything else and developing a truly Biblical worldview. A person's view on this subject is a window by which we view the world. Dr. Soloman gives us a verifiable frame to evaluate whether his findings are flawed or something that is understandably sensible to merit serious consideration.

When we tell others our view on evolution is based on scientific evidence rather than solely on the book of Genesis, we have a much stronger argument. Dr. Soloman builds the case that evolution, with his accompanying philosophy, is identified with a worldview at such a level that readers must sincerely struggle how the theory could possibly go against the evidence.

This book has the capacity to train a whole new generation to confront and survive the worldview challenges they encounter all their life. To those of you who read this book in its entirety will be able to survive in a post-modern culture without being brainwashed by prejudice, faulty thinking, and preconceived assumptions.

Dr. Soloman has made exciting reading for the ordinary person, but which is thoroughly researched by scholarship to present a deep understanding of the subject. The situation we often find ourselves in is flawed thinking and lazy scholarship. To that end, I invite you to read Dr. Soloman's book which will prove to not only be enjoyable reading, but one of the most prolific and insightful tomes you will ever read on the origins of the universe.

Dr. Peter W. Teague,

President Emeritus

Lancaster Bible College | Capital Seminary & Graduate School

PREFACE

The Universe Had a Beginning

The scientific consent 100 years ago was that the universe was everlasting. This idea started to unravel with the inferences of Albert Einstein's Theory of Relativity back in 1916, where his equations pointed to an expanding universe. Yet he didn't like that conclusion and so supplemented a constant to his equation that invalidated the expansion. Later, he admitted it had been the major mathematical blunder in his life.

Then in 1929 astronomer Edwin Hubble acknowledged he observed galaxies expanding outward, which meant they had been much closer together in the past. Einstein, fascinated, wanted to see the indication for himself and in 1931 he went to the Mt. Wilson Observatory in Los Angeles, California. Einstein examined through the telescope, observed the evidence and then concluded, "I now see *the necessity* of a beginning." This began a change in the scientific boldness toward the cosmos.

Years after, in 1965, two U.S. scientists distinguished the leftovers of the original burst of energy of the creation occurrence characteristically so-called the "Big Bang." They both won a Nobel Prize in physics. One of them, Arno Penzias, later professed, "The best data we have [about the beginning of the universe] are exactly what I would have predicted had I nothing to go on but the first five books of Moses, the Psalms and the Bible as a whole" ("Clues to Universe Origin Expected," *The New York Times*, March 12, 1978, p. 1).

With the indication at hand, what was written in Genesis 1:1 truly surprised many scientists by its precision: "In the beginning, God created the heavens and the earth." Here it says the universe of matter and energy *appeared at a certain point in time* and was all created by an Ultimate Creator who existed before all of this occurred. It was an enormous proof of God's existence, with no real substitute enlightenments for a universe that, according to modern physics, *appeared out of nothing.*

The Universe is Fine-Tuned for Life

Almost 50 years ago, in 1973, cosmologist Brandon Carter instituted that the independent constants or laws in physics have one exceedingly rare characteristic in common—*they are precisely the values needed to establish and sustain a universe capable of producing life.* This is another vast and virtually accepted proof for a universe that has been prudently designed.

Scientists have instituted some 30 constants or laws of physics that rule the universe. All are dissimilar to each other and yet are finely tuned to unbelievable proportions to make life possible. The indication points to "Someone" spending much too much time tuning all of these laws so they would work in unison.

Amazingly, the Bible exposed this truth long before any scientist revealed these facts. As Jeremiah 33:25 states, "But I, the Lord, have a covenant with day and night, and *I have made the laws* that *control* earth and sky" (*Good News Translation*).

The Origin of Life – The Genetic Code

Opposing to what many have been led to trust, scientists have no genuine explanation for how life arose.

Even the well-known atheist and evolutionist Richard Dawkins acknowledged concerning the appearance of life, "*Nobody knows how it happened*" (*Climbing Mount Improbable,* 1996, p. 282). Moreover, one of the pioneers of the DNA code, the atheist Francis Crick, determined, "An honest man, armed with all the knowledge available to us now, could only state that in some sense, *the origin of life* appears at the moment to be *almost a miracle,* so many are the environments which would have had to have been gratified to get it going" (*Life Itself: Its Origin and Nature,* 1981, p. 88).

In the past 60 years, biologists have found that life started with a massive amount of accurate information *already embedded* in the cell. The human genome unaided is a molecule with almost 3 billion genetic letters, all precisely well-organized to give instructions to the cell. Furthermore, scientists have never instituted inorganic matter to generate a coded system of information and the machinery to interpret it. From the most primitive cells to human beings, all have the same elementary operating system of mind-boggling intricacy, with codes, transmitters and receivers all working together.

Furthermore, the origin of life aenigma has a "chicken-and-egg question"—which came first, the chicken or the egg? In this case, to get life to transpire, you need *both* the complete genetic code and the proteins—the machine parts—that read the code and build new proteins. Deprived of the code, you can't shape proteins. And without proteins, you can't process the code. So how could both have risen at the same time?

Biological life Exquisitely Programmed - "Robotic Machines"

In order to comprehend what is fashioning inside a cell, a good illustration is picturing a large city swarming with life and movement.

Biochemist Michael Denton defines the cell this way: "To clutch the reality of life as it has been discovered by molecular biology, we must amplify a cell a billion times until it is twenty kilometers in diameter and looks like a giant airship large enough to cover a great city like London or New York. What we would then see would be an entity of unmatched intricacy and adaptive design . . .

"We would see around us, in every direction we looked, all sorts of robot-like machines. We would notice that the simplest of the functional components of the cell, the protein molecules, were astonishingly complex pieces of molecular machinery, each one consisting of about three thousand atoms arranged in highly organized 3-D spatial conformation.

"We would wonder even more as we watched the strangely purposeful activities of these weird molecular machines, particularly when we realized that, despite all our accumulated knowledge of physics and chemistry, the task of designing one such molecular machine—that is one single functional protein molecule—would be completely beyond our capacity at present" (*Evolution: A Theory in Crisis,* 1986, p. 329).

This is the reason biochemists have a hard time have faith in explaining that blind evolution can build such machinery—and obtain all the parts to work together from the start. Furthermore, to keep the human body operational, biologists calculate that "about *330 billion cells* are substituted *daily,* equivalent to about 1 percent of all our cells" (Mark Fischetti, "Our Bodies Replace Billions of Cells Every Day," *Scientific American,* April 1, 2021).

"We take life for granted," adds Douglas Ell, "because it is everywhere. Our planet is overrun by *biological machines.* There are at least 10 million different types (species) of machines; some estimate that tens of millions of other types (species) have not yet been discovered . . .

"Coordinated systems permit blue whales to dive thousands of feet below sea level without being crumpled and sing complex songs that travel across oceans. Other systems guide bees to do a dance that tells other bees where to find the best sources of pollen. There are systems for hiding, systems for fighting, systems for reproducing, systems for obtaining food, systems for communicating, and so on" (*Counting to God,* p. 110).

Such findings show that everything about life is programmed to the last detail and that virtually nothing has been left to chance. Does this superb design point to evolution or to God? The answer is clear.

The Earliest Evidence of Life

Though Darwin titled his book *On the Origin of Species by Means of Natural Selection,* he was never able to substantiate that assumption. Many people adopt that the theory of evolution, with its countless mutations and natural selection as the means of change, can account for the origin and development of all the living things on this planet.

Yet this is sleight of hand, since evolution can account for *micro*evolution, or changes *within* the species (such as dogs of varying sizes, shapes and colors), but not *macro*evolution, or changes from one kind of creature to another. Natural selection can tell you something about the *survival* of the species, but nothing about the *arrival* of the species. It undoubtedly cannot trace the origin of the approximately *10 million species* on earth. These are classified into some 33 main body types or phyla, such as sponges, worms, insects and mammals.

Darwin foretold that as more of the fossil record was revealed, it would display types of species gradually appearing, beginning with one or a few, and then multiplying from simple to more complex life forms. He wrote, "If numerous species . . . have really *started into life at once,* the fact would be *fatal* to the theory of evolution through natural selection" (*Origin of Species,* 1859, p. 305). Yet that is precisely what has been found—major body types appearing at what's considered the *beginning* of the fossil record rather than in deposits laid down later.

Scientists call this "the Cambrian Explosion," referring to major types of plants and animals suddenly appearing *fully formed* in that fossil layer. This is the *opposite* of what Darwin and evolutionists had claimed would be found—and they have no real explanation or answers. Of the 33 main body types, 23 of them (or 70%) appear at the recognized *beginning* stage of the fossil record.

The issue in question here, by analogy, would be as discovering together such diverse inventions as a washing machine, a refrigerator, a bicycle, a car and an airplane. While they do have some shared features, they have very discrete purposes and resolutions. Likewise, the major types of creatures found in the Cambrian layer, such as sponges, worms, trilobites and jawless fish, are very diverse, complex and appear suddenly, with no evidence of these main body types evolving from other creatures.

As paleontologist Niles Eldredge admitted: "If life had evolved into the wondrous profusion of creatures little by little, then there should be some fossiliferous record of those changes . . . But no one has found any evidence of such in-between creatures . . . All of the fossil evidence to date has failed to turn up any such missing links" (George Alexander, "Alternate Theory of Evolution Considered," *Los Angeles Times,* Nov. 19, 1978). Indeed, Darwin has been let down by the fossil record!

Earth Has "Just Right" Conditions to Sustain Life

In 1966, Carl Sagan presented the famous TV documentary series *Cosmos.* He believed in order to have life you just needed two conditions—a right kind of star and a planet at the right distance. This supposition showed to be totally baseless.

Now, more than 50 years later, scientists have come to the comprehension that *more than 200 conditions* have to be "just right" for life to exist and thrive. The probabilities are about 1 in $10^{2,685,000}$. The probability of that happening comes out at about **1 in $10^{2,685,000}$**, or 10 followed by **2,685,000** zeros. For comparison, the Universe only has 10^{80} atoms. The infographic finishes by letting you know that the probability of you existing as you is pretty much zero.

As author Eric Metaxas explains: "Today there are more than 200 known parameters necessary for a planet to support life—*every single one of which must be perfectly met, or the whole thing falls apart.* Without a massive planet like Jupiter nearby, whose gravity will draw away asteroids, a thousand times as many would hit Earth's surface. The odds against life in the universe are simply astonishing" ("Science Increasingly Makes the Case for God," *The Wall Street Journal,* Dec. 25, 2014).

The Bible tells us: "The Lord is God. He made the skies and the earth. He put the earth in its place. He did not want the earth to be empty when he made it. He created it *to be lived on.* I am the Lord. There is no other God" (*Isaiah 45:18, Easy-to-Read Version*).

The Universe Mathematically Designed

Amazingly, the universe has been discovered to be mathematically designed. It follows orderly laws that can be described in mathematical terms. Sir James Jeans, one of the great astronomers of the 20th century, remarked: "From the intrinsic evidence of his creation, the Great Architect of the Universe *now begins to appear as a pure mathematician . . .* The universe begins to look more like a great thought than like a great machine" (*The Mysterious Universe,* 1930, pp. 134, 137).

A substantial problem for evolutionists and atheists is this: *Evolution can't do math,* since it is founded on random variations and mutations, and math necessitates an intelligent agent who can first formulate a mathematical blueprint of laws before creating things so they will be logical. This is why the present cosmos can be traced back to mathematical rules.

As Einstein distinguished, "The most incomprehensible thing about the universe is that it is comprehensible." He meant that it could be understood in *mathematical* terms but that an explanation for that was outside math.

As far back as the early 1900s, scientists were discovering the laws that govern the subatomic realm, the tiny microcosm designated by quantum mechanics. It has very diverse rules than our macro world and appears to make room for such things as free will to arise.

Many scientists came to comprehend that *not all is determined by matter and energy.* Experiments illustrate that an observer can change a particle through means of observing it. The consequences are that we can determine the outcome of our lives by the choices we make.

It brings to mind what God said: "*Today I have given you a choice between life and death, success and disaster. I command you today to love the Lord your God. I command you to follow him and to obey his commands, laws, and rules. Then you will live . . .*" (*Deuteronomy 30:15-16, ERV*).

Science Points to The Existence of God

Cutting-edge science is continually showing more intricacy and deeper design, not only in the cosmos, but also in all living things.

The biblical patriarch Job once challenged skeptics to look at the design of the creatures around them and notice they witness to a Supreme Designer and Creator. He stated: "Even birds and animals have much they could *teach* you; ask the creatures of earth and sea for their wisdom. All of them know that *the Lord's hand made them*" (*Job 12:7-9, GNT*).

Therefore, by exploring all the evidence and grasping where it leads, we hope you will believe in God, the Creator and the "Origin of the Universe." *Keep* believing in Him, and earnestly seek His will for your life!

Sabrie Soloman

ACKNOWLEDGMENTS

The Author is extremely grateful for the opportunity to express his heartfelt acknowledgments to the individuals who have played a vital role in the successful publication of "The Truth of The Origin of The Universe" book by Khanna Book Publishing Company.

First and foremost, the author extends his deepest gratitude to the Director and Editor-in-Chief, Mr. Mukul Seth, for his unwavering dedication and expertise in shaping the book's narrative. His keen eye for detail and impeccable editing skills have significantly enriched the content and ensured its quality.

Also, the author expresses his sincere appreciation to the Editorial Team, comprising of Mr. Mukesh Sharma and Mr. Yogesh Kumar. Their valuable feedback, insightful suggestions, and meticulous attention to detail have greatly enhanced the overall coherence and readability of the manuscript.

A special mention goes out to the Marketing Team, led by Mr. Rakshit Khanna for his tireless efforts in promoting the book and reaching a wider audience. Their innovative marketing strategies and relentless pursuit of excellence have played a crucial role in generating interest and enthusiasm among readers.

Furthermore, I am truly grateful for the dedication and hard work of the Sales Team, consisting of Mr. Nand Lal, Mr. Surinder Kumar, and Mr. Roshan Lal. Their exceptional sales acumen and customer-centric approach have boosted sales and fostered positive relationships with clients and partners.

The author would also like to extend his heartfelt thanks to Creation Ministries International for its invaluable contributions to the research and development of the book. Their pioneering work in the field of creation science has laid the foundation for a deeper understanding of the origins of the universe.

Sabrie Soloman

ABOUT THE AUTHOR

Dr. Sabrie Soloman, Ph.D., Sc.D., MBA, PE – He is the Chairman & CEO of American SensoRx, Inc., USA; Founder of Advanced Manufacturing Technology Post Graduate Studies at Columbia University, NY, USA; Professor of Advanced Technology at Columbia University. Dr. Soloman authored numbers of technical books published and translated worldwide: Sensors Handbook (2 editions), Sensors and Control Systems in Manufacturing (2 editions); Affordable Automation; Introduction to Electromechanical Engineering; Modern Welding Technology; 3D Printing Technology; 3D Bioprinting Technology & Design to name a few. Dr. Soloman holds numerous Patents, Technical Awards, and several US Product Registrations. Dr. Soloman is considered an international authority on advanced manufacturing technology, robotics, biomedical engineering, pharmaceuticals, and automation in the microelectronic, automotive, beef, pork, poultry industries. He has been and continues to be instrumental in developing and implementing several industrial and modernization programs through the United Nations to Europe, Asia, and African Countries. He is the first to introduce and implement unmanned flexible synchronous/asynchronous manufacturing systems in the microelectronic and the meat industries, and the first to incorporate advanced vision technology in wide array of robot/micro-robot manipulators. Dr. Soloman was selected to deliver the US presidential closing address "Innovative Remote Sensors Technology," at the Universal Design Conference," New York, USA. Dr. Soloman was the President of the International Christian Union at New Castle-Upon-Tyne University. He debated the "Origin of the Universe" before numerous attendees presenting his case against intellectuals and proponents of Big Bang's Singularity in Great Britain. .

CONTENTS

CHAPTER 15: UNDERSTANDING TIME DILATION 87-126

CHAPTER 17: THE STRING THEORY AND THE ORIGIN OF THE UNIVERSE 183-224

INTRODUCTION

HYPES BIG BANG

The Big Bang theory is the atheist prevailing scientific explanation for the origin of the universe. According to this theory, the universe began approximately 13.8 billion years ago when allegedly a massive explosion caused the expansion of space. While the theory is supported by a wide range of evidence, including cosmic microwave background radiation and the abundance of elements in the universe, it is often the subject of controversy and debate. Some individuals may hype the Big Bang theory to generate interest and controversy, while others may express skepticism or alternative theories.

Blackhole

Blackholes are highly dense celestial objects that possess such strong gravitational pull that nothing, even light, can escape their grasp. They are believed to exist at the center of most galaxies, including our own Milky Way. While black holes have captivated the imagination of scientists and the public alike, there are also some misconceptions and hype surrounding these objects. Some individuals may exaggerate the characteristics and capabilities of black holes, such as the suggestion that they can swallow up the entire universe or that they are portals to other dimensions. It is important to approach information about black holes critically and rely on scientific data rather than speculation or sensationalism

Time Dilation

Time dilation is a phenomenon predicted by Einstein's theory of general relativity. It states that time passes more quickly for objects moving relative to an observer. For example, a clock on an airplane moving at a high speed would appear to run slower than a clock on the ground. While this phenomenon is fundamental to our understanding of the universe, it is often exaggerated or misrepresented in literature and popular culture. Some individuals may claim that time dilation allows for time travel or that it causes significant changes in a person's lifespan. However, these assertions are not supported by any scientific evidence.

Space-Time Warp

Space-time warp is another phenomenon predicted by Einstein's theory of general relativity. It refers to bending or distortion of the fabric of space-time due to the presence of massive objects or gravitational effects. While space-time warp has been observed in various astronomical phenomena, such as gravitational lensing and black holes, it is often exaggerated or misrepresented in popular culture. Some individuals may claim that space-time warp allows for faster-than-light travel or other fantastical possibilities. However, these assertions are not based on scientific evidence and should be considered speculative.

String Theory

String theory is a theoretical framework in physics that attempts to unify quantum mechanics and general relativity. It suggests that the fundamental building blocks of the universe are not particles but tiny vibrating strings. While string theory has gained attention due to its potential to explain fundamental mysteries in physics, it is often the subject of hype and speculation. Some individuals may overstate its predictive power or suggest that it can solve all fundamental questions about the universe. It is important to approach information about string theory critically and rely on scientific data rather than speculation or sensationalism.

UFO`s

Unidentified Flying Objects (UFOs) have gained attention in the media and popular culture due to their mysterious nature and the possibility of extraterrestrial life. While the existence of UFOs is a matter of ongoing scientific investigation, it is important to separate fact from fiction. Some individuals may exaggerate the number of UFO sightings or suggest that aliens are already interacting with humans. However, there is no concrete evidence to support these claims and it is important to approach UFO-related information with caution.

Carbon Dating

Carbon dating is a technique used to estimate the age of the Earth, living organisms, and even objects of historical interest. It is based on the decay rate of carbon-14, a radioactive isotope of carbon. While carbon dating is a powerful tool, it is often misrepresented or exaggerated in popular culture. Some individuals may claim that carbon dating is inaccurate or that it can be easily manipulated. However, scientific evidence supports the accuracy and reliability of carbon dating techniques. It is important to approach information about carbon dating critically and rely on scientific data rather than speculation or sensationalism.

Remember, scientific theories and concepts should always be evaluated critically and based on scientific evidence. While these topics may generate interest and debate, it is important to avoid spreading misinformation or exaggerations.

The Genesis of the Cosmos: Different Perspectives with the Creator at the Helm

The nature of our universe's origin remains one of the most profound mysteries, compelling us to explore various theories that range from the purely scientific to those involving a divine creator. Each perspective offers a unique lens through which to comprehend the inception of existence, acknowledging the possibility of an overarching intelligence that might have orchestrated the universe's creation. The harmony between the complexities of the cosmos and the structured laws governing it may suggest a deliberate design, captivating the human imagination and fueling both scientific inquiry and spiritual speculation.

Assessment of Radiocarbon Dating Reliability

As one of the primary methods used to determine the age of ancient artifacts and geological samples, radiocarbon dating has been instrumental in reconstructing historical timelines. However, the technique is not without its limitations and sources of error. This critique delves into the accuracy of radiocarbon dating, assessing factors that can lead to discrepancies such as contamination, calibration curves, and the assumption of constant atmospheric C-14 levels. By understanding these challenges, researchers can refine dating methods to provide more accurate historical data. It is essential to continuously evaluate and improve radiocarbon dating techniques to reconcile anomalies and enhance our comprehension of Earth's past.

The Mystery of Space-Time Curvature

General relativity has transformed our understanding of space and time by introducing the concept of a malleable space-time fabric. This curvature of space-time, a fundamental aspect of gravitational phenomena, gives rise to the enigmatic effects of time dilation and the possibility of black holes—regions from which nothing can escape, not even light. The mystery of space-time curvature has implications not only for the behavior of massive celestial objects but also for the potential of future technologies and the very nature of reality. Revolutionary as it remains, our continued exploration into the depths of this spatial-temporal warping holds the promise of uncovering new truths about the universe.

Sabrie Soloman

13

THE HYPES ABOUT BIG BANG THEORY

PROBLEMS FOR THE BIG BANG UNIVERSE

No doubt surprisingly for many people the idea that the universe had a beginning in the big bang is in some ways unsatisfactory. It raises a series of valid questions, which are listed below, Figure 13.1.

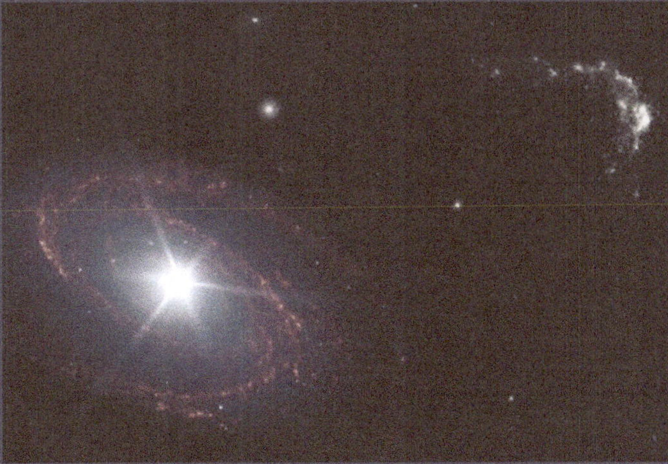

Fig.13.1: *Cosmology has some Big Problems with Big Bang – Curtsy Thanapol Sisrang Getty Images*

- "If the universe had a beginning, then what happened before the beginning?"[1]
- "What caused it?"
- "If time emerged at the big bang, then was there a time before time?"
- "How can the universe appear spontaneously out of nothing at all?"
- Equally valid concerns with respect to the horizon problem
- The flatness problem.

THE HORIZON PROBLEM

This is the fact that light has not had enough time to travel from a point on one extreme edge of an expanding universe to a point on the diametrically opposite extreme edge. Nevertheless, the temperature (i.e., the CMB) is the same (to one part in 100,000) for both points, as well as being the same in all directions of our universe, Figure 13.2.

These two points on the sky are separated today by 90 billion light-years. That means if you've got a universe that's been expanding sedately and is only 13.8 billion years old, those two points could never have been in contact with each other, which means there's no explanation for how they could be so precisely the same."

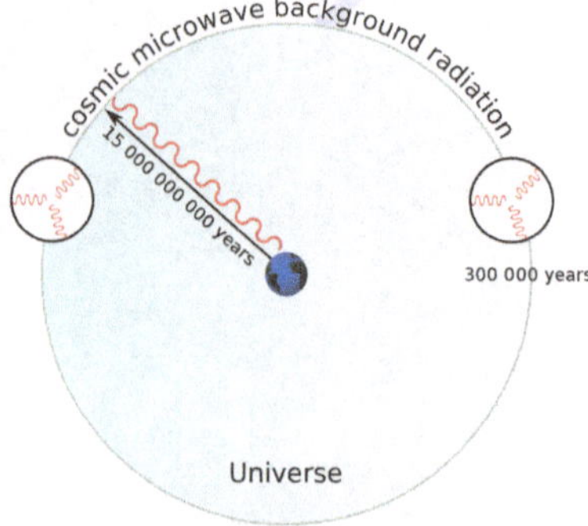

Fig.13.2: Problem with the Universe Horizon – Curtsy Wikipedia

LIGHT-TRAVEL TIME PROBLEM

The 'distant starlight problem' is sometimes used as an argument against biblical creation. People who believe in billions of years often claim that light from the most distant galaxies could not possibly reach earth in only 6,000 years. However, the light-travel–time argument cannot be used to reject the Bible in favor of the big bang, with its billions of years. This is because the big bang model *also* has a light-travel–time problem, Figure 13.3.

Fig.13.3: Light-Time Problem with Bib Bang - Curtsy NASA

THE BACKGROUND

In 1964/5, Penzias and Wilson discovered that the earth was bathed in a faint microwave radiation, apparently coming from the most distant observable regions of the universe, and this earned them the Nobel Prize for Physics in 1978. This Cosmic Microwave Background (CMB) comes from all directions in space and has a characteristic temperature. While the discovery of the CMB has been called a successful prediction of the big bang model, it is actually a ***problem*** for the big bang. This is because the precisely uniform temperature of the CMB creates a light-travel-time problem for big bang models of the origin of the universe, Figure 13.4.

Fig.13.4: *Cosmic Microwave Background Experiments - Curtsy Wikipedia*

UNDERSTANDING THE PROBLEM

The temperature of the CMB is essentially the same everywhere—in all directions (to a precision of 1 part in 100,000). However (according to big bang theorists), in the early universe, the temperature of the CMB ***would have been very different at different places in space due to the random nature of the initial conditions***. These different regions could come to the same temperature if they were in close contact. More distant regions would come to equilibrium by exchanging radiation (i.e., light). The radiation would carry energy from warmer regions to cooler ones until they had the same temperature.

The problem is this: even assuming the big bang timescale, there has not been enough time for light to travel between widely separated regions of space. Accordingly, how can the different regions of the current CMB have such precisely uniform temperatures if they have never communicated with each other? ***This is a light-travel-time problem.***

The big bang model assumes that the universe is many billions of years old. While this timescale is sufficient for light to travel from distant galaxies to earth, it does not provide enough time for light to travel from one side of the visible universe to the other. At the time the light was emitted, supposedly 300,000 years after the big bang, space already had a uniform temperature over a range at least ten times larger than the distance that light could have travelled (called the 'horizon'). So, how can these regions look the same, i.e., have the same temperature? How can one side of the visible universe 'know' about the other side if there has not been enough time for the information to be exchanged? This is called the 'horizon problem'. Secular astronomers have proposed many possible solutions to it, but no satisfactory one has emerged to date.

CREATION SCIENTISTS

In the Beginning created the universe simultaneously, in a steady state by the word of His mouth. All the corners of the universe were consistent in temperature and environment. The universe was created in a perfect order as He said it was good, it was very good.

The big bang requires that opposite regions of the visible universe must have exchanged energy by radiation, since these regions of space look the same in CMB maps. But there has not been enough time for light to travel this distance. Both biblical creationists and big bang supporters have proposed a variety of possible solutions to light-travel–time difficulties in their respective models. So big-bangers should not criticize creationists for hypothesizing potential solutions, since they do the same thing with their own model. The horizon problem remains a serious difficulty for big bang supporters, as evidenced by their many competing conjectures that attempt to solve it. Therefore, it is inconsistent for supporters of the big bang model to use light-travel time as an argument against biblical creation, since their own notion has an equivalent problem. Attempts to overcome the big bang's 'light-travel–time problem'

Currently, the most popular idea proposed by Big bangers is called 'inflation'—a conjecture invented by "Alan Guth" in 1981. In this scenario, the expansion rate of the universe (i.e., space itself) was vastly accelerated in an 'inflation phase' early in the big bang. The different regions of the universe were in very close contact before this inflation took place. Thus, they were able to come to the same temperature by exchanging radiation *before* they were rapidly (faster than the speed of light) pushed apart. According to inflation, even though distant regions of the universe are not in contact today, they were in contact before the inflation phase when the universe was small.

However, the inflation scenario is far from certain, Figure 13.5. There are many different inflation models, each with its set of difficulties. Moreover, there is no consensus on which (if any) inflation model is correct. A physical mechanism that could cause the inflation is not known, though there are many speculations There are also difficulties on how to turn off the inflation once it starts—the 'graceful exit' problem. Many

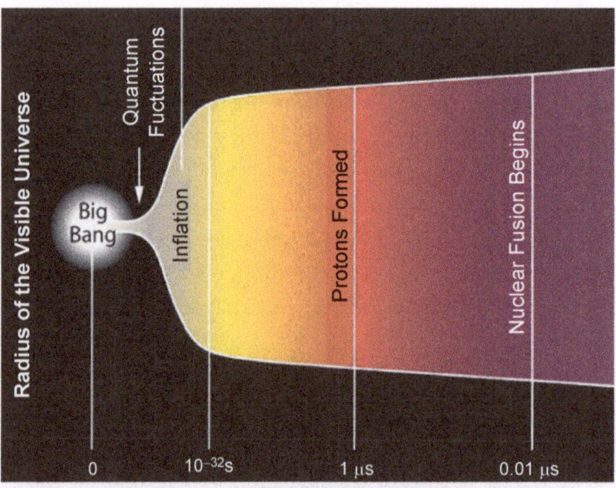

Fig.13.5: Big Bang Cosmic Inflation – Curtsy - Wikimedia Commons.

inflation models are known to be wrong—making predictions that are not consistent with observations, such as Guth's original model. Also, many aspects of inflation models are currently unable to be tested.

Some astronomers do not accept inflationary models and have proposed other possible solutions to the horizon problem. These include: scenarios in which the gravitational constant varies with time, the 'ekpyrotic model' which involves a cyclic universe, scenarios in which light takes 'shortcuts' through extra (hypothetical) dimensions, 'null-singularity' models, and models in which the speed of light was much greater in the past. (Creation scientists have also pointed out that a changing speed of light may solve light-travel–time difficulties for biblical creation.) Nonetheless, the author strongly objects this proposal offered by some creation scientist as the speed of light is an absolute value and according to Einstein Theory of Relativity the light can only be bent due to strong gravitational force.

In light of this disagreement, it is safe to say that the horizon problem has not been solved

UNDERSTANDING THE FLATNESS PROBLEM

This is the fact that all measurements we ever make in space are straight (meaning they conform to Euclidian geometry). Our universe appears to be completely flat, which seems very strange, because it could be curved like the surface of a sphere or curved like the surface of a saddle.

FASTER-THAN-LIGHT INFLATION

The theory of 'Faster-Than-Light Inflation,' meaning that the universe was once expanding incredibly fast, doubling in size every 10^{-37} seconds. That's one ten-million-million-million-million-million-millionths of a second." This allegedly solves the horizon and flatness problems because: "it suggests the universe has to be extremely big. And that means that it's always going to look flat."

Big Bang proponents made an analogy with the little piece of the earth one is standing on. "[It] looks flat, even though we know the Earth is curved, because it's very small compared to the size of the Earth." Applying this 'logic' to the universe, they said that "the two pieces of the sky that are so far away from each other were once in contact with each other. They could jiggle around and get to the same temperature, but then they were ripped apart."

Many people did not realize that what is being suggested—inflation—is a postulate to try to solve many intractable problems with the big bang hypothesis; but one for which there is absolutely no experimental basis or even a deducible mechanism.

The proposal is that the universe expanded rapidly at many times the speed of light. And then just as suddenly this super-expansion came to a screeching halt—also for no known reason.

If proponents of Genesis creation were to seek to solve a scientific conundrum in a similar way, inventing new scientific laws and processes, secularists would definitely cry foul at such a proposition. They would be justified in labelling it a convenient miracle invented for the sake of solving the problems of one's model, Figure 13.6.

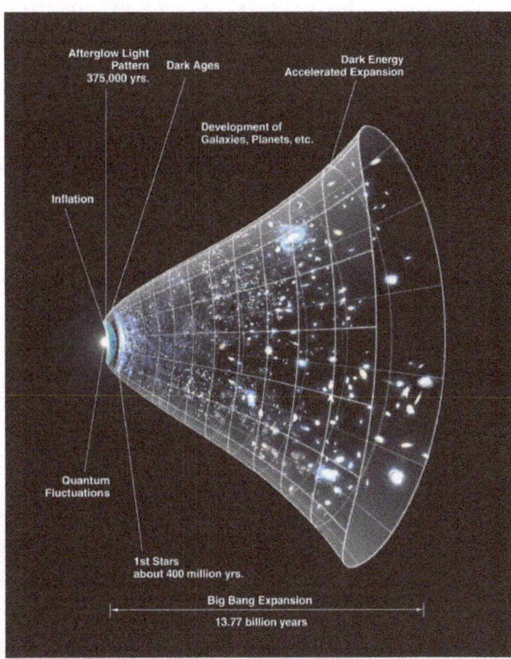

Fig.13.6: Universe Expansion – Curtsy Wikimedia Commons

Before the Bib Bang! **How could that be?**

Big Bang proponents sated "If we are right about inflation, then this rapid expansion must have occurred *before* the thing we used to call 'the big bang.'" This leads to a discussion on the need for a mechanism, which theoretical physicist, my colleague Professor "Brian Greene" (Physics & Mathematics, Columbia University) is enlisted to supply.

Greene tells us: "If you have an energy that is uniformly spread out through a region of space, it can yield a new kind of gravity, repulsive gravity—gravity that doesn't pull things together, but pushes things apart." He goes on to say: "This kind of fuel, if you would, called the inflation field, it's like a fuel that generates this repulsive gravity, is what drove the universe to start expanding in the first place."

ETERNAL UNIVERSE

Many people may well wonder what the purpose of all this is, until Big Bang opponent stated "A more speculative addition to the theory exists, and it opens the doors to an intriguing possibility—ours may not be the only universe." they posed two questions:

1. "How long was inflation going on for before the big bang?"
2. "Could it have been going on for an indefinite period of time? Could you push the origin of the universe back and back and back into the infinite past, so we have an eternal universe?"

Multiverse?

This is followed by another conundrum for creation scientists, when when big bang proponent stated: "There are other possibilities" and they elaborated: "The other possibilities suggest that we are one of a grand collection of universes—we are part of a multiverse. ... You have this repulsive gravity coming from the inflation field ... it's such an efficient process that you can virtually never fully use up the fuel that generated our expansion, so our big bang happens but there is some fuel left over. What does it do? It can generate *another* big bang, so you get this wonderful process of big bang after big bang after big bang, yielding universe after universe, after universe." Figure 13.7

Not so according to Stephen Hawking and the 2nd Law!

Fig.13.7: *Multiverse Hypothesis – Curtsy Wikimedia Commons*

No less a big-bang promoter than "Stephen Hawking" disagreeing with this theory because it is contrary to a scientific law to which there are no exceptions. He says: "If your theory disagrees with the Second Law of Thermodynamics, it is in bad trouble. In fact, the theory that the universe has existed forever is in serious difficulty with the Second Law of Thermodynamics. The Second Law states that disorder always increases with time. Like the argument about human progress, it indicates that there must have been a beginning. Otherwise, the universe would be in a state of complete disorder by now, and everything would be at the same temperature."

SECULAR ALTERNATIVE TO CREATION

The big bang is the current secular alternative to creation by God. The latter is objected to so strongly by some that they reject even a beginning to the universe. This means that they have to concoct alternative theories, no matter how unscientific, illogical, or intellectually unsatisfying these may be. There is, in fact no better explanation for the universe than the one which God has given us in Genesis, that He created all things by His Word.

COSMIC INFLATION: DID IT REALLY HAPPEN?

Built on a Shifting Sand

Astrophysicists have measured the temperature of the Cosmic Microwave Background (CMB) radiation and its small variations (anisotropies). Also, they have found it is *partially polarized*. They make the claim that,

The largest contribution to the polarization was imprinted *during the epoch of recombination*, when local quadrupole intensity fluctuations, incident on free electrons, created linear polarization via Thomson scattering [emphasis added].

Fig.13.8: *Building on Shifting Sand*

This is a key element in the alleged evolution of the big bang universe. The big bang supposedly produced a super-hot plasma of electrons, protons, and photons, and this plasma was opaque. The "epoch of recombination" is assumed to have occurred about 380,000 years after the bang, when it was cool enough for electrons to combine with protons to become neutral hydrogen atoms. This made space transparent to photons, so the CMB radiation separated from matter in the big bang fireball, called 'photon decoupling'.

Once radiation decoupled from matter it travelled freely throughout the Universe, no longer interacting with matter. Thus, it should carry information of the physics from the early universe, Figure 13.8. This radiation, allegedly, after it cooled by about a factor of 1100, is observed at the earth as the CMB radiation.

The epoch of recombination produced the largest scalar E-mode polarization on the CMB radiation and it allegedly occurred long after any inflationary epoch. It is also assumed that there were gravitational waves, which generated both E-mode and B-mode polarizations as well as gravitational lensing, which generated B modes by distorting the scalar-induced E-mode pattern in the CMB radiation.

Prior to the epoch of recombination, primordial gravitational waves from an alleged cosmic inflation are said to have left their imprint on the CMB radiation in the form of B-mode polarization. By attempting to distinguish the various contributions, at different scale sizes, it is hoped to detect some evidence of the era of inflation. All of this hinges integrally on the standard ΛCDM ('big bang') cosmology being correct. If the model is wrong—remember also dark matter and dark energy are essential components here—then the house of cards crumbles.

Credit: NASA/ESA

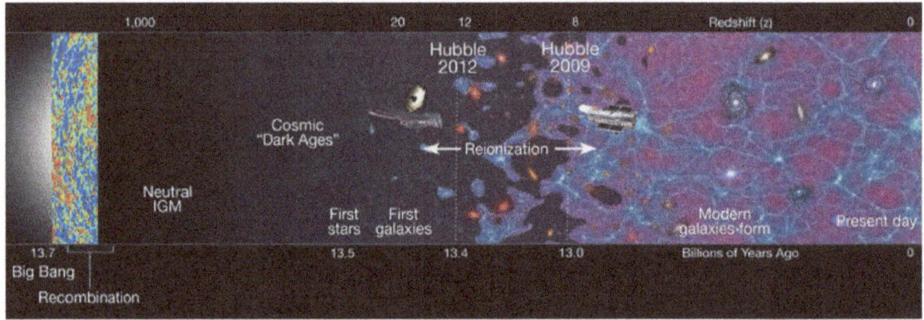

Fig.13.9: Cosmic Timeline Illustrated Showing Redshift Across the Top, and 'lookback' Time to the Big Bang along the Bottom.

NEW CMB E-MODE POLARIZATION DATA

Does newly published polarization data of the CMB radiation, from the South Pole Telescope (SPT), really show that *cosmic inflation really happened*? According to "Hugh Ross" it really did happen! He stated in his recent article, "Cosmic Inflation: It Really Happened":

"*... we now can be 99.9999999 percent certain that the universe indeed went through an inflation event when the universe was only about* ***10^{-34} seconds*** *old. To put it another way, there is less than 1 chance in 900,000,000 that the universe did not experience an inflation event very early in its history.*"

Why was that not news on all the science/boffin news websites? The last time this claim was made, in March 2014, by the telescope BICEP2 team, also from the South Pole, it was splashed around on all the news websites and published in all the news media. But it was a big flop. The BICEP2 team was heavily criticized for not more carefully looking for other sources, in particular foreground sources in the galaxy that could also result in the same signal they claimed was evidence of the inflation-era primordial gravitational waves that were their claimed 'smoking gun' evidence. In the final analysis they admitted they were wrong, Figure 13.9.

Last year, with the grandiose claim there was also mention of a Nobel Prize for a 'proof' of the big bang inflation epoch. *Why is there no such story now? Why aren't the news sites full of this story?*

Just for a sanity check, and to be certain, I googled the words "the South Pole Telescope (SPT), cosmic inflation really happened". All the 'hits' I got related to the 2014 BICEP2 claimed discovery, except for one, this article by "Hue Ross."

On top of that, according to Ross:

"This is good news for Christians because for more than 2,500 years the Bible and biblical commentaries were the only books proclaiming the fundamental features of big bang cosmology."

I am almost lost for words! For all history, and certainly the last 2000 years, except for about the past 200 years, Jewish and Christian scholars have believed in fiat creation, which occurred about 4000 BC. That could hardly be construed as "biblical commentaries were the only books" teaching a big-bang-type cosmology.

As for the Bible itself, the main texts in relation to the origin of the Universe are in Genesis, which are much older than 2500 years. The South Pole Telescope team has now published their latest measurements. But their paper was submitted to the preprint archive (which is freely available) on 4 November 2014. *Therefore, this is not new news*! At least Ross waited until it was published in a peer-reviewed journal, something the BICEP2 team did not do. But wait! The South Pole Telescope team did not make any claim of such a discovery or 'proof' of inflation in the very paper Hue Ross gets his evidence from. Strange?

In cosmology, how can you be sure that you have a unique proof? Could some other effect explain the same result? For example, it was shown that dust emission from the galaxy was the cause of the effect the BICEP2 Telescope team saw in the polarization of the CMB radiation. It was not due to primordial gravitational waves after all. The believers of course hold onto hope that they will indeed find such an effect, but is not that an indication that the model or worldview determines what *can* be looked for?

EXPANDING UNIVERSE

Ross assumes the validity of an expanding universe is strong despite the unverifiable assumed existence of the dark entities needed for this to be so. He even claims it a successful prediction of biblical texts. He cites several OT scriptures in support of cosmological expansion of the Universe. But careful examination of the meaning and intent of the Hebrew words used in those scriptures indicates that the ancients had no such concept of the standard rubber balloon analogy of cosmological expansion. At best those scriptures describe no more expansion than the stretching of tent or curtain material and, in some cases, beaten brass. Tent material may stretch a few percent of its size, but most certainly there is no concept of linear expansion by a factor of 1100, as required for the CMB radiation. If you include the inflation stage *the linear expansion factor* increases to more than 10^{29}.

Ross wrote "ongoing expansion since that beginning" was a biblical prediction, Figure 13.10. But to say that the Bible predicted the expanding universe, including cosmic inflation, because it is in reference to "since the beginning," then that prediction must be interpreted as involving a gargantuan 'stretching' and 'spreading'. These words are used in the biblical verses he cites, one of which is Isaiah 40:22, which uses the expressions "stretches out the heavens as a curtain" and "spreads them out as a tent" 10^{29} times to falsely justify the hypothesis of Cosmic Microwave Background. These are similes, and as such should carry some correspondence over to the intended meaning.

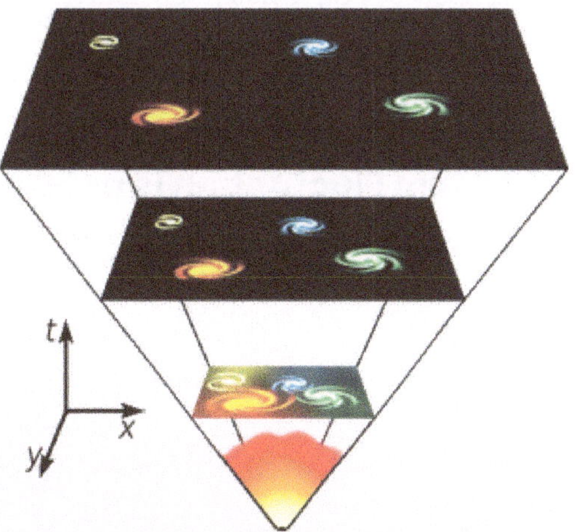

Fig.13.10: *Expanding Universe – Curtsy – Wikimedia Commons*

The cooling of the CMB radiation is apparently also a prediction of the Bible according to Ross. He states: "A pervasive law of decay that implies a continuously cooling cosmos." In regards to this point, he cited Romans 8:18–22.

The key verse is Romans 8:22, stating that the whole creation suffers in agony. This verse describes the effects of the ***Curse of sin*** that entails the entry of death and suffering into God's "very good" creation (something that Ross says predated sin). It is somewhat of a stretch to get ***the second law of thermodynamics***, ***the law of decay***, from that, but that is what he means.

There is no doubt that we see decay and we also see ***isolated systems headed to equilibrium***. This results in cooling. These are experimental facts. But what is not proven is that the universe is expanding and as a result the expansion induces adiabatic cooling in the CMB radiation. To say this is what the scripture meant is clearly reading into it more than the Author intended.

It is true that the Bible describes a "singular beginning of matter, energy, space, and time", which are Ross's words. Genesis 1:1 state God created the Universe. However, Genesis 1 and Exodus 20:11 also state He did it in ***six approximately 24-hour days***, not in a big bang nor with the sequence of events the big bang evolution posits. To imply that any scriptures describe a big-bang-inflationary origin is really stretching the truth.

Finally, Dr. Hue Ross claims the Bible supports "constant laws of physics," which I agree with. The laws of physics are part of God's created order and we know we can trust in Him, who remains constant throughout all time (Hebrews 1:11, 13:8). This biblical perspective was foundational to the development of modern science, spurred on by the Reformation, when the authority of the Bible was rediscovered. However, to suggest that those laws are the product of the inflationary hot big bang early universe as many secular (or pagan) theoretical cosmologists assert, and not the product of a Creator, is what Ross must inherit as he believes the big bang is what the Bible describes. Where does he draw a line? Alternatively, there are those theorists, such as "Paul Davies," who believe not that God created the laws and the Universe, but that the laws created the Universe without any creator.

In Ross's writings, he over-reached when he said the Bible predicted an inflationary-big-bang-expanding universe. He claimed that cosmic inflation has now been proven, even when the secular cosmologists have not claimed that. **Quite obviously Ross does not understand cosmology**. A few secular theorists do when they say: "***Cosmology ... isn't a science***".

The Universe began in time. That is the only similarity between the Bible's account and the Big Bang's. That similarity, though, stems from the big bang requiring an expanding universe to have an origin in time. The Bible however explains the order by which the universe is expanding. However, it does not support the Big Bang Chaotic expansion. Nonetheless, even that origin in time is now under attack from the secularists who don't like any origin in time. What will Ross do if and when the secular cosmologists adopt an eternal big bang universe What will the theology of "Hugh Ross," "William Lane Craig" and others like them become then?

Quantum Fluctuation is not God's Creation

Big Bang scientists believe that the universe expanded due to fluctuation hypothesis, figure 15.11.

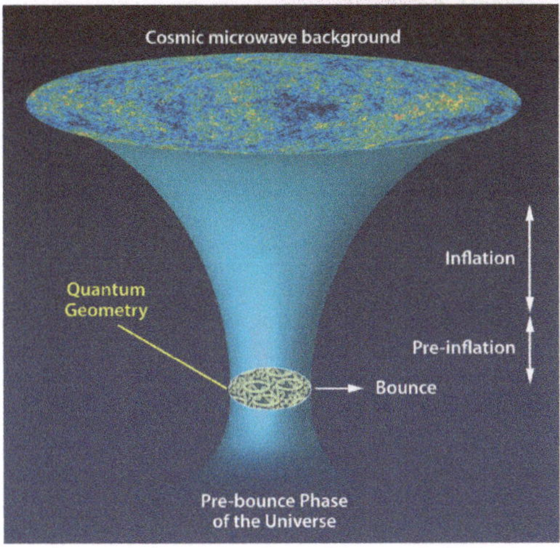

Fig.13.11: *The Hypothesis of Quantum Fluctuation of Big Bang*

In a sense, *Genesis 1:1* is the most important verse in the Bible: "In the beginning, God created the heavens and the earth." If we can believe this verse, no other verse in the Bible should be a problem, Figure 13.12. For example, if God can create the whole universe, then raising people from the dead, walking on water, feeding thousands with two fish and five loaves, raising Lazarus from the Dead, and causing a virgin to conceive would be easy beyond words.

Fig.13.12: *Biblical Creation Vs Big Bang Hypothetical Universe*

Also, in this one verse, all other false religions are rejected, Table 13.1:

Table 13.1

World Religions	Christianity
Evolutionism: that goo became you via the zoo.	• God created all things.
Humanism: man is the measure of all things.	• God is the ultimate reality. • Man is part of the created order. • God created us so He is the measure of all things.

Materialism: Matter (or mass-energy) is the only reality. This is a synonym of Naturalism: natural laws describe all things.	• God created matter (and mass-energy); or, God created nature. • God is thus sovereign over the natural world. • Thus matter (mass-energy) are not eternal or self-existent.
Pantheism: all is god; god and creation are the same thing.	• God created the universe. • Thus, God is distinct from His creation.

Conversely, if we can't trust this verse, then nothing else in the Bible makes sense. Since this verse is so foundational, it is not surprising that atheists have feverishly attacked this concept. Some of the attacks are childish, while others have the veneer of philosophy or advanced science.

Who Created God?

The Bible doesn't attempt to prove that God exists—it proclaims this truth as obvious. But a common question from little children (and not-so-little atheists) is: "If God created the universe, then who created God?" Or, "If everything has a cause, then who caused God?" But no serious apologist ever argued that way. As we have pointed out in several sections in this book, one of the main *real* arguments is:

1. Everything **which has a beginning** has a cause.
2. The universe has a beginning.
3. Therefore, the universe has a cause.

It is not everything that has a cause, but only everything which begins to exist. The universe requires a cause because it had a beginning. This can be shown by the *Laws of Thermodynamics*. The *First Law of Thermodynamics* states that natural processes can neither create nor destroy mass-energy (mass-energy interchange can occur according to

$$E = mc^2,$$

but the total remains the same). But the *Second Law* states that the amount of energy *available for work* is running out, or entropy is increasing to a maximum. If the total amount of mass-energy is limited, and the amount of usable energy is decreasing, then the universe cannot have existed forever, Figure 13.13.

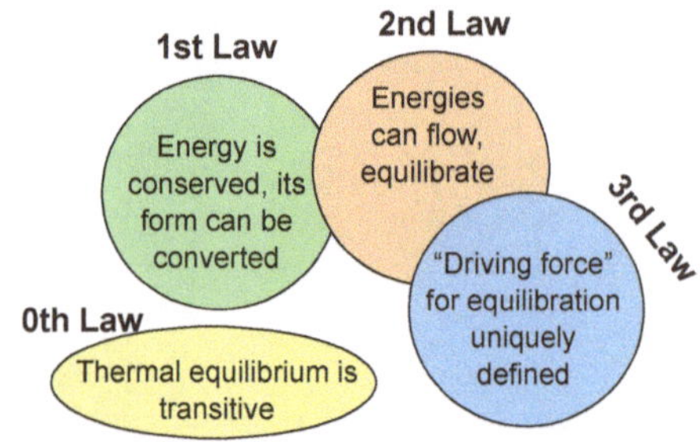

Fig.13.13: Laws of Thermodynamics

The Thermodynamic Laws

Otherwise, it would *already* have exhausted all usable energy—the 'heat death' of the universe. For example, all radioactive atoms would have decayed, every part of the universe would be the same temperature, and no further work would be possible. Accordingly, the obvious corollary is that the universe began a finite time ago with a lot of usable energy, and is now running down.

In addition, Einstein's general relativity, which has much experimental support, shows that time is linked to matter and space. Therefore, time itself would have begun along with matter and space; an insight first pointed out by Augustine in the fourth century. Since God, by definition, is the Creator of the whole universe, he is the Creator of time. Therefore, He is not limited by the time dimension He created, so has no beginning in time—God is "the One who is high and lifted up, who *inhabits eternity*, whose name is Holy" (*Isaiah 57:15*). Therefore, He doesn't have a cause.

CAUSE AND EFFECT

It is a metaphysical principle that things which begin have a cause, but it is also self-evident—no-one really denies it in his heart, Figure 13.14.

All science and history would collapse if this law of cause and effect were denied. Accordingly, would all law enforcement, if the police didn't think they needed to find a cause for a stabbed body or a burgled house. Also, the universe cannot be self-caused—nothing can create itself, because that would mean that it existed before it came into existence, which is a logical absurdity.

Fig.13.14: *Cause and Effect Illustration*

Despite this, the favorite philosopher of modern atheists, the Scotsman David Hume (1711–1776), disagreed. He taught that one might conceive of something coming into being without a cause.

However, British analytic philosopher (and conservative Roman Catholic) G.E.M. (Elizabeth) Anscombe (1919–2001) argued cogently that no one really conceives of any such thing. To paraphrase one of her points, suppose that a banana suddenly appeared on your plate. You would not think, "Hume was right after all—this banana really did come into being without a cause." No, you would think, "How did that banana get there?" and look for the likely cause. Maybe there was a hole in the ceiling above it, or in the plate below it. If that were ruled out, then maybe you were temporarily unaware of your surroundings, and in that time, someone placed the banana there without your noticing. Failing that, maybe a magician's trick, or even a miracle, was the cause. Regardless, even an *unknown cause* would be more likely than *no cause*.

Further, Anscombe pointed out, we would be less likely to think that this banana came into being than that it *already existed* and was somehow *moved* to the place. I.e., the cause was in *transportation* not in creation out of nothing.

So even though Hume claimed that one could easily conceive of something coming into being without a cause, in reality, he likely never really conceived any such thing. Indeed, it seems impossible to conceive. Hume himself, in more lucid moments, even admitted as much:

But allow me to tell you that I never asserted so absurd a Proposition as that anything might arise without a cause: I only maintained, that our Certainty of the Falsehood of that Proposition proceeded neither from Intuition nor Demonstration; but from another Source.

Universe from Nothing?

Despite the above, a number of atheists have claimed that the universe really came from 'nothing.' The universe burst into something from absolutely nothing. And as it became bigger, it became filled with even more stuff that came from absolutely nowhere. How is that possible? The claim is rendered to the "theory of inflation" to explain the claim.

More recently other atheists have promoted the same notion. However, some non-creation scientists have vehemently disagreed.

They were disdained of cosmologists discussing about universes being created out of nothing, particles coming out of nothing, different types of nothing, nothing being unstable. This is nonsense. The word nothing is often used loosely—I have nothing in my hand, there's nothing in the fridge etc. But the proper definition of nothing is "not anything". Nothing is not a type of something, not a kind of thing. It is the absence of anything, Figure 13.15.

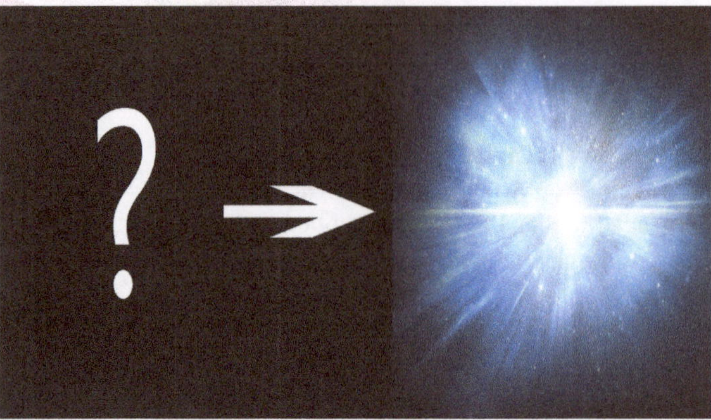

Fig.13.15: Universe from Nothing

Understanding Quantum Fluctuation

Some physicists assert that quantum mechanics violates this cause/effect principle and can produce something from nothing. For instance, "Paul Davies" writes:

... spacetime could appear out of nothingness as a result of a quantum transition. ... Particles can appear out of nowhere without specific causation ... the world of quantum mechanics routinely produces something out of nothing.

Fig.13.16: Implication of Quantum Fluctuation – Curtsy - NeoLeo/Shutterstock.com

Nonetheless, this is a gross misapplication of quantum mechanics. Quantum mechanics never produces something out of nothing. Davies himself admitted on the previous writing that his scenario 'should not be taken too seriously.'

Also, theories that the universe is a quantum fluctuation must presuppose that there was *something* to fluctuate—their 'quantum vacuum' is a lot of matter-antimatter potential—not 'nothing'. So, this is another equivocation.

Now let's examine the claim that something comes out of nothing. Does it make sense to say that there are different types of not anything? That not anything is not stable? Is likely that they are talking about is the ***quantum vacuum***.

The quantum vacuum is a type of something. It has properties. It has energy, it fluctuates, it can cause the expansion of the universe to accelerate, it obeys the (highly non-trivial) equations of quantum field theory. We can describe it. We can calculate, predict and falsify its properties. ***The quantum vacuum is not nothing.***

This suggests a very simple test for those who wish to talk about nothing: if what you are talking about has properties, then it is not nothing. It is pure equivocation to refer to the quantum vacuum as nothing when a philosopher starts asking the question "why is there something rather than nothing?" Scientists are not asking "why are there particles rather than just a quantum vacuum?" Figure 13.17. They are asking "why does anything exist at all?" As "Stephen Hawking" once asked, why does the universe go to all the bother of existing?

Fig.13.17: Quantum Vacuum – Curtsy Anne Baring

We can now see that this question cannot be answered by any of the methods we normally call scientific. Scientific theories are necessarily theories of something, some physical reality. Equations describe properties, and thus describe something. There cannot be equations that describe not-anything. Write down any equation you like—you will not be able to deduce from that equation that the thing that it describes must exist in the real world. Existence is not a predicate.

Where, for starters, are the laws of quantum mechanics themselves supposed to have come from? atheist scientists are more or less upfront, as it turns out, about not having a clue about that. They are simply takes the basic principles of quantum mechanics for granted. They seem to be thinking that these vacuum states amount to the relativistic-quantum-field-theoretical version of there not being any physical stuff at all.

The atheistic scientists think that the laws of relativistic quantum field theories entail that vacuum states are unstable. And that, in a nutshell, is the account they propose of why there should be something rather than nothing.

Nonetheless, that is totally wrong. Relativistic-quantum-field-theoretical vacuum states—no less than giraffes or refrigerators or solar systems—are particular arrangements of elementary physical stuff. The true relativistic-quantum-field-theoretical equivalent to there not being any physical stuff at all isn't this or that particular arrangement of the fields—what it is (obviously, and ineluctably, and on the contrary) is the simple absence of the fields!

The fact that some arrangements of fields happen to correspond to the existence of particles and some don't is not a whit more mysterious than the fact that some of the possible arrangements of my fingers happen to correspond to the existence of a fist and some don't. And the fact that particles can pop in and out of existence, over time, as those fields rearrange themselves, is not a whit more mysterious than the fact that fists can pop in and out of existence, over time, as my fingers rearrange themselves. And none of these poppings—if you look at them aright—amount to anything even remotely in the neighborhood of a creation from nothing.

The atheistic scientists' writings are just the latest in a series of philosophically inept writings by the *soi-disant* 'new atheists.'

The spate of bad books on philosophy and religion by prominent scientists—Dawkins' *The God Delusion*, Hawking and Mlodinow's "*The Grand Design*," and Atkins' *On Being*, among others—are notable not only for the sophomoric philosophical and theological errors they contain but also for their sheer *repetitiveness*.

Atheistic scientists' fallacious writings account of how something can come from nothing, though presented as great breakthrough, and praised as such by Dawkins in his afterword, is largely a rehash of ideas already put forward by Hawking, Mlodinow, and some less eminent physics popularizers. Dawkins has been peddling the "Who created the creator?" meme since the eighties.

Critics have exposed their errors and fallacies again and again. Yet these writers keep repeating them anyway, for the most part simply ignoring the critics. What accounts for this? To paraphrase a famous remark of Ludwig Wittgenstein's, I would suggest that a picture holds these thinkers' captive, a picture of the quantitative methods of modern science that have made possible breathtaking predictive and technological successes.

The Bible presupposes that God began the universe. The fact of the universe's beginning points strongly to a Creator consistent with the biblical God. Some atheists, following Hume, have asserted that something can begin without a cause, but this is not only unreasonable, it is arguably inconceivable. The 'New Atheists' have resorted to quantum bluffing to claim that something really can come from nothing. But they must equivocate about the word 'nothing'. This really should mean *nothing—no* properties. However, their proposed quantum vacuum is not nothing; it must be *something, with* properties—e.g., the quantum vacuum, which is being bound by the laws of quantum physics, so that it can 'fluctuate.'

"In the beginning God created the heavens and the earth." It stands to reason, Figure 13.18.

Fig.1.18: *In the Beginning God Created the Heaven and the Earth – Curtsy Randal S. Chase*

Simultaneous Causation and the Beginning of time

Have you ever asked yourself the following questions:

- Do we need simultaneous causation?
- If God is outside of time, can He create time and have a motive to do so without time passing?

Do we need simultaneous causation? For time's beginning, yes. There's no other logical option. Clearly, causes cannot happen after their effects in time, even when the effect is time's beginning. But could the cause of time's beginning occur *before* time's beginning *in time*? Of course not! Nothing can occur before *time's* beginning *in time*; such an idea is self-contradictory.

Thus, we have only one option left: the cause of time's beginning occurred *when* time began. In other words, cause and effect - since the effect is time's beginning.

The time's beginning, not necessarily the *universe's* beginning. We can conceive of the two being distinct without contradiction. For instance, say that God counted to three before He created the universe. In this scenario, the universe began *after* time began. If God counted in sequence before He created the universe, time would've existed before the universe began.

However, as one of God's attributes is "infinity," He consequently, sees the past, present and future simultaneously. And since Time/Space/Matter are intwined and inseparable, it is very likely that the time and the universe were created simultaneously. Also, as He is the God of Infinite order and knowledge, we may never know what was in His mind at the point of creation! Nonetheless, either creating simultaneous or sequential, He created them all and in All!

Moreover, the relevance of simultaneous causation isn't affected by how long-ago time began. Of course, the Bible teaches that it began around 6,000 years ago. But even if time began with the supposed 'big bang' 13.7 billion years ago, simultaneous causation remains the only way for time to begin as an effect. The biggest problem with using the big bang as evidence for God is that it contradicts the biblical time frame and event order.

Nonetheless, is simultaneous causation a coherent concept? Since effects cannot precede their causes in time, it could only be incoherent if effects *must* follow their causes in time. But why think this is true?

First, there's no obvious incoherence in the term 'simultaneous cause.' When we look at the term 'married bachelor,' a basic understanding of both words quickly shows that it's an impossible idea. Bachelors are *unmarried* men. Not so with 'simultaneous cause,' A cause is in some sense 'responsible' for an effect; there's nothing about this that immediately tells us cause and effect can't happen at the same time. Thus, the burden of proof lies with those who would say simultaneous causation is impossible.

Could it be that we only ever experience effects following causes in time? Even if that were true (which is debatable), it wouldn't show that effects *must always* follow their causes in time. A prince of a tropical nation who only ever experiences liquid water can't thereby argue that ice is impossible. Indeed, if time began and had a cause, it *must* be an exception to this.

Consider the theological consequences of saying an effect *must* follow its cause in time. If so, then time must be uncaused. This is true whether time began or not, though a beginningless past runs into philosophical problems—Doubt your doubts! Without a moment preceding time's beginning in which a cause could operate, time must be uncaused if time began. Similarly, without a beginning, time cannot have a cause preceding it. Thus, time can't be an effect if it has no beginning. And if time can't be an effect, then it must be uncaused.

But if time must be uncaused, *not even God* can be time's cause. Can this be avoided by positing time as a divine attribute? No, that's a category mistake. *Eternity* is a divine attribute, which describes God's *relation to* time. However, the Bible teaches that God is the sole source of all things, and it strongly implies that time began. Indeed, any theism worthy of the name must insist that God is the sole source of all being.

But if time is uncaused, God is not responsible for time, and so is not the sole source of all being. Such a 'god' is no God at all. Thus, without any obvious incoherence in the idea, or any solid evidence against it, the theist has no reason to abandon simultaneous causation, and plenty of reason to embrace it.

Nonetheless, we can give examples of simultaneous causation: e.g., a ball sitting on a cushion. This example goes back to "Immanuel Kant." In this example, the depression in the cushion (the effect) lasts as long as the ball is sitting on the cushion (the cause). This would be true even if the ball had been sitting on the cushion forever, or if God had created them *ex nihilo* so that they began to exist simultaneously. As such, at any given time cause and effect are both occurring; i.e., they are occurring simultaneously. Thus, far from being "a philosophical fudge factor," simultaneous causation is both coherent and even something we regularly experience.

Is Time's Beginning an Uncaused Cause?

Of course, if time began, then the atheist's only way out is to say that time began uncaused. But if time began uncaused, it would be an inexplicable brute fact. And why think time's beginning is inexplicable? Nothing comes from nothing. If it did, then anything could come from nothing, not just time or universes. Balls, angels, ducks, unicorns, and Sherlock Holmes could all just pop into being without cause. Even worse, saying that causes are not always needed ruins the potential for any explanatory discipline, including science. After all, how could we justify trying to explain anything if anything could pop into being inexplicably?

Thus, something had to be responsible for time's beginning. God is the sole source of all being, exists necessarily, and is the greatest conceivable being. Therefore, God is the best candidate cause for time's beginning. And since simultaneous causation is a coherent idea, there's no causal problem with God doing so.

God - 'Before Time'

Now, you may ask: "If God is outside of time, he can create time and have a motive to do so without time passing,? Atheists would then be right however that the concept of 'before time' is meaningless if that's true."

SIMULTANEOUS CREATION

First, if God is outside time, can He be in time? If He can't, can He create time? How you answer these questions will determine whether God can have a timeless intention to create time. However, your answer will depend crucially on how you view God's relation to time. This is a difficult subject on which Bible-believing Christians disagree.

Time Before Time!

Still, is it true that "the concept of 'before time' is meaningless"? It depends on what we mean by "before time." Obviously, if we mean by it 'temporally before time', yes, that's a meaningless self-contradiction. Theists have said the same (e.g., Augustine and Leibniz) as much as atheists. But we don't have to read it that way, since 'before' can have connotations other than temporal priority. For instance, it can be stipulated to mean something like 'logically prior' or 'explanatorily prior' without temporal connotations. And God is certainly 'ontologically prior' to time in that He is ultimately responsible for its existing.

God is responsible for time's beginning. But that means at least one effect occurred *when* its cause occurred. It could be no other way for time's beginning. And there's no logical problem with saying so. We thus can't say that effects *must* follow their causes in time. As such, we can use causality and time's beginning as arguments for God. God is the only plausible candidate for causing time to begin, Figure 13.19.

Fig.1.19: *Time Caused Time Creation*

GOD AND TIME

God exists without beginning or end; that much the Bible is clear on. But what does that mean about the nature of time and how God relates to it? Is He essentially timeless? Must He be in time? Or can God choose between being timeless and in time?

God *cannot* have either beginning or end; He *must* be eternal.

There are different ways of construing God's eternity available to the Christian, because the Bible doesn't fully explain the nature of God's relation to time. One is called *omni-temporality*, which means that God has existed for an infinite duration of objectively temporal moments. This idea has become somewhat fashionable among some modern Christian philosophers. However, I see a deep incoherence in the idea. It seems impossible to count sequentially from $-\infty$ to 0, much less $-\infty$ to ∞. This means there can't be an infinite series of moments. Even God's omnipotence is irrelevant here, because not even omnipotence can do the logically impossible (such as make 2+2 equal 5, or make a married bachelor).

Another way of thinking about God's eternity is to say that He is essentially timeless. The common statement 'God is outside of time' reflects this general view of God's eternity. This means that God is completely *static*; even His thinking does not change *at all*. This has been the traditional view of the church ever since at least the 4th and 5th centuries, largely due to the church father Augustine.

However, this view of God is hard to reconcile with the *dynamic* relational depiction of God in the Bible, and especially Christ's Incarnation. It also means that our experience of time is completely subjective, since on this view the only truly objective perspective is God's *timeless* perspective. It also means that God is not really related to His creation in an objective temporal sense. The only way He is said to '*change*' and '*relate*' with respect to creation is as other beings around Him have a subjective experience of change. This is like how a father might appear to his son to get shorter through the years, when in fact the father's height has remained constant, and only the son's height has changed. The 'change' in the father is merely apparent and subjective; it's not objectively real. Unlike omni-temporality, this view is not logically impossible. Moreover, it fits well with the depiction of God's changelessness in Scripture (e.g., *Malachi 3:6*). And, since God is perfect, it might be argued that any change in God would necessarily be a change for the *worse*, which would of course be impossible, making God essentially unchangeable (and therefore timeless).

A third view is the idea that God can choose whether to be timeless or temporal. Or, perhaps more correctly, He can choose to stay static, or become temporal. This avoids the metaphysical problems of omni-temporality. It also fits well with the dynamic relational depiction of God in Scripture, and Christ's Incarnation. But it does mean we have to adopt a 'softer' form of changelessness to explain passages like *Malachi 3:6*. Moreover, the idea that God can 'switch' from a timeless to a temporal mode of existence is rather counterintuitive. And since God is perfect, may not *any* change in God be a change for the worse? This view would say that not every sort of change in a perfect being entails a change for the worse; there may be *value-neutral* changes that God can undergo (an example of such a value-neutral change might be God's knowledge that it's 3 pm changing to knowledge that it's 3:01 pm a minute later). But again, I don't think this view is demonstrably incoherent.

As you can see, all the views on offer have their difficulties. And none of them are uniquely taught in or directly derivable from Scripture. This is an issue over which Bible believing Christians can disagree.

Now, both the second and third views have God existing in a timeless state apart from creation. However, the insightful question is: how can a timeless being be personal?

In our experience, thinking, willing, relating, and acting, all things that *persons necessarily do*, all *take time* to do,! But *must* they? I don't see why they must. For instance, why does God need *time* to think about e.g., His own goodness? Such a thought could simply be in His mind changelessly, and thus *timelessly*. God doesn't *learn* anything through any sort of sequential process; God knows all truths innately and immediately. And while there are some things that God could not *objectively* do unless He were objectively 'in time' (such as relate to *objectively* temporal creatures), even that doesn't mean He would be less than personal. It just means there are certain things He can only do *in time.*

Even 'activity' need not take time. For instance, if God's action is 'each of the divine persons relating perfectly to each other,' would that not be a *changeless* activity? After all, the very act of being God necessarily *and changelessly* entails the perfect love relation of the three divine persons. There is no reason to think that must involve any change, and thus sequence, at least apart from creation. 'God transcends sequence.' If we were to say that in a more fruitful way, we might say: 'if God is essentially timeless, then our experience of time is purely subjective.' What this shows is that 'God transcending time' is not so much a statement about God, but a statement about *the nature of time.* Is time a subjective or objective reality? If time is a subjective reality, God is timeless. If time is an objective reality, God is in time.

THE HYPES ABOUT BIG BANG THEORY

Big Bang is Not God's Method of Creation

Some prominent Christian apologists claim that the big bang was God's method of creation. Another common view is that the big bang is an apologetic for biblical creation. By this reasoning, Genesis 1:1 says that there was a beginning, and the big bang was also the beginning of the universe. Thus, the big bang is evidence for creation, not evolution. This is a mistaken conclusion. The ministries of the Christian apologists, as well as others, generally take a high view of Scripture which strengthens Christian faith. The critique of big-bang advocaçy in this chapter should in no way be taken as a broad criticism of any of these aplogist.

Christian apologists advocating the big bang include William Lane Craig, Norman Geisler, Hugh Ross, David Noebel and Lee Strobel. William Lane Craig, Research Professor of Philosophy at Talbot School of Theology, was interviewed by Lee Strobel. Strobel asked, "And the universe came into being in what has been called the Big Bang?" Craig answered: "Exactly. As [astrophysicist] Stephen Hawking said, 'Almost everyone now believes that the universe, and time itself, had a beginning at the Big Bang.'"

In an interview with Norman Geisler, at the time Research Professor at Southern Evangelical Seminary in Charlotte, N.C., Geisler referred to agnostic Robert Jastrow as claiming that "the Big Bang points to God". On the basis that the Bible and big-bang theory both posit a beginning for the cosmos, Geisler also accepts that "modern astrophysics ... affirms" that there was a big bang. Strobel himself concludes, on the strength of big-bang advocacy going back to secular scientists such as Hawking and Jastrow, that "atheism cannot credibly account for the Big Bang."

Apologist David Noebel criticizes secular humanist Paul Kurtz for failing to reckon with the big bang, which, according to Noebel, is a metaphor for creation suggesting "a creative point like that in Genesis 1:1, which is outside the purview of Secular Humanist cosmology." Noebel further claims that there is still controversy among Christians "about the age of the universe, [but] not whether a Big Bang occurred ..." Noebel thus implies that the big bang is above questioning in the Christian community. But this is not true.

In Dismantling the Big Bang: God's Universe Rediscovered, Christian coauthors Alex Williams and John Hartnett begin with "Four Reasons to Reject the Big-Bang Theory". Their first reason is, "It doesn't work." Aside from the scientific difficulties with this theory, the simple fact is that not all Christians accept it. Some of these Christians are highly credentialed in physics and astrophysics. Hartnett has a Ph.D. in physics from the University of Western Australia, where he worked in the Frequency Standard and Metrology research department, as well as working with the European Space Agency's Atomic Clock Ensemble in Space and is a tenured research professor. It is a short step from claiming that virtually all Christians accept the big bang to the equally fallacious assertion that the big bang is incompatible with atheism. Yet Hartnett reflects that at a young age when he was attracted to a career in cosmology, "I would have described myself as an atheist, believing that the big bang had all the answers ..." There are atheists as well as Christians who believe the big bang.

Astronomer "Hugh Ross," founder and director of the apologetics ministry Reasons to Believe, has written that Christian resistance to big-bang theory is from "a failure to understand the biblical roots of big-bang cosmology." This claim is part of a chapter entitled "The Big Bang: The Bible Said It First". To Ross, the big bang is in the Bible because it teaches that in the beginning God "stretched out the heavens". But God's stretching out the heavens signifies the big bang only in the minds of those who want to believe it. Instead, this phrase connotes not a chaotic event like the big bang, but an orderly process consistent with the highly structured origin of the cosmos in Genesis 1.

Further, Genesis 1 teaches that the creation was a fiat via the spoken word of God, not a process such as the big bang. Seeing the big bang in Scripture is therefore a reading-in of extra-biblical beliefs—an eisegesis—and not an exegesis. Probably at one time or another almost every manmade idea has been "seen" in Scripture, including the justification of slavery before the American Civil War. The Bible nowhere explicitly teaches a big bang, but does explicitly teach recent fiat creation. The big bang as a naturalistic process occupying billions of years, and recent fiat creation, are not compatible. There are in fact many variants of big-bang theory. In this paper, "big-bang theory" means all variants collectively; the phrase "big-bang model" refers to a specific variant.

SCIENTIFIC DIFFICULTIES WITH BIG-BANG HYPOTHESIS

Belief that big-bang theory is a suitable Christian apologetic is flawed because it overlooks serious scientific weaknesses. Big-bang theory has in fact grown more problematic with time. In the late 1940s, it appeared that the big bang could explain the synthesis of virtually all elements and their relative abundances in the universe. In the 1950s, it became apparent the big bang could account for at most the synthesis of only the first few elements, and the rest must have been synthesized in stars. But stellar synthesis of elements ("stellar nucleosynthesis") has also turned out to be problematic. In his Nobel lecture, William Fowler, a pioneer in big bang and stellar nucleosynthesis theory, acknowledged:

"*In spite of the past and current research in experimental and theoretical nuclear astrophysics ... . Hoyle's grand concept of element synthesis in the stars [is not] fully established, ... It is not just a matter of filling in the details. There are puzzles and problems in each part of the cycle that challenge the basic ideas underlying nucleosynthesis in stars* [emphasis added]."

Over the years, theorists have found it necessary to add various "un-observables" to the big bang to bring it into line with observations. Dark matter, dark energy, the missing mass, and a hypothetical super-fast expansion called cosmic inflation happening right after the big bang itself are all by nature unobservable—i.e., they are defined in such a way that they are not observable but can be inferred only indirectly. Scientists cannot observe dark matter, for example. Thus big-bang theorists have to believe in these concepts by [blind] faith. Christian apologists embracing the big bang as if Scripture revealed it appear to be unaware of this subjectivity.

These unobserved entities are needed to control the universe's expansion rate so the big-bang universe matches the real universe in properties like its size. Without these entities, big-bang theory fails. The emphasis of this chapter is not on the big bang's scientific problems—these have been discussed in the reference just cited—but it is fitting to be reminded that non-Christian big-bang advocates continue to express severe doubts about the theory. One has written: "*Theorists ... invented the concepts of inflation and cold dark matter to augment the big bang paradigm and keep it viable, but they, too, have come into increasing conflict with observations. In the light of all these problems, it is astounding that the big bang hypothesis is the only cosmological model that physicists have taken seriously*."

Astronomer John Fix has commented on the idea that primordial cosmic inflation validates the big bang: "[T]his explanation for the period of inflation may sound like a fairy tale ... . It seems unlikely that people will ever be able to confirm the validity of these theories by means of experiments [emphasis added]." As for dark matter, "its existence must remain an article of faith for the true believer in the standard [big bang] model." In other words, big-bang hypothesis will remain a faith-based construct.

Fig.13.20: *The WMAP Satellite (NASA)*

None of this means that big bang researchers have ceased looking for observational confirmation of the hypothesis. However, one of the latest examinations of radiation in space done to confirm the big bang has instead pointed away from it. This radiation is a very weak microwave radiation that fills all the background of space, so is called the Cosmic Microwave Background (CMB). Big-bang theory predicts that CMB should have a certain type of fluctuation imposed on a smooth background.

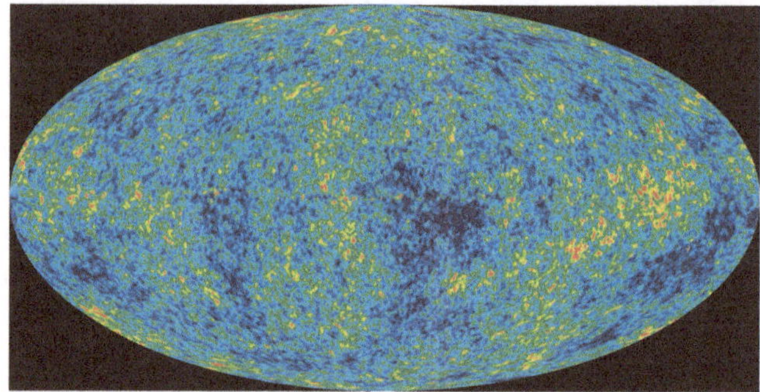

Fig.13.21: *The Cosmic Microwave Background as Detected by WMAP (2008).*

The Wilkinson Microwave Anisotropy Probe (WMAP; Figure13.20) was sent into space to confirm the big-bang prediction, but it did not. Instead, the WMAP data Figure 13.21, are "far outside the current expectations" of big-bang cosmology, and have added "to the anomalies seen in the CMB", so that numerous free parameters (currently about a dozen) are required to produce a precision cosmology in which big-bang models agree with observation. This development is the opposite of Christian apologetics claims that the big bang has been overwhelmingly confirmed. But didn't the big bang correctly predict the hydrogen-helium abundance ratio (the H/He ratio) of the cosmos, and didn't it predict the existence of the CMB? Various Christian apologists seem to think so, but in fact the big-bang model has been adjusted backwards to fit the observed H/He ratio. It did not make a prediction:

"It is commonly supposed that the so-called primordial abundances [of the elements] provide strong evidence for Big Bang cosmology. However, a particular value for the ... ratio needs to be assumed ad hoc to obtain the [observed] abundances [emphasis added]."

Big-bang hypothesis predicted the existence of the CMB but not the correct temperature. On the eve of the first CMB detection in 1965, the predicted temperature was over fifteen times too large. Thus the big bang has not been a successful predictive theory, but one that has needed continuous patching to agree with new observations. Aside from scientific difficulties, big-bang hypothesis is actually an evolutionary one which many have come to perceive as a creationary one because certain theorists and others have used creation to mean evolution, a topic discussed in more detail below. But using the big bang as an apologetic means that one is building a doctrinal foundation on the shifting sands of manmade ideas.

Could the Big Bang have been God's Method of Creation?

People commonly believe that the universe is expanding, but before the 1920s the reigning cosmic model described a static universe, not an expanding one, (Figure 13.3. Data emerging in the 1920s were interpreted to mean that the universe was not static after all. If the universe were expanding, some said, then running the expansion backwards would mean that the universe began at a single point in time. Today this cosmic singularity is commonly believed to have contained all the mass/energy of the universe when it exploded some 14 billion years ago.

On the other hand, the concept of an expanding universe is not anti-biblical, since there is no requirement that the expansion must have originated with a cosmic singularity. God could have initiated a cosmic expansion of the just-created universe. Eventually, in the minds of many, the evidence for cosmic expansion was equated with evidence for a Big Bang. But in the 1920s, the advent of the cosmic expansion concept only opened the door to big-bang hypothesis. Modern big-bang hypothesis was not devised until the 1940s, and the big bang did not become the dominant model of cosmic evolution until the mid-1960s with the discovery of the CMB.

In 1920, the expanding universe concept was still in the future. Debate revolved around whether our galaxy, the Milky Way, includes all of the observable universe (the small universe view), or whether there are galaxies beyond the Milky Way (the large universe). Astronomer "D.H. Curtis" held out for a small universe. Astronomer "Harlow Shapley" advocated a large one. The two scientists came head-to-head in one of the most famous debates in the history of science. The Curtis–Shapley debate took place at the Smithsonian Institute on 26 April 1920. There was as yet little direct evidence of galaxies outside our own. Even so, the large universe view won the day. Verification of the large-universe view opened the door to the possibility of cosmic expansion, first proposed in the early 1900s. Acceptance of an expanding universe, as noted above, in turn opened the door to the formulation and eventual dominance of the big bang hypothesis.

The Curtis–Shapley debate was commemorated in 1998 at the same location and was called the "Nature of the Universe Debate: Cosmology Solved?" "Jim Peebles" of Princeton and "Michael Turner" of the University of Chicago argued for different versions of the big bang. The moderator was "Margaret Geller." Concluding, Geller asked, "How many think that neither of these models will be represented in such a future debate in 80 years?" About 500 were present. The room was filled with hands in the air. So much for scientists' confidence in the leading cosmological models. How can anyone be certain that God used the big bang to create when widespread doubts persist about the long-term survival of the hypothesis itself?

Does the Big Bang Imply a Beginning?

Prominent Christian apologist and big bang advocate "Hugh Ross" claims that the big bang implies a beginning. According to Ross, since Genesis also teaches a beginning, the big bang must therefore be a valid biblical apologetic for creation. Apologist "Lee Strobel" quotes William Lane Craig as saying,

"Atheists themselves used to be very comfortable in maintaining that the universe is eternal and uncaused. ... The problem is that they can no longer hold that position because of modern evidence that the universe started with the Big Bang."

By this line of reasoning, the big bang must be a valid Christian apologetic because it has confounded the atheists.

Unfortunately, such claims are little more than wishful thinking, because:

(1) prominent non-Christians have asserted that the big bang is in fact no evidence for God or a beginning, and

(2) the desire for a cyclic universe is still very much alive among non-Christian believers in the big bang. They simply see the current big bang as one part of a longer, possibly eternal series of cycles.

Carl Sagan, a believer in the big bang, famously asserted that, "The Cosmos is all that is or ever was or ever will be." In Sagan's view therefore, only the material is real. Even though he explicitly denied the existence of God, he did not deny that the universe could be open, i.e.. a non-cyclic cosmos with a beginning, saying "*Very likely, the universe has been expanding since the Big bang ...*" But he left the door open to the possibility of a cyclic universe: "*it is by no means clear that it will continue to expand forever.*" "Gamow" was an atheist and preferred a cyclic universe with no beginning:

"*We can now ask ourselves two important questions: why was our universe in such a highly compressed state, and why did it start expanding? The simplest, and mathematically most consistent, way of answering these questions would be to say that the Big Squeeze which took place in the early history of our universe was the result of a collapse which took place at a still earlier era, and that the present expansion is simply an 'elastic' rebound which started as soon as the maximum permissible squeezing density was reached.*"

Fig.13.22: Artistic Rendering of the Universe Expanding After the Big Bang

But then he denied the possibility of cyclic behavior: "*There is no chance that the present expansion will ever stop or turn into a collapse.*"

Astronomer "James Jeans" was not a Christian, yet he proposed that the universe had a beginning. He did this before the big bang was fashionable, so the big bang is not necessary to conclude that there must have been a beginning. Jeans' argument was based on the fact that the entropy (i.e., disorder) of the cosmos is constantly increasing:

"The more orthodox scientific view is that the entropy of the universe must forever increase to its final maximum value, Figure 13.22. It has not yet reached this: we should not be thinking about it if it had. It is still increasing rapidly, and so must have had a beginning; there must have been what we may describe as a 'creation' at a time not infinitely remote [emphasis added]."

Jeans could be described as a Deist who saw God as having no relationship to the physical universe. For Jeans, the universe was self-existent: "We can only think of [the solar system] as something continually changing and evolving, working out its own future from its past." In fact, no one really knows where the big bang came from.

"Nobody has the foggiest idea what happened the Tuesday before the Big Bang. Who can say whether there was a previously collapsing universe or an incipient quantum fluctuation. That whole domain is part of Bubble-land."

In other words, the Bible teaches a beginning, but there is nothing in big bang hypothesis demanding a beginning.

"Paul Davies," a big-bang cosmologist who believes in a beginning, denies that an external supreme eternal being could have brought matter into existence. He asks, "*Did God Cause the Big Bang*?" and answers that God causing the big bang:

"... makes 'little sense' because a supernatural creation cannot be a causative act in time, for the coming-into-being of time is part of what we are trying to explain. Therefore, such an explanation cannot be a case of cause and effect."

Cambridge University astrophysicist "Stephen Hawking" is arguably one of the most recognized names in modern evolutionary (i.e., big bang) cosmology. Yet he believes in no beginning at all: "[The universe] had no beginning, no moment of Creation." Further, he makes no explicit profession of atheism, saying,

"Science seems to have uncovered a set of laws that ... tell us how the universe will develop with time .... These laws may have originally been decreed by God, but it appears that he has since left the universe to evolve according to them and does not now intervene in it" [emphasis added]." But he does not regard God as Creator:

"But if the universe is really completely self-contained, having no boundary or edge, it would have neither beginning nor end; it would simply be. What place, then, for a creator?"

Clearly there is nothing in big-bang hypothesis which inherently points to God or creation.

CONFUSING CREATION WITH EVOLUTION

"George Gamow," arguably the prime mover behind modern big-bang hypothesis, was one of the first to use the word creation when he really meant evolution. This was many decades ago, and Gamow's linguistic conflation met public resistance, which he addressed in the second printing of his (misnamed) book The Creation of the Universe. In a note for the second printing, Gamow wrote,

"In view of the objections raised by some reviewers concerning the use of the word 'creation,' it should be explained that the author understands this term, not in the sense of 'making something out of nothing,' but rather as 'making something shapely out of shapelessness,' as, for example, in the phrase 'the latest creation of Parisian fashion.'"

Thus, to Gamow, the big-bang creation was not the fiat creation that the Bible teaches. The re-definition of creation which Gamow espoused had long been fashionable in the liberal/modernist community. "John Gibson" is a theological liberal who asserts that a Genesis 1 creation and the Fall lack "any hint elsewhere" in Scripture. He equates the origins account in Genesis with the Babylonian belief that creation was bringing order out of chaos, similar to Gamow's definition. Gibson also writes that maybe God created the chaos, but then maybe "it was there in the beginning, independent of him."

"John H. Walton," an evangelical, notes that, "In the ancient world something came into existence when it was separated out as a distinct entity, given a function, and given a name." This is also similar to Gamow's creation concept. Walton notes that ancient Israel's cosmogony was different from this, for it posited that God is eternally existing but the creation is not. Gamow had—possibly unknowingly— imported into scientific discussions the pagan concept of creation from antiquity, and tantamount to the modern liberal definition. Henceforth in scientific discussions, evolution would increasingly be called creation.

But outside of the Christian community, the dominant view is that the big-bang beginning was merely a quantum mechanical fluctuation. Cosmologists "John Barrow" and "Frank Tipler," in The Anthropic Cosmological Principle, have a section entitled "Creation Ex Nihilo." But echoing Gamow, their creation does not involve God. According to them, "These ideas envision the whole universe to be a giant, quantum mechanical virtual fluctuation of the vacuum." Yet how can a vacuum fluctuation give rise to the very existence of the vacuum that fluctuated?

The existence of the vacuum has to be assumed before one can posit such a fluctuation.

Nowadays one can assume almost as a matter of course in cosmology that when creation is used, evolution is meant, as in the following:

"Each planet seems to provide another set of essential clues for unraveling the mystery of creation. Every planet has proceeded along some peculiar path of evolution all its own—yet each still defines a certain stage in a general process."

The author of these sentences is in fact expressing his hope that new discoveries will solve the mystery of evolution, as the second sentence makes clear. Christian apologists who advocate the big bang as a creation model have been led astray by this long-standing conflation of evolution with creation. This conflation was promoted most forcefully by Gamow, a professed atheist.

GIVING THE BIG BANG CREDIT IT DOESN'T DESERVE

Does teaching the big bang in an old universe lead people to Christ? Hugh Ross tells of a visitor to the Sunday school class he teaches who heard class members telling of people "who came to faith in Christ as a result of my ministry." "That day," Ross writes, the visitor "relinquished his belief in the 'evils' of young-earth creationism." This story illustrates Ross' conflation of old-age advocacy with evangelism. If a long chronology and its attendant evolutionary models such as the big bang are leading people to Christ, then teaching young-earth creation—without the big bang—must be a hindrance to the gospel. But the Bible teaches that even those teaching or behaving erroneously can lead people to Christ (Philippians 1:15–18). Evangelism happens because of gospel words that go out (Romans 10:15), not because of the human error mixed with those words.

Space scientist "Robert Jastrow" claims to have found God (though not Christianity) in his study of big-bang cosmology. In this context, he famously wrote:

"For the scientist who has lived by his faith in the power of reason, the story ends like a bad dream. He has scaled the mountains of ignorance; he is about to conquer the highest peak; as he pulls himself over the final rock, he is greeted by a band of theologians who have been sitting there for centuries."

In the mercy of God toward fallen man, His creation is pervaded with His own self-revelation (Romans 1:20). This revelation is sufficiently pronounced that none has an excuse for failing to know Him. It would not be surprising, therefore, if even the flawed theories of men may contain elements of truth. Indeed, there is a design apologetic based on the big bang. Big-bang advocates, both Christian and non-Christian, have noticed that the parameters in the big-bang hypothesis must be finely adjusted. One such parameter is the mass of the universe:

"The universe ... is either barely open or barely closed. In the language of cosmology, we say that the universe is 'very nearly flat,' a flat universe being one in which there is just enough mass to bring the expansion to a stop ... . Given all of the infinite possible masses that the universe could have, why does it have a mass so close to this critical value? Why is the universe almost flat?"

The universe is so intricately constructed that any hypothesis will discern design, even a false one. Darwinists, for example, routinely discover intricate designs in living things. Their belief in the rise of life by chance does not prevent their observing designed structures. The cosmos also exhibits structures whose workings are described by natural law. But efforts to apply these laws to the origins of these structures via big-bang hypothesis have failed, unless one invokes constructs outside these laws such as dark matter, dark energy and inflation. This failure suggests that the laws of nature which describe how the cosmos functions cannot be responsible for its origin.

Putting the big bang on a pedestal because it can be taken to imply a biblical truth—that there was a beginning, or that the universe is designed—is like claiming that all cults and false religions must lead to God and salvation—for each one also contains elements of truth.

Christians can and should be grateful for each person saved through a Christian apologetic outreach, even one advocating the big bang. But Christian apologetics needs to discard its reliance on manmade philosophies, and should be based on a cosmology starting with the recent fiat creation of Genesis 1. This is the cosmology that reveals the power of God like no other. It is the cosmology that dominated before the advent of big-bang hypothesis and its precursor, the Darwinian Revolution. It is especially ironic for a Christian to teach the power of God in future events, yet to fall back on the claim that in the past God was constrained to use a process over billions of years. After all, the Bible teaches that Jesus Christ, the Creator, is "the same yesterday, and today, and forever" (Hebrews 13:8).

The story is told of Francis Deak, a European statesman prior to the catastrophic political Revolution of 1848 and the host of changes it brought about. After that Revolution, "an Austrian official remarked, 'Deak cannot demand after so many accomplished facts that we should begin affairs all over again.'"

Overhearing the conversation, Deak responded, "Why not? ... . If a man has buttoned one button of his coat wrong, it must be undone again from the top."

"The button might be cut off."

"Then the coat could never be buttoned right at all." Neither can the cosmos be understood aright by big bang hypothesis. Christian apologists should abandon big bang hypothesis, and build cosmological hypothesis again from the top by giving the Word of God its rightful authority in the matter of cosmic origins.

DISCOVER GOD'S UNIVERSE

Disregards Big Bang

Big bang thinking is permeating Christianity, undermining the Gospel and leading countless souls astray. Many prominent church figures are urging people to abandon the fundamental doctrines of creation and the authority of the Bible. Leading Church of England scientist/priests, Rev. Dr "Arthur Peacocke" and Rev. Dr Sir "John Polkinghorne" KBE, FRS, are both urging the church to replace Genesis creation with big bang theory and cosmic evolution.

Fig.13.23: Ant Nebula - NASA, ESA and The Hubble Heritage Team (STScl/AURA)

The Ant Nebula is approximately 1.6 light years long and part of a Milky Way constellation called Norma, some 3,000–6,000 light years away from Earth, Figure 13.23. The curious pattern of gases being ejected from the dying star results in the visible ant-like appearance though exactly how this has occurred is not yet fully understood.

According to Peacocke, the Creator is the author of processes whereby things make themselves. This very likely includes life elsewhere in the universe. Jesus Christ 'represents the consummation of the evolutionary creative process that God has been affecting in and through the world ... What was revealed in Jesus the Christ could also, in principle, be manifest both in other human beings and indeed also on other planets, in any sentient, self-conscious, non-human persons ...' (p. 114). Entirely gone is the concept of man the sinner and Christ the Savior. Jesus is thus just a highly evolved human being, rather than God incarnate, and we can evolve as He did and become like Him.

Polkinghorne takes an approach much like this, too, and reaches similar conclusions. In a public lecture entitled 'Is there a destiny beyond death?' he proposed a view of eternal life in which there is no cross. Instead of the cross, he inserted purgatory as a means of making us fit to enter into the presence of God.

I imagine that if you wanted to insult Jesus then you could hardly do better than to say that his death on the cross did *not* restore us to full fellowship with God and thus, we will need to go through a process of purgation (cleansing) in order to achieve that end for ourselves. Yet this is the logical consequence of Polkinghorne's big bang beginning followed by cosmic evolution.

The big bang's millions of years, with Earth cooling slowly and geological processes forming fossils, mean that death must be a natural part of life and have been around before sin. Thus, death is not the penalty for sin. Therefore, Christ's death on our behalf means nothing.

The Reflection Nebula in the Orion constellation. It is about 1,500 light years from Earth and was discovered first by creationist Sir "William Herschel" and his wife "Caroline," some two centuries ago, Figure 13.24.

Fig.13.24: NASA, The Hubble Heritage Team

Tragically, all this heresy is based on a mirage. In a powerful new book, *Dismantling the Big Bang: God's Universe Rediscovered*, (Disregard Big Bang) Australian authors *Alex Williams* and Dr *John Hartnett* show us (in easy-to-read layman's language) that the standard big bang models (there are several varieties) are simply not credible. They do not enable us to explain the origin of galaxies, stars, or planets—which is astonishing, given that the very purpose of the big bang is to explain the origin of the material universe. (Without stars and galaxies, there's not much left of that universe, to put it mildly). In short, Peacocke and Polkinghorne are urging Christians to relinquish the genuine history of the universe and exchange it for a counterfeit.

On the other hand, the Bible does have a credible and consistent cause for all that we see around us. 'In the beginning was the Word, the Word was with God and the Word was God. He was in the beginning with God. All things were made through him ... The Word became flesh and dwelt amongst us and we beheld his glory, the glory as of the only-begotten Son of the Father' (*John 1:1-3, 14*).

These familiar words from John's Gospel constitute the most powerful explanation for the origin, nature and destiny of the universe that the world has ever heard.

COSMOLOGY BELONGS TO THE CHRISTIAN

Cosmology is the study of the universe (cosmos). We humans are stuck in both space and time in a very small part of a very vast universe, Figure 13.25. We cannot traverse the universe to study its size or shape, nor can we go backwards or forwards in time to see how it changes. All we can do is contemplate it through our telescopes—and make up and compare (and try to test) stories that try to explain where it came from and how it got to be the way it is.

The standard big bang models are the most popular stories going around now, but most people don't realize that they are just one class of many possible different kinds of narratives.

Fig.13.25: *Antonella Nota (ESA/STScI) et al., ESA, NASA.*

As "Disregards Big Bang" book shows, some stories about the universe are better than others because they explain more. In ancient times, some thought that spirit beings moved the planets around the solar system—until the 17th century when Sir "Isaac Newton" discovered the laws of motion and the force of gravity.

In the 20th century, when Edwin Hubble discovered that the universe was expanding, these same laws of gravity and motion led to the idea that in the past the galaxies must have all been together in one place, in which case gravity would have sucked everything into a state of infinitely dense and hot energy. To get out of that state, something incomprehensible must have happened. An eruption of unimaginable magnitude must have occurred—hence the name 'big bang'.

But this 'story' is riddled with insoluble conundrums. For example, there is no known force in physics that could have caused the expansion of the universe out of such an initial state. Then many other things which likewise have no known cause (such as 'inflation') must follow later in the process. One event does not lead naturally to the next, so a multitude of incomprehensible causes have to be cobbled together to get the expansion going and to make it produce a universe. The Disregard Big Banf book shows that in this and other aspects, big bang theory is internally inconsistent.

In dramatic contrast, biblical cosmology *is* internally consistent. Its proposed cause—God—*can* create a universe. We know this because God the Son, Jesus Christ, came to Earth and did miracles of creation such as turning water into wine, and multiplying loaves and fishes. (Even a dead fish is far more complex than an entire galaxy.)

Not only does the Bible provide us with a ***credible cause*** (God) but this same cause also provides a *consistent* explanation for the nature and development of the world we live in. For the Word of God not only created the universe, but also upholds the universe and guides the universe to its destiny in that same Word, the person of Jesus Christ (*Colossians 1:16–17*).

Cosmologists have discovered that the laws of physics and chemistry and the structure of the galaxy and the solar system are very finely tuned to support life on Earth. This is consistent with the idea of intelligent design.

Also, the Bible in Genesis explains the 'fallen' state of the world with its suffering and death—a result of God's curse upon man's sin. Therefore, a straight forward reading of the Bible explains the origin, nature and destiny of the universe in a scientifically credible and consistent manner. The authors also show that the Bible's timescale is more credible than the billions of years of standard cosmology.

The great news is that the Bible is not only a source of salvation to satisfy our souls, it is a goldmine of intellectual understanding to satisfy our minds. Jesus Christ is not just an immense figure in history, he is the Alpha and the Omega, the Beginning and the End (*Revelation 22:13*), and that is the greatest cosmological statement humanity is ever likely to hear.

SECULAR SCIENTISTS REFUTE THE BIG BANG

Naïve Apologetics

It's amazing to recognize how many Christian leaders have not merely tolerated the 'big bang' idea, but embraced it wholeheartedly. To hear their pronouncements, believers should welcome it as a major plank in our defense of the faith. 'At last, we can use science to prove there's a creator of the universe.'

However, the price of succumbing to the lure of secular acceptability, at least in physics and astronomy, has been heavy. We have long warned that adopting the big bang into Christian thought is like bringing the wooden horse within the walls of Troy. The followings describe the reasons:

- The big bang forces acceptance of a sequence of events totally incompatible with the Bible (e.g. earth after sun instead of earth before sun—The big bang's billions of years of astronomical evolution are not only based on naturalistic assumptions, they are contrary to the words of Jesus Himself, who said people were there from the beginning, not towards the end of an interminably long 'creation' process (*Mark 10:6*).

- The slow evolution of the stars, then solar system and planets (including earth) in big bang thinking means that 'big bang Christians' are invariably dragged into accepting 'geological evolution' (millions of years for the earth's fossil-bearing rocks to be laid down). Accordingly, they end up denying the global Flood, and accepting death, bloodshed and disease (as seen in the fossils) before Adam. This removes the Fall and the Curse on creation from any effect on the real world, as well as removing *the* biblical answer Christians have always had to the problem of suffering and evil (God made a perfect world, ruined by sin).
- Marrying one's theology to today's science means that one is likely to be widowed tomorrow.

In fact, the signs are strong that exactly that is happening, and that those who have 'bought' the big bang for its allegedly irrefutable science have been 'sold a pup.' A bombshell 'Open Letter to the Scientific Community' by 33 leading scientists has been published on the internet (Cosmology statement) and in *New Scientist* (Lerner, E., Bucking the big bang, *New Scientist* **182**(2448)20, 22 May 2004). An article on www.rense.com titled 'Big bang theory busted by 33 top scientists' (27 May 2004) says, 'Our ideas about the history of the universe are dominated by big bang theory. But its dominance rests more on funding decisions than on the scientific method, according to Eric Lerner, mathematician Michael Ibison of Earthtech.org, and dozens of other scientists from around the world.'

The open letter includes statements such as:

- The big bang today relies on a growing number of hypothetical entities, things that we have never observed—inflation, dark matter and dark energy are the most prominent examples. Without them, there would be a fatal contradiction between the observations made by astronomers and the predictions of the big bang theory.'
- 'But the big bang theory can't survive without these fudge factors. Without the hypothetical inflation field, the big bang does not predict the smooth, isotropic cosmic background radiation that is observed, because there would be no way for parts of the universe that are now more than a few degrees away in the sky to come to the same temperature and thus emit the same amount of microwave radiation. … Inflation requires a density 20 times larger than that implied by big bang nucleosynthesis, the theory's explanation of the origin of the light elements.' [This refers to the *horizon problem*, and supports what we say in Light-travel time: a problem for the big bang.]
- 'In no other field of physics would this continual recourse to new hypothetical objects be accepted as a way of bridging the gap between theory and observation. It would, at the least, *raise serious questions about the validity of the underlying theory* [emphasis in original].'
- 'What is more, the big bang theory can boast of no quantitative predictions that have subsequently been validated by observation. The successes claimed by the theory's supporters consist of its ability to retrospectively fit observations with a steadily increasing array of adjustable parameters, just as the old Earth-centered cosmology of Ptolemy needed layer upon layer of epicycles.'

The dissidents say that there are other explanations of cosmology that do make some successful predictions. These other models don't have all the answers to objections, but, they say, 'That is scarcely surprising, as their development has been severely hampered by a complete lack of funding. Indeed, such questions and alternatives cannot even now be freely discussed and examined.'

Those who urge Christians to accept the big bang as a 'science fact' point to its near-universal acceptance by the scientific community. However, the 33 dissidents describe a situation familiar to many creation scientists: 'An open exchange of ideas is lacking in most mainstream conferences ... doubt and dissent are not tolerated, and young scientists learn to remain silent if they have something negative to say about the standard big bang model. Those who doubt the big bang fear that saying so will cost them their funding.'

Evolutionist and historian of science, Evelleen Richards, has noticed that it's hard even for rival *evolutionary* theories to get a hearing when challenging the ruling paradigm— This should give some idea of the difficulties biblical scientists face.

But don't we read, even in the daily newspapers, about many 'observations' that only ever seem to support the big bang? In fact, these prominent secular scientists say:

'Even observations are now interpreted through this biased filter, judged right or wrong depending on whether or not they support the big bang. Accordingly, discordant data on red shifts, lithium and helium abundances, and galaxy distribution, among other topics, are ignored or ridiculed.'

Science is a wonderful human tool, but it needs to be understood, not worshipped. It is fallible, changing, and is severely limited as to what it can and cannot determine. As creation scientists have often pointed out, instead of a scientific concept, the big-bang idea is more a dogmatic religious one—based on the religion of humanism. As these big-bang opposers point out:

'Giving support only to projects within the big bang framework undermines a fundamental element of the scientific method—the constant testing of theory against observation. Such a restriction makes unbiased discussion and research impossible.'

Furthermore, contrary to the naïve pronouncements of many who should know better, it is not in any sense a matter of 'looking into a telescope and "seeing"? the big bang billions of years ago.' As always, observations are interpreted and filtered through worldview lenses. Those who developed the big bang were guided by secular worldview filters just as much as those who are now crying that the emperor has no clothes. They wanted a universe that created itself; their opponents want an eternal, uncreated universe. From a Christian perspective, both are in open defiance of their Creator's account of what really happened.

With Darwinism on the run, the Enemy of souls is seeking to seduce believers into embracing a more subtle, yet far deadlier way of evading the authority of the Bible. With progressive creationism/big-bangery rampaging through the evangelical community, he must think he is on a winner.

BIG BANG UNIVERSE IS HYPOTHETICAL UNIVERSE

Missing Antimatter

Fig.13.36: A Balanced Universe

The currently leading evolutionary cosmogony 'birth of the universe' is the big bang hypothesis. This basically says: nothing exploded and became everything, Figure 13.26. One part of this is energy turning into matter, as per the famous Einstein formula

$$E = mc^2$$

However, when this occurs, the standard laws of particle physics state that an *equal amount of matter and antimatter must be produced.* Yet our universe comprises overwhelmingly of matter, with only rare and fleetingly short-lived antimatter particles, produced for example in experimental high energy collisions. These are of the same mass but opposite charge (if the particle is charged) and magnetic moment as the corresponding matter particle, which they soon interact with and are mutually annihilated—antielectron (positron) with electron, antiproton with proton, antineutron with neutron, etc. How did matter survive annihilation from an equal amount of antimatter?

According to big bang proponents, most matter *was* annihilated by antimatter, forming the photons of the cosmic background radiation. And there are a billion of those photons for every proton. Therefore, they believe that the early universe contained a billion *and one* protons for every billion antiprotons—leaving just that one proton surviving.

But for energy to produce this slight imbalance, there would need to be asymmetry in the fundamental makeup of the universe. Accordingly, researchers at the European Organization for Nuclear Research, or CERN, in Meyrin, Switzerland, have been trying to find imbalances between protons and antiprotons.

The latest experiment has measured the magnetic moment of the antiproton 350 times more precisely than previously—to 1.5 parts per billion. This was a enormously difficult experiment, because these antiprotons must be generated, then trapped by a powerful magnetic field to prevent them annihilating with a proton; and also because the magnetic moment is so *tiny*!

The experiment showed that the *antiproton has an equal and opposite magnetic moment to the proton.* I.e., not enough asymmetry to explain the overabundance of matter! This has so undermined the big bang cosmology that the lead researcher, Christian Smorra, said:

"*All of our observations find a complete symmetry between matter and antimatter, which is why the universe should not actually exist.*"

Well, not a universe that formed in some big bang, anyway. Discoveries like this and others which shake confidence in the big bang might well lead to the secular world abandoning this belief in future, in favor of some other naturalistic origins myth. This should therefore be a lesson to those misguided Christian apologists who 'marry' Genesis to the big bang—they might well find themselves 'widowed' in the future. Therefore, they will need to reinterpret their reinterpretations, which were in any case biblically untenable.

APPENDIX

i. As is often the case in cosmology, it was not actually the data that suggested expansion but a certain bias for interpreting the data. One of the earliest framers of the "expanding universe" concept wrote: "The expansion of the universe is a matter of astronomical facts interpreted by the hypothesis of relativity." Lemaître, G., Contributions to a British Association discussion on the evolution of the universe, Nature 128, Supplement, p. 704, 1931.

ii. Alpher, R., Bethe, H. and Gamow, G., The origin of the chemical elements, Physical Review 74:1198–1199; p. 1198, 1948. Other papers leading to modern big-bang hypothesis are discussed in Henry, ref. 18, pp. 55–56. George Gamow was the driving force behind early modern big-bang hypothesis and was also known for his sense of humor. For the paper referenced above, Gamow added Bethe's name as a coauthor. This made the authors' last names start with (and sound like) the first three letters of the

Greek alphabet, alpha, beta, gamma. This was Gamow's alpha-beta-gamma paper. Center for History of Physics, "A Gamow joke", , paragraph 1, accessed 26 February 2009.

iii. Trimble, V., The 1920 Shapley-Curtis discussion: background, issues, and aftermath, Publications of the A stronomical Society of the Pacific 107:1133–1144, 1995; pp. 1140–1142. The format of this meeting was actually that of a discussion, not a debate, as there was no time for rebuttal. Even so, the encounter is now known as the Great Debate. Nemiroff, R.J., The 75th anniversary astronomical debate on the distance scale to gamma-ray bursters: an introduction, Publications of the Astronomical Society of the Pacific 107:1131–1132, 1995; p. 1131.

iv. Robertson, P., Revolutions of 1848, Harper, New York, p. 307, 1960

v. Jonathan Henry earned his doctorate from the University of Kentucky in Chemical Engineering. He is now Chairman of the Science Division and Professor of Natural Science at Clearwater Christian College in Florida. In 1987 he began speaking and writing in defence of "recent creation" when his teaching schedule permitted. He has authored The Astronomy Book published by Master Books

vi. A powerful recent book, Gary Bates's *Alien Intrusion: UFOs and the Evolution Connection*, Master Books, Arkansas, USA, 2004, highlights the absence of scientific evidence for the existence of extra-terrestrial life, and gives massive evidence for a satanic effort to deceive mankind into thinking aliens exist and have arrived on earth.

vii. Other apologists have used different arguments. For example, the leading medieval theologian and apologist *Thomas Aquinas* (c. 1225–1274), while he believed that the universe began in time because the Bible said so, didn't think this could be proved philosophically. Instead, in his famous 'five ways' (*Summa Theologiae*, Question **2**, The existence of God) he argued that even if the universe had no beginning, then it would *still* not exist or undergo change *here and now* unless there were a necessary being, 'prime mover', and 'first cause', who upholds the universe in its *moment-by-moment existence*. "This all men speak of as God." That is, while the *Kalām* argument concerns God's *creative* work, Aquinas's arguments are for God's *sustaining* work since He finished His creation on Day 7 (Genesis 2:1–3). Then Aquinas spends hundreds of pages arguing that the 'God' with these properties would be perfectly good, all-powerful, and all-knowing.

viii. Readers should note English physicist and big-bang-proponent Paul Davies' comment: "According to modern physics, the big bang represented the origin of space and time, as well as of matter and energy. This means that *time itself came into existence with the big bang*. Questions like: What happened before the big bang? or What caused the big bang? are therefore meaningless. There was no before." [Emphasis in the original.] Source: Davies, P., 'Science, God and the Laws of the Universe', *ABC Radio 24 Hours*, August 1992, p. 36–39.

ix. This depends on the mass density of the universe. Too much and gravity would cause the universe to collapse. Too little and the universe would expand forever. Observations show 'flatness', which means the density is minutely below the threshold required for collapse. This is a cosmological fine-tuning problem, where the force of the expansion matches the force of gravity to one part in 10^{60}.

x. Thomson scattering occurs when low-energy photons are scattered elastically by photons. High-energy photons often give energy to the electrons, so the scattered photons have lower energy (so longer wavelength), called Compton scattering.

xi. E and B are common symbols in physics for the electric and magnetic fields, respectively. The polarization of radiation can be decomposed into two components—E-mode (radial, curl free) and B-mode (circular, divergence free). In big-bang cosmology, E-mode polarization of the CMB radiation is said to result from Thomson scattering, while B-modes are not from magnetic fields but are said to result from gravitational lensing effects.

xii. Jesus would not have created a new star to demonstrate His power—to be appreciable, it would have had to be close enough to endanger people's lives.

xiii. Dr Ross promotes the idea that 'creation' happened progressively over billions of years, which entails having death and suffering before the Fall of Adam, which undermines the Gospel, the goodness of God and eschatology. He also asserts that Noah's Flood was a local event. His views are critiqued in articles listed here: Progressive Creationism.

xiv. Magnetic moment is the torque produced by interaction with a magnetic field. Despite having no charge, neutrons, too, have a magnetic moment. This is because a neutron comprises three charged quarks: two down quarks (charge −⅓) and one up quark (+⅔); the antineutron comprises two down antiquarks (+⅓) and one up antiquark (−⅔).

xv. The Cryogenic Dark Matter Search collaborators once claimed some bumps in their data might be a weakly interacting massive particle (WIMP) signal at the 3σ level. But it was not a detection. And a faster-than-light neutrinos claim was found to be wrong even though it was from a 6σ result. For more information see Hartnett, J.G., Our galaxy near the centre of concentric shells of galaxies?, 26 May 2014; biblescienceforum.com.

xvi. The proposition is often justified by showing how neatly it solves the problems of the big bang. But since it was designed to do that, this is hardly evidence for it. The obvious circularity has been pointed out by others. See Has the 'smoking gun' of the 'big bang' been found?

xvii. Actually, the word 'cause' has several different meanings in philosophy. But in this section, we are referring to the *efficient cause*, the chief agent causing something to happen.

xviii. Other apologists have used different arguments. For example, the leading medieval theologian and apologist Thomas Aquinas (c. 1225–1274), while he believed that the universe began in time because the Bible said so, didn't think this could be proved philosophically. Instead, in his famous 'five ways' (*Summa Theologiae*, Question **2**, The existence of God) he argued that even if the universe had no beginning, then it would *still* not exist or undergo change *here and now* unless there were a necessary being, 'prime mover', and 'first cause', who upholds the universe in its *moment-by-moment existence*. "This all men speak of as God." That is, while the *Kalām* argument concerns God's *creative* work, Aquinas's arguments are for God's *sustaining* work since He finished His creation on Day 7 (Genesis 3–2:1). Then Aquinas spends hundreds of pages arguing that the 'God' with these properties would be perfectly good, all-powerful, and all-knowing.

xix. The word 'cause' has several different meanings in philosophy. But in this section, we are referring to the *efficient cause*, the chief agent causing something to happen.

REFERENCES

1. Strobel, L., The Case for Faith, Zondervan, Grand Rapids, MI, pp. 80–118, 2000.
2. Quoting from Hawking, S.W. and Penrose, R., The Nature of Space and Time, Princeton University, Princeton, NJ, p. 20, 1996.
3. Geisler, N., Creation and the Courts: Eighty Years of Conflict in the Classroom and the Courtroom, Crossway, Wheaton, IL, p. 257, 2007.
4. Noebel, D., Understanding the Times: The Collision of the World's Competing Worldviews, Summit Press, Manitou Springs, CO, pp. 102–103, 2006.

5. Henry, J., A critique of progressive creationism in the writings of Hugh Ross, Creation Research Society Quarterly 43:16–24, 2006; pp.
6. Fowler, W., The quest for the origin of the elements, Science 226:922–935, 1984; p. 934.
7. Henry, J., The elements of the universe point to creation: introduction to a critique of nucleosynthesis hypothesis, Journal of Creation 29:(2):53–60, 2006; pp. 55–58.
8. Oldershaw, R., What's wrong with the new physics? New Scientist 127(1748):56–59, 1980; p. 59.
9. Fix, J., Astronomy, WCB/McGraw-Hill, Boston, MA, p. 616, 1999.
10. Sandage, A., Observational tests of world models, Annual Review of Astronomy and Astrophysics 26:561–630, 1988; p. 623.
11. McGaugh, S., Confrontation of modified newtonian dynamics predictions with Wilkinson Microwave Anisotropy Probe first year data, Astrophysical Journal 611:26–39, 2004; p. 26.
12. Rudnick, L., Brown, S. and Williams, L.R., Extragalactic radio sources and the WMAP cold spot, , 3 August 2007; accessed 27 February 2009.
13. Arp, H., Burbidge, G., Hoyle, F., Narlikar, J. and Wickramasinghe, N., The extragalactic universe: an alternative view, Nature 346:807–812, 1990; p. 811.
14. Penzias, A., and Wilson, R., A measurement of excess antenna temperature at 4080 Mc/s, Astrophysical Journal 142:419–421; p. 420, 1965. Dicke, R., Peebles, P., Roll, P. and Wilkinson, D., Cosmic black-body radiation, Astrophysical Journal 142:414–419, 1965; p. 415.
15. Nemiroff, R.J. and Bonnell, J.T., The nature of the universe debate in 1998, Publications of the Astronomical Society of the Pacific 111:285–287, 1999; p. 285.
16. Van Flandern, T., Big bang alternatives, Biblical Astronomer 11(Spring):57–58; p. 58, 2001. Van Flandern, a big bang critic, was present at this event. But Nemiroff and Bonnell also note that "Many raised their hands". Nemiroff and Bonnell, ref. 30, p. 286.
17. Sagan, C., Cosmos, Random House, New York, p. 4, 1980.
18. Charles Naeser, one of Gamow's colleagues, stated "... Gamow ... never did hide the fact that he was an atheist ..." Anderson, D., Interview with Dr Charles Naeser, , paragraph 43, accessed 26 February 2009.
19. Gamow, G., The Creation of the Universe, Mentor, New York, p. 37, 1952.
20. 39. Jeans, J., The Mysterious Universe, Dutton, New York, pp. 176–177, 1932.
21. Berman, B., Strange universe: bubbleland, Astronomy 28(6):106, 2000.
22. Davies, P., The Mind of God, Penguin, New York, pp. 39–72, 1992.
23. Davies, ref. 42, p. 69. But see also Craig, W.L., God, creation & Mr. Davies, Brit. J. Phil. Sci. 37:163–175, 1986 which address the same kind of claim.
24. Hawking, S.W., A Brief History of Time, Bantam Books, New York, p. 116, 1988
25. Gibson, J., Genesis, Westminster, Philadelphia, PA, p. 27, 1981.
26. Walton, J., Ancient Near Eastern Thought and the Old Testament, Baker, Grand Rapids, MI,.
27. Barrow, J.D. and Tipler, F.J., The Anthropic Cosmological Principle, New York, p. 440, 1986.
28. Short, N.M., Planetary Geology, Prentice-Hall, Englewood Cliffs, N.J., p. 334, 1975.
29. Jastrow, R., God and the Astronomers, Norton, New York, p. 116, 1978.

30. Trefil, J.S., The Moment of Creation, Scribners, New York, p. 47, 1983.

31. Peacocke, A., Humanity's Place in Cosmic Evolution, in Dick, S.J., (ed.) *Many Worlds: The New Universe, Extra-Terrestrial Life and the Theological Implications*, Templeton Foundation Press, Philadelphia, USA, pp. 89–118, 2000. Quoted in an online review at: www.faithnet.org.uk/AS%20Subjects/Philosophyofreligion/peacocke.htm, 11 October 2005.

32. Polkinghorne, J., *Is there a destiny beyond death?*, St George's Anglican Cathedral Perth, Lecture No. 10, Western Australia, 2003. Return to text.

33. Williams, A. and Hartnett, J.,*Dismantling the Big Bang: God's Universe Rediscovered*, Master Books, Arkansas , USA, 2005.

34. Lewis, G. and Barnes, L., *A Fortunate Universe: Life in a Finely Tuned Cosmos*, ch. 6, Cambridge University Press, 2016.

35. From the old name in French, Conseil Européen pour la Recherche Nucléaire (European Council for Nuclear Research).

36. Smorra, C. and 16 others, A parts-per-billion measurement of the antiproton magnetic moment, *Nature* **550**(7676):371–374, 19 October 2017 | doi:10.1038/nature24048.

37. Smorra, C., cited in Osborne, H., The universe should not actually exist, CERN scientists discover, newsweek.com, 25 October 2017.

38. Wieland, C., Secular scientists blast the big bang: what now for naïve apologetics? *Creation* **27**(2):23–25, 2005; creation.com/bigbangblast.

39. Sarfati, J., Evolution/long ages contradicts Genesis order of Creation, *Creation* **37**(3):52–54, 2015; creation.com/evolution-v-genesis-order.

40. Hawking, S., The Beginning of Time, hawking.org.uk, accessed March 2017.

41. Crites, A.T., *et al.*, Measurements of E-Mode Polarization and Temperature-E-Mode Correlation in the Cosmic Microwave Background from 100 Square Degrees of SPTpol Data, *Astrophysical Journal* **805**(1):36, 2015 | doi: 10.1088/0004-637X/805/1/36.

42. Hartnett, J.G., Has the 'smoking gun' of the 'big bang' been found?, 20 March 2014; creation.com/bbgun.

43. Ross, H., Cosmic Inflation: It Really Happened, reasons.org, 3 August 2015.

44. Hartnett, J.G., New study confirms BICEP2 detection of cosmic inflation wrong, 5 February 2015; creation.com/inflation-wrong.

45. Hartnett, J.G., Hey, not so fast with the Nobel Prize!, 3 April 2014; creation.com/inflation-doubt.

46. This amounts to saying that an inflation event happened to within more than 6 standard deviations (6σ).

47. Planck Collaboration (2013), Planck 2013 results XVI, Cosmological parameters, *Astronomy & Astrophysics* 571: (A16), See Table 2, p. 12, November 2014 | http://dx.doi.org/10.1051/0004-6361/201321591. Return to text.

48. In 2013 WMAP nine-year data analysis produced a similar result ($n_s = 0.013 \pm 0.972$), See Nine-year Wilkinson Microwave Anisotropy Probe (WMAP) observations: final maps and results, *The Astrophysical Journal Supplement Series* 208(2), September 2013. And an online review paper, Lahav, O. and Liddle, A.R., The Cosmological Parameters, September 2011, quotes $n_s = 0.014 \pm 0.967$. These

results have double the errors of the latest paper yet still $n_s < 1.0$. However, no claimed proof of inflation was made back as far as 2011.

49. Hartnett, J.G., 'Dark photons': another cosmic fudge factor, 18 August 2015; creation.com/dark-photons.
50. Hartnett, J.G., Does the Bible really describe expansion of the universe? *J. Creation* 25(2):125–127, August 2011; creation.com/expansion?
51. Hartnett, J.G., The big bang is not a Reason to Believe, 20 May 2014; creation.com/bang-not-reason.
52. Similar lists are in Morris, *The Genesis Record, A scientific and devotional commentary on the Book of Beginnings*, p. 38, Baker Book House, Grand Rapid, MI, 1976, p. 38; and Fruchtenbaum,A.G., *The Book of Genesis*, p. 35, Ariel's Bible Commentary, Ariel Ministries, San Antonio, TX, 2009.
53. See this comprehensive analysis and refutation: Ammi, K., Atheism, creation.com/atheism, 11 June 2009; and *Christianity for Skeptics*, ch. 3.
54. This is the 'hard' form of 'agnosticism', as coined by 'Darwin's Bulldog', T.H. Huxley (see also Grigg, R., Darwin's bulldog—Thomas H. Huxley, *Creation*, 31(3):39–41, 2009; creation.com/huxley). A 'soft agnostic' merely claims himself not to know that there is a God.
55. Theologians first used the word 'omnipotent' (all-powerful) to mean that *God has no limitations from **outside** Himself.* See Sarfati, J., If God can do anything, then can He make a being more powerful than Himself? What does God's omnipotence really mean? creation.com/omnipotence, 12 January 2008.
56. See also *Christianity for Skeptics*, ch. 1. This is called the *Kalām* Cosmological Argument. It goes back to the church theologian Bonaventure (1274–1221), and was also advocated by medieval Arabic philosophers. The word *kalām* is the Arabic word for 'speech', but its broader semantic range includes 'philosophical theism' or 'natural theology'. The *kalām* argument's most prominent modern defender is the philosopher and apologist Dr William Lane Craig (1949–): *The Kalām Cosmological Argument*, Barnes & Noble, New York, 1979. Unfortunately, Dr Craig compromises on the plain meaning of Genesis—see William Lane Craig's intellectually dishonest attack on biblical creationists.
57. Anscombe, G.E.M., "Whatever has a beginning of existence must have a cause": Hume's argument exposed, *Collected Philosophical Papers*, Volume 1, Basil Blackwell, 1981.
58. Anscombe, G.E.M., Times, beginnings and causes, in her *Collected Philosophical Papers*, Volume 2, Basil Blackwell, 1981.
59. Feser, E., *The Last Superstition: A Refutation of the New Atheism*, Kindle Locations 5253-5254, St. Augustine's Press. Kindle Edition, 2012.
60. Lemley, B., Guth's grand guess, *Discover* 23(4):32–39, April 2002; discovermagazine.com.
61. Krauss, M., *A Universe from Nothing: Why There is Something Rather than Nothing*, Free Press, 2012. See also review by Reynolds, D.W., *J. Creation*, 27(1):30–35, 2013.
62. Barnes, L., Out of nothing, *letterstonature.wordpress.com*, 1 April 2011.
63. Davies, P., *God and the New Physics*, p. 215, Simon & Schuster, 1983.
64. Albert, D., On the Origin of Everything: review of *A Universe From Nothing*, by Lawrence M. Krauss, *New York Times*, 23 March 2012; nytimes.com.
65. Feser, E., Not Understanding Nothing, A review of *A Universe from Nothing, First Things*, June/July 2012; firstthings.com.

66. Similar lists are in Morris, *The Genesis Record, A scientific and devotional commentary on the Book of Beginnings*, p. 38, Baker Book House, Grand Rapid, MI, 1976, p. 38; and Fruchtenbaum,A.G., *The Book of Genesis*, p. 35, Ariel's Bible Commentary, Ariel Ministries, San Antonio, TX, 2009.
67. See this comprehensive analysis and refutation: Ammi, K., Atheism, creation.com/atheism, 11 June 2009; and *Christianity for Skeptics*, ch. 3.
68. This is the 'hard' form of 'agnosticism', as coined by 'Darwin's Bulldog', T.H. Huxley (see also Grigg, R., Darwin's bulldog—Thomas H. Huxley, *Creation*, 31(3):39–41, 2009; creation.com/huxley). A 'soft agnostic' merely claims himself not to know that there is a God.
69. Theologians first used the word 'omnipotent' (all-powerful) to mean that *God has no limitations from* ***outside*** *Himself.* See Sarfati, J., If God can do anything, then can He make a being more powerful than Himself? What does God's omnipotence really mean? creation.com/omnipotence, 12 January 2008.
70. See also *Christianity for Skeptics*, ch. 1. This is called the *Kalām* Cosmological Argument. It goes back to the church theologian Bonaventure (1274–1221), and was also advocated by medieval Arabic philosophers. The word *kalām* is the Arabic word for 'speech', but its broader semantic range includes 'philosophical theism' or 'natural theology'. The *kalām* argument's most prominent modern defender is the philosopher and apologist Dr William Lane Craig (1949–): *The Kalām Cosmological Argument*, Barnes & Noble, New York, 1979. Unfortunately, Dr Craig compromises on the plain meaning of Genesis—see William Lane Craig's intellectually dishonest attack on biblical creationists.
71. Anscombe, G.E.M., "Whatever has a beginning of existence must have a cause": Hume's argument exposed, *Collected Philosophical Papers*, Volume 1, Basil Blackwell, 1981. Return to text.
72. Anscombe, G.E.M., Times, beginnings and causes, in her *Collected Philosophical Papers*, Volume 2, Basil Blackwell, 1981.
73. Feser, E., *The Last Superstition: A Refutation of the New Atheism*, Kindle Locations 5253-5254, St. Augustine's Press. Kindle Edition, 2012.
74. Lemley, B., Guth's grand guess, *Discover* 23(4):32–39, April 2002; discovermagazine.com.
75. Krauss, M., *A Universe from Nothing: Why There is Something Rather than Nothing*, Free Press, 2012. See also review by Reynolds, D.W., *J. Creation*, 27(1):30–35, 2013.
76. Albert, D., On the Origin of Everything: review of *A Universe From Nothing*, by Lawrence M. Krauss, *New York Times*, 23 March 2012; nytimes.com.
77. Feser, E., Not Understanding Nothing, A review of *A Universe from Nothing, First Things*, June/July 2012; firstthings.com.

BLACKHOLE – SPACE – TIME WARP

UNDERSTANDING BLACK HOLE

When a particularly large star nears the end of its life, it often becomes a black hole.

A combined array of planet-scale of 8-ground-based radio telescopes forged through international collaboration — was designed to capture images of a black hole.

In April, 2019, EHT researchers unveiled the first direct visual evidence of the supermassive black hole in the center of Galaxy-Messier 87 and its shadow, Figure 14.1.

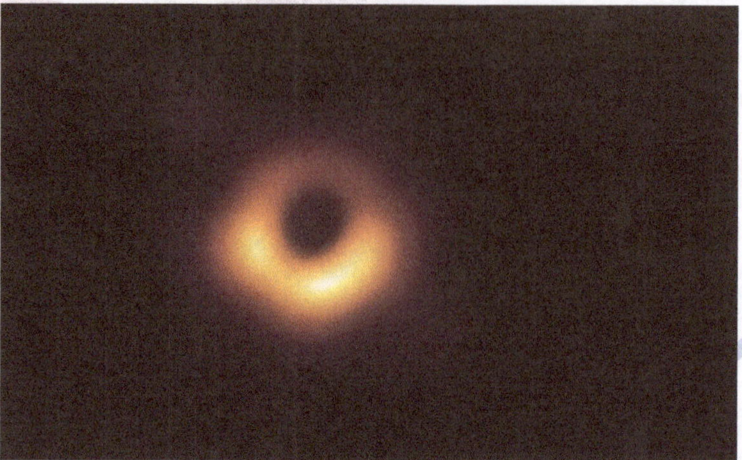

Fig.14.1: First – Image of a Black Hole. The Event Horizon Telescope (EHT) — Curtsy - Event Horizon Telescope Collaboration, European Southern Observatory (ESO) press release, April 2019.

Physicists and astronomers have been theorizing and studying black holes for over 200 years. These scientists can help us better understand the stellar evolution of black holes, which in turn can help us understand the dynamics and mechanics of our universe. On April 10, 2019, a network of radio-telescopes called the "Event Horizon Telescope (EHT)" captured and delivered our first image of a black hole! This significant event opens exciting new paths and fields to explore in the study of black holes.

Black Holes were once massive stars. Stars emit heat and light because of nuclear reactions that take place in their cores. Hydrogen atoms are converted into helium atoms, and the energy that results from the reaction causes the star to shine. The energy also helps the star keep its spherical shape. While gravity works to collapse a star onto itself, the nuclear energy counteracts the gravitational force and maintains the shape of the star.

After a lifetime spent fusing elemental components into helium and heavier atoms (all the way up to iron), the star will eventually run low on fuel. This means that fewer nuclear reactions take place in the star and there is too little energy to fight against gravity. When this happens, smaller stars like the Sun turn into white dwarves – incredibly dense stars that concentrate all of the star's mass into a volume about the size of the Earth. This type of stellar evolution hypothetically can take place over billions of years.

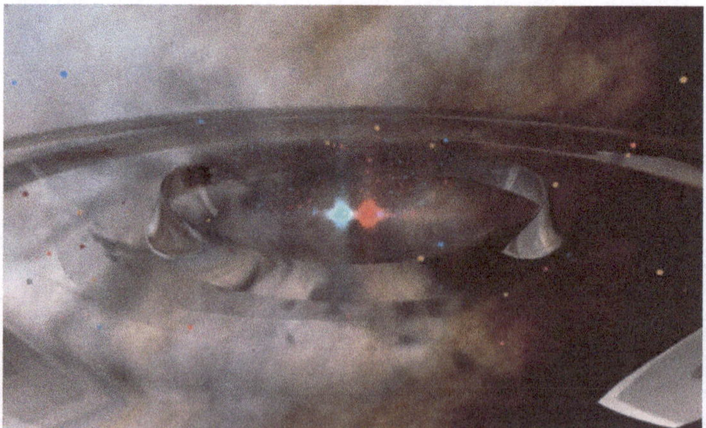

Fig.14.2: Two Black Holes Collide. Curtsy - 2017. NASA Goddard Space Flight Center

When two black holes collide, Figure 14.2, they release massive amounts of energy in the form of gravitational waves that last a fraction of a second and can be heard throughout the universe – if you have the right acoustic instruments.

Our Sun is not big enough to become a black hole, even after a few billion years. Much larger stars that range from being four to a billion times more massive than the Sun can turn into black holes at the end of their lives. Bigger stars "burn" their hydrogen fuel much more quickly, leading to stellar lifespans of only tens of millions of years, Figure 14.2.

These larger stars have so much material in them that when they burn out, their intense gravity will collapse them into black holes. The strong gravity just keeps pulling and pulling, compressing matter into a smaller and smaller volume. At some point, the gravity field compresses the stellar remnant into a "black hole." The name comes from the fact that even light cannot escape the pull of this intense gravity field if it passes within the distance from the black hole known as the "event horizon."

Fig.14.3: *History of Chandra X-Ray Observatory. A 2-week Observation Through the Optic Eye of the Chandra X-Ray Observatory – Curtsy 2003. NASA Marshall Space Flight Center revealed this stunning explosion occurring in the super massive black hole at the Milky Way's center, known as Sagittarius A or Sgr A**

Black holes can be observed and detected in several ways. They exert a strong gravitational effect on their surroundings, and we can measure the motions of objects that orbit them, like other stars. Also, the black hole's gravity is so strong that light that passes near their "event horizon" can get refracted, as though by a lens, altering the positions or appearance of stars and galaxies farther away from Earth than the black hole, Figure 14.3.

Black holes are often surrounded by other material, such as dust or other stars. The black hole's intense gravity can pull apart this material, generating intense radio, x-ray, and gamma radiation, and some of this occurs above the event horizon, where it can escape and be detected and measured.

In addition, observations have recently demonstrated the occurrence of "gravitational radiation," first predicted by Einstein, occurring when black holes pass near one another or collide, essentially pushing a wave of stretched and compressed space-time through the universe.

For many years, black holes could be detected, and their effects measured, but they were not directly observed. But in 2019 a black hole and its shadow were captured in an image for the first time. This historic feat involved a network of eight ground-based radio telescopes around the globe, operating together as if they were one telescope the size of our entire planet. This network is known as the Event Horizon Telescope.

Fig.14.4: NASA Visualization Shows a Black Hole's Warped World. Curtsy - 2019. Jeremy Schnittman, NASA Goddard Space Flight Center

This visualization of a black hole illustrates how its gravity distorts our view, warping its surroundings as if seen in a carnival mirror. The visualization simulates the appearance of a black hole where infalling matter has collected into a thin, hot structure called an accretion disk. The black hole's extreme gravity skews light emitted by different regions of the disk, producing the misshapen appearance, Figure 14.4.

Despite everything that has been discovered about black holes, scientists still have many exciting and unanswered questions - what are conditions like inside the event horizon, especially near the "singularity" at the black hole's center? In this region, the high density of material creates conditions unlike anywhere else in the known universe. Scientists have also theorized that black holes may "evaporate" over time, slowly dissipating and ultimately vanishing, but this phenomenon has never been observed directly. In addition, scientists are studying whether information can be destroyed by entering a black hole, or if there are mechanisms by which the information can be "recycled" and retained in the visible universe. These questions, and many more, will keep observers and scientists occupied for decades to come.

A 2-week observation through the optic eye of the Chandra X-Ray Observatory revealed this stunning explosion occurring in the Ultra-massive black hole at the Milky Way's center, known as ESA/Hubble Figure 14.5.

Black holes can be observed and detected in several ways. They exert a strong gravitational effect on their surroundings, and we can measure the motions of objects that orbit them, like other stars. Also, the black hole's gravity is so strong that light that passes near their "event horizon" can get refracted, as though by a lens, altering the positions or appearance of stars and galaxies farther away from Earth than the black hole.

Black holes are often surrounded by other material, such as dust or other stars. The black hole's intense gravity can pull apart this material, generating intense radio, x-ray, and gamma radiation, and some of this occurs above the event horizon, where it can escape and be detected and measured.

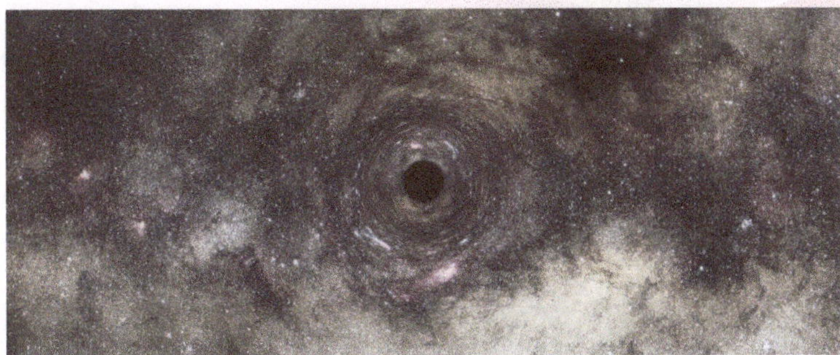

Fig.14.5: Ultra-massive Black Hole Discovered to Be 33 billion Times More Massive Than the Sun showing how the gravitational field warps light traveling from behind the black hole.– Curtsy (ESA/ Hubble, Digitized Sky Survey, Nick Risinger/skysurvey.org, N. Bartmann)

In addition, observations have recently demonstrated the occurrence of "gravitational radiation," first predicted by Einstein, occurring when black holes pass near one another or collide, essentially pushing a wave of stretched and compressed space-time through the universe.

For many years, black holes could be detected, and their effects measured, but they were not directly observed. But in 2019 a black hole and its shadow were captured in an image for the first time. This historic feat involved a network of eight ground-based radio telescopes around the globe, operating together as if they were one telescope the size of our entire planet. This network is known as the Event Horizon Telescope.

Fig.14.6: NASA Visualization Shows a Black Hole's Warped World. Curtsy - 2019. Jeremy Schnittman, NASA Goddard Space Flight Center.

This visualization of a black hole illustrates how its gravity distorts our view, warping its surroundings as if seen in a carnival mirror, Figure 14.6. The visualization simulates the appearance of a black hole where infalling matter has collected into a thin, hot structure called an accretion disk. The black hole's extreme gravity skews light emitted by different regions of the disk, producing the misshapen appearance.

Despite everything that has been discovered about black holes, scientists still have many exciting and unanswered questions – what are conditions like inside the event horizon, especially near the "singularity" at the black hole's center? In this region, the high density of material creates conditions unlike anywhere else in the known universe. Scientists have also theorized that black holes may "evaporate" over time, slowly dissipating and ultimately vanishing, but this phenomenon has never been observed directly. In addition, scientists are studying whether information can be destroyed by entering a black hole, or if there are mechanisms by which the information can be "recycled" and retained in the visible universe. These questions, and many more, will keep observers and scientists occupied for decades to come.

BACK HOLE – SPACE/TIME WARP

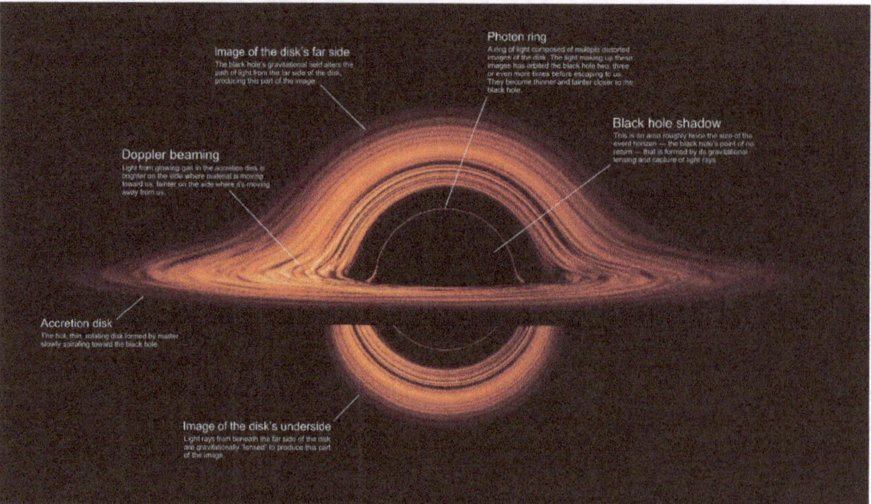

Fig.14.7: NASA Visualization of a Black Hole's Warped World – Curtsy NASA's Goddard Space Flight

This NASA's new visualization of a black hole illustrates how its gravity distorts our view, warping its surroundings as if seen in a carnival mirror. The visualization simulates the appearance of a black hole where infalling matter has collected into a thin, hot structure called an accretion disk. The black hole's extreme gravity skews light emitted by different regions of the disk, producing the misshapen appearance, Figure 14.7.

Bright knots constantly form and dissipate in the disk as magnetic fields wind and twist through the churning gas. Nearest the black hole, the gas orbits at close to the speed of light, while the outer portions spin a bit more slowly. This difference stretches and shears the bright knots, producing light and dark lanes in the disk.

Seen nearly edgewise, the turbulent disk of gas churning around a black hole takes on a crazy double-humped appearance. The black hole's extreme gravity alters the paths of light coming from different parts of the disk, producing the warped image. The black hole's extreme gravitational field redirects and distorts light coming from different parts of the disk, but exactly what we see depends on our viewing angle. The greatest distortion occurs when viewing the system nearly edgewise.

Viewed from the side, the disk looks brighter on the left than it does on the right. Glowing gas on the left side of the disk moves toward us so fast that the effects of Einstein's relativity give it a boost in brightness; the opposite happens on the right side, where gas moving away from us becomes slightly dimmer. This asymmetry disappears when we see the disk exactly face on because, from that perspective, none of the material is moving along our line of sight.

Closest to the black hole, the gravitational light-bending becomes so excessive that we can see the underside of the disk as a bright ring of light seemingly outlining the black hole. This so-called "photon ring" is composed of multiple rings, which grow progressively fainter and thinner, from light that has circled the black hole two, three, or even more times before escaping to reach our eyes. Because the black hole modeled in this visualization is spherical, the photon ring looks nearly circular and identical from any viewing angle. Inside the photon ring is the black hole's shadow, an area roughly twice the size of the event horizon — its point of no return.

GRAVITY WARPS SPACE AND TIME

"Simulations and movies help us visualize what Einstein meant when he said that gravity warps the fabric of space and time." Until very recently, these visualizations were limited to our imagination and computer programs. I never thought that it would be possible to see a real black hole. Yet on April 10, the Event Horizon Telescope team released the first-ever image of a black hole's shadow using radio observations of the heart of the galaxy M87.

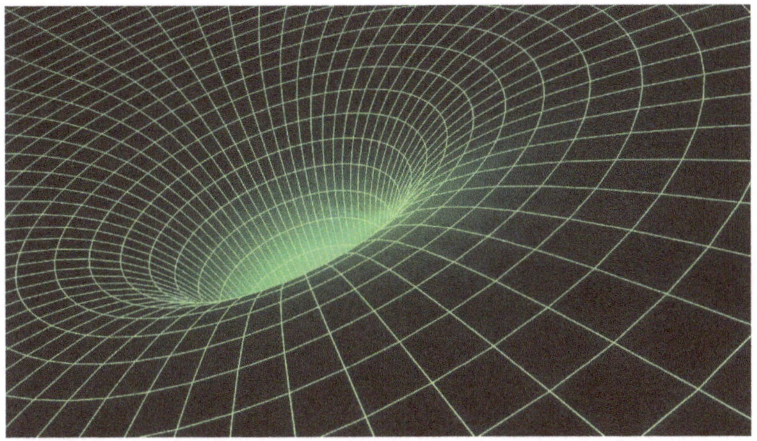

Fig.14.8: An Illustration of Heavily Curved Spacetime, Outside the Event Horizon of a Black Hole. Curtsy - JohnsonMartin/Pixabay

As you get closer and closer to the mass's location, space becomes more severely curved, eventually leading to a location from within which even light cannot escape: the event horizon. At large distances, the spatial curvature is indistinguishable for equal mass black holes, neutron stars, white dwarfs, or any other comparably massed object, Figure 14.8.

One of the most mind-bending concepts about the Universe itself is that gravity isn't due to some unseen, invisible force, but comes about because the matter and energy in the Universe bends and distorts the very fabric of space itself. Matter and energy tell space how to curve, and that curved space lays out the path upon which matter and energy move. The distance between two points isn't a straight line, but a curve determined by the fabric of space itself.

So where would you go if you wanted to find the regions of space that had the greatest amount of curvature? You'd pick the locations where you had the most mass concentrated into the smallest volumes: black holes. But not all black holes are created equal. Paradoxically, it's the smallest, lowest-mass black holes that create the most severely curved space of all. Here's the surprising science behind why.

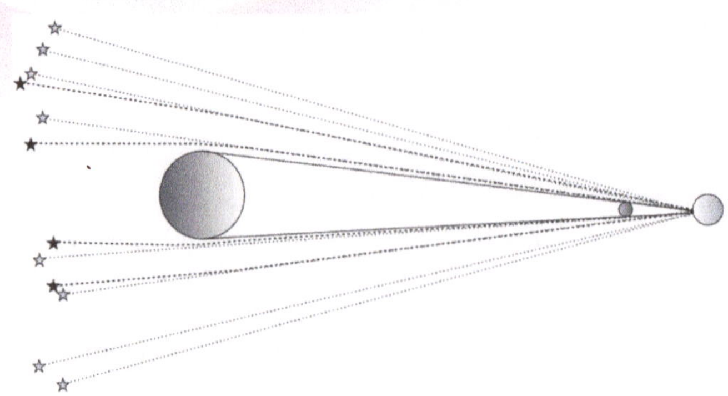

Fig.14.9: During a Total Eclipse, Stars would Appear to be in a Different Position than their Actual Locations. Curtsy E. Siegel/Beyond the Galaxy

This appearance is due to the bending of light from an intervening mass: the Sun. The magnitude of the deflection would be determined by the strength of the gravitational effects at the locations in space through which the light rays passed, Figure 14.9.

When we look out at the Universe, particularly on large cosmic scales, it behaves as though space were virtually indistinguishable from flat. Masses curve space, and that curved space deflects light, but the amount of deflection is minuscule even for the most concentrated amounts of mass we know of.

The solar eclipse of 1919, observed by Albert Einstein and a team of observing scientists, where the light from distant stars was deflected by the Sun, caused the path of light to bend by less than a thousandth-of-a-degree. This was the first observational confirmation of General Relativity, caused by the largest mass in our Solar System.

Gravitational lensing goes a step beyond that, where a very large mass (like a quasar or galaxy cluster) bends space so severely that the background light gets distorted, magnified, and stretched into multiple images. Yet even trillions of solar masses cause effects on scales of tiny fractions-of-a-degree.

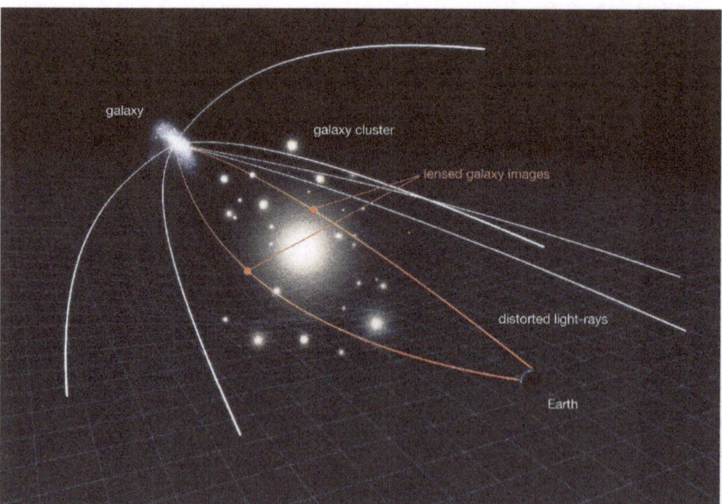

Fig.14.10: An illustration of Gravitational Lensing Illustrating How Background Galaxies—or any Light Path is Distorted —Curtsy - NASA, ESA & L. Calçada

The distortion is by the presence of an intervening mass, but it also shows how space itself is bent and distorted by the presence of the foreground mass itself. When multiple background objects are aligned with the same foreground lens, multiple sets of multiple images can be seen by a properly-aligned observer, Figure 14.10.

But it's neither our proximity to a mass nor the total amount of mass that determines how severely space is curved. Rather, it's the total amount of mass that's present within a given volume of space. The best way to visualize this is to think about our Sun: a 1 solar-mass object with a radius of about 700,000 kilometers. At the very limb of the Sun, 700,000 km from its center, light deflects by about 0.0005 degrees.

You could compress the Sun into about the size of Earth (similar to a white dwarf): about 6,400 km in radius. Light grazing this object's limb would deflect by about 100 times as much: 0.05 degrees.

You could compress the Sun into about a ~35 km radius (similar to a neutron star). Light grazing its limb would deflect a lot: by about a dozen degrees.

Or you could compress the Sun so much it becomes a black hole: with a radius of about 3 km. Light grazing its limb would be swallowed, while light just outside it could deflect by 180° or even more.

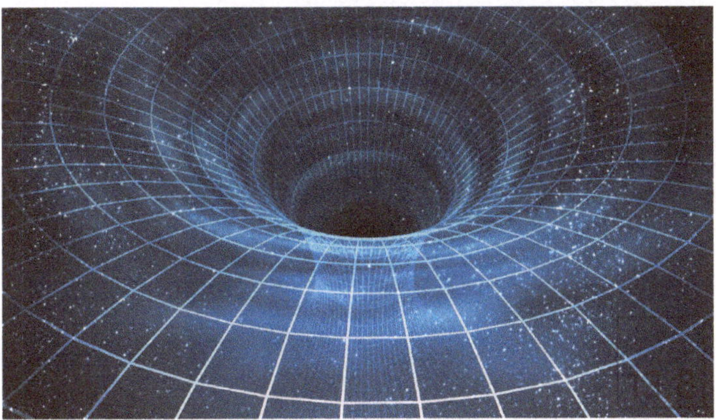

Fig.14.11: Once You Cross the Threshold to Form a Black Hole, Everything Inside the Event Horizon Crunches Down to a Singularity That is, at Most, One-dimensional. No 3D Structures can Survive Intact. Curtsy vchalup / Adobe Stock

But there's something important to think about in all of these scenarios. The total amount of mass—whether you have a Sun-like star, a white dwarf, a neutron star, or a black hole—is the same in each problem. The reason that space is more severely curved is because the mass is more concentrated, and you're able to approach it much more closely, Figure 14.11.

If you instead stayed at the same distance from the center-of-mass in each scenario, 700,000 km away from a 1 solar mass object regardless of how compact it was, you'd see the exact same deflection: about 0.0005 degrees. It's only because we can get very close to the most compact masses of all, i.e., black holes, that light deflects by such a severe amount as it grazes its limb.

This is a universal property of all black holes. When light just barely grazes the outside of the event horizon, it's right on the border of getting swallowed, and it will maximally be bent around the outskirts of the black hole.

The gravitational bending and capture of light by the event horizon is the cause of the shadow captured by the Event Horizon Telescope. The photons that aren't captured create a characteristic sphere, and that helps us confirm General Relativity's validity in this newly-tested regime, Figure 14.12.

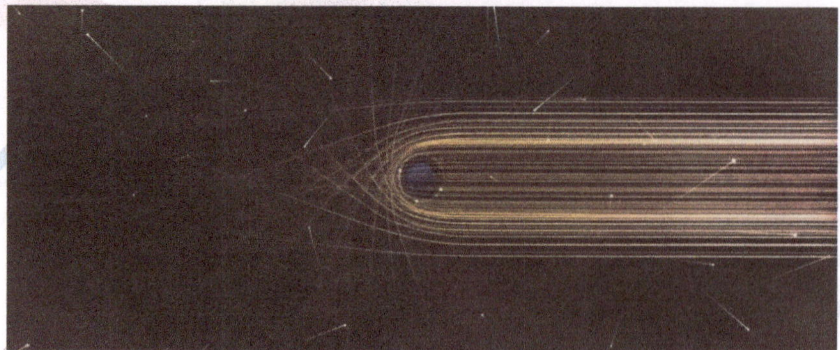

Fig.14.12: This Artist's impression Depicts the Paths of Photons in the Vicinity of a Black Hole. Credit - Nicolle R. Fuller/NSF

But not all black holes are created equal. Sure, there are some metrics by which every black hole looks the same, and those are important. Every black hole has an event horizon, and that horizon is defined by the location where the speed you'd need to travel in order to escape from its gravitational pull exceeds the speed of light. From outside the horizon, light can still make it to locations in the outside Universe; inside the horizon, that light (or any particle) gets swallowed by the black hole.

But the more massive your black hole is, the larger in radius its event horizon is. Double the mass and the radius of the event horizon doubles. Sure, much of things will scale the same way:

- the escape velocity at the horizon is still the speed of light,
- the amount of light deflection follows the same mass-and-radius relationship,
- and—if we could image them all directly—they'd all exhibit that same donut-like shape we saw from the Event Horizon Telescope's first image.

They are revealed by the Event Horizon Telescope in a galaxy some 60 million light-years away. The dotted line represents the edge of the photon sphere, while the event horizon itself is interior even to that, Figure 14.13.

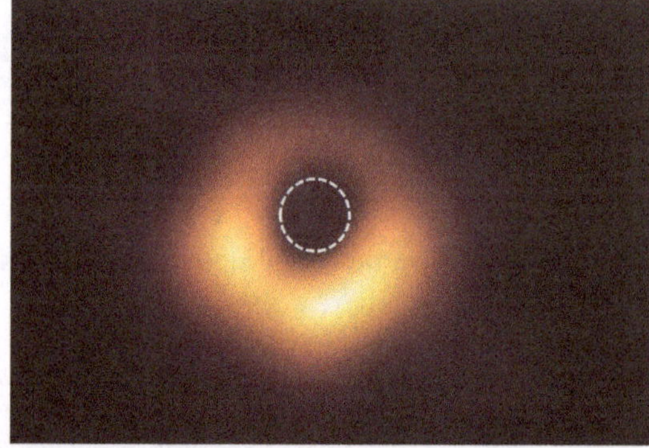

Fig.14.13: The features of the Event Horizon itself, Silhouetted Against the Backdrop of the Radio Emissions from Behind it. Curtsy - Event Horizon Telescope Collaboration et al.; Annotation: E. Siegel

But there are a few properties that aren't comparable for black holes of different masses. Tidal forces, for example, are a case where the differences are enormous. If you were to fall toward the event horizon of a black hole, you'd experience forces that would attempt to tear you apart by stretching you in the direction of the black hole's center while simultaneously compressing you in the perpendicular direction: spaghettification.

If you fell into the black hole at the center of the galaxy M87 (the one imaged by the Event Horizon Telescope), the difference between the force on your head and the force on your toes would be tiny, less than 0.1% of the force of Earth's gravity. But if you fell into a black hole with the mass of the Sun, the force would be many quintillions of times as great: enough to tear your individual atoms apart.

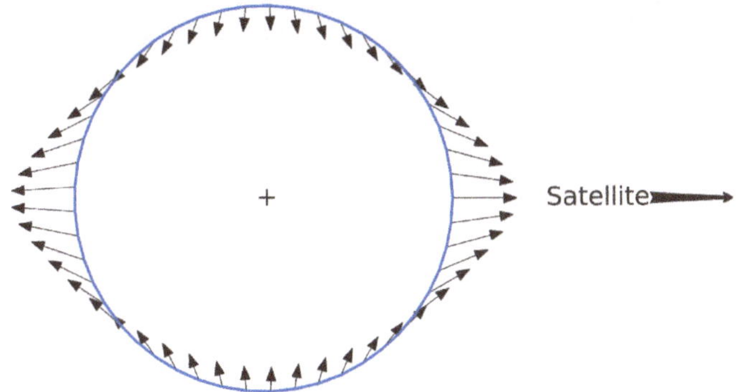

Fig.14.14: When Two Objects of a Finite Size Gravitationally Attract One another, the Gravitational Force on Different Parts of the Object are Different from the Average Value. Curtsy - Krishnavedala/ Wikimedia Commons

This effect causes what we see and experience as tidal forces, which can get extremely large at short distances, Figure 14.14.

Perhaps the most striking difference between black holes of different masses, however, comes about from a phenomenon we've never actually observed: Hawking radiation. Wherever you have a black hole, you have a very small amount of low-energy radiation being emitted from it.

HAWKING RADIATION

Although we've concocted some very pretty visualizations of what causes it—we typically talk about the spontaneous creation of particle-antiparticle pairs where one falls into the black hole and one escapes—that's not what's really going on. It is true that radiation is escaping from the black hole, and it's also true that the energy from that radiation has to come from the mass of the black hole itself. But this naive picture of particle-antiparticle pairs popping into existence and one member escaping is grossly oversimplified.

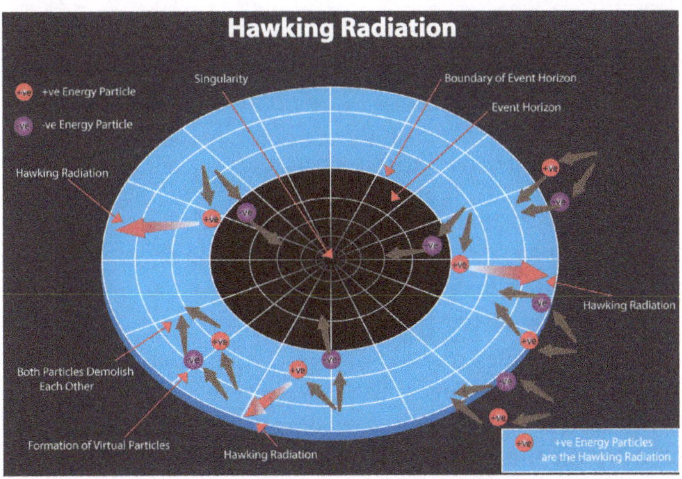

Fig.14.15: The Most Common, and Incorrect, Explanation for How Hawking Radiation Arises is an Analogy with Particle-Antiparticle Pairs. Curtsy - Physics Feed

If one member with negative energy falls into the black hole's event horizon, while the other member with positive energy escapes, the black hole loses mass and outgoing radiation departs the black hole. This explanation has misinformed generations of physicists, and came from Hawking himself, Figure 14.15.

QUANTUM VACUUM

The real story is a little more complicated, but vastly more illuminating. Wherever you have space itself, you also have the laws of physics that exist in our Universe, which includes all the quantum fields that underlie reality. These fields all exist in their lowest-energy state when they permeate empty space, a state known as "the quantum vacuum."

The quantum vacuum is the same for everyone so long as they're in empty, uncurved space. But that lowest-energy state is different in places where the spatial curvature is different, and that's where Hawking Radiation actually comes from: from the physics of quantum field theory in curved space. Far enough away from anything, even a black hole, the quantum vacuum looks like it does in flat space. But the quantum vacuum differs in curved space, and differs more dramatically where space is more severely curved.

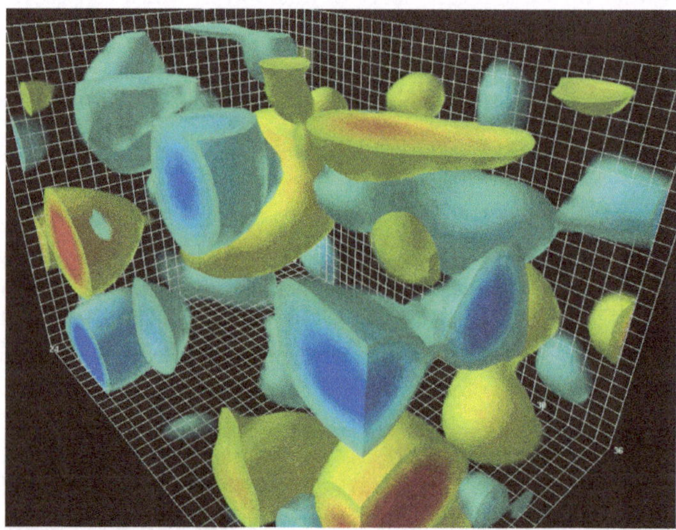

Fig.14.16: Visualization of a Quantum Field Theory Calculation Showing Virtual Particles in the Quantum Vacuum. Curtsy - Derek Leinweber

(Specifically, for the strong interactions.) Even in empty space, this vacuum energy is non-zero, and what appears to be the 'ground state' in one region of curved space will look different from the perspective of an observer where the spatial curvature differs. As long as quantum fields are present, this vacuum energy (or a cosmological constant) must be present, too, Figure 14.16.

That means, if we want the brightest, most luminous, most energetic Hawking radiation to come from our black hole, we'd want to go to the lowest-mass black holes we can find: the ones where the spatial curvature at their event horizon is the strongest. If we were to compare a black hole like the one at the center of M87 with the imaginary one we'd have if the Sun became a black hole, we'd find:

- the more massive black hole has a temperature that's billions of times lower,
- has a luminosity that's ~20 orders of magnitude lower,
- and will evaporate on timescales that are ~30 orders of magnitude longer.

This means that it's the lowest-mass black holes of all that are the locations where space is the most strongly-curved out of all the places in the Universe, and—in many ways—make for the most sensitive natural laboratory to test the limits of Einstein's General Relativity.

This event followed by a neutron star product that then collapses into a black hole, a direct-to-black-hole merger may have occurred on April 25, 2019. The only two surefire neutron star-neutron star mergers both produced black holes in the end: one of about 2.7 solar masses and one of about 3.5 solar masses. They are the lowest-mass black holes to date in the known Universe, Figure 14.17.

Fig.14.17: *Instead of Two Neutron Stars Merging to Produce a Gamma Ray Burst and a Rich Plethora of Heavy Elements. Curtsy - National Science Foundation/LIGO/Sonoma State University/A. Simonnet*

It might seem counterintuitive to think that the lowest-mass black holes in the Universe curve space more severely than the supermassive behemoths that populate the centers of galaxies, but it's true. Curved space isn't just about how much mass you have in one place, because what you can observe is limited by the presence of an event horizon. The smallest event horizons are found around the lowest-mass black holes. For metrics like tidal forces or black hole decay, being close to the central singularity is even more important than your overall mass.

This means that the best laboratories for testing many aspects of General Relativity—and to search for the first subtle effects of quantum gravity—will be around the smallest black holes of all. The lowest-mass ones we know come from neutron stars that merge to form black holes, just 2.5-to-3 times the Sun's mass. The smallest black holes are where space is bent the most, and may yet hold the key to the next great breakthrough in our understanding of the Universe.

GRAVITY

The Mystery Force

Gravity holds us firmly on the ground and keeps the earth circling the sun. This invisible force also draws down rain from the sky and causes the daily ocean tides.

It keeps the earth in a spherical shape, and prevents our atmosphere from escaping into space.

It would seem that this everyday gravity force should be one of the best understood concepts in science. However, just the opposite is true. In many ways, gravity remains a profound mystery. Gravity provides a stunning example of the limits of current scientific knowledge.

What is gravity?

Isaac Newton asked this question in 1686, and concluded that gravity was an attractive force between all objects. He realized that the same force that causes an apple to fall to the ground also holds the moon in its orbit. Earth's gravity actually causes the moon to fall about one millimeter away from a straight-line path, each second, as it orbits the earth (Figure 14.18). Newton's universal Law of Gravity is one of the great science discoveries of all time.

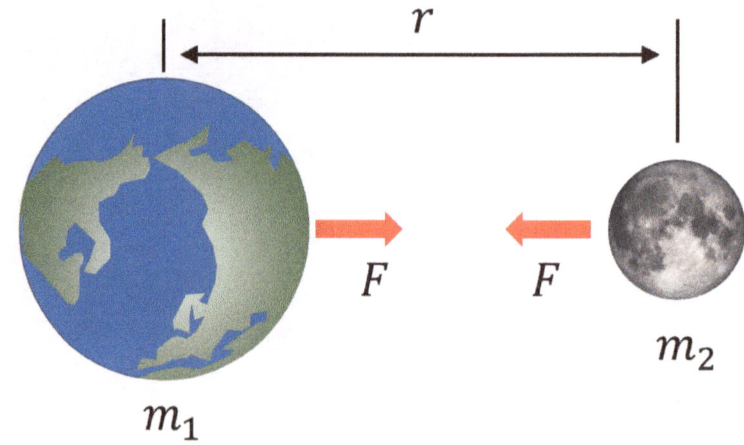

Fig.14.18: Earth Gravity Effect

Gravity is one of four known fundamental forces of nature (Table 14.1). Note that gravity is by far the weakest of the four, yet it dominates on the scale of large space objects. As Newton showed, the attractive gravity between any two masses gets less and less as the distance between them gets greater, but it never quite reaches zero.

Therefore, every particle in the entire universe actually attracts every other particle. Gravity is a long-range force in contrast to the strong and weak forces (Table 14.1). Magnetic and electric forces are also long-range, but gravity is unique in being both long-range, and always attractive, thus never cancelling out (unlike electromagnetism, where the forces may either attract or repel).

Table 14.1: The Four Fundamental Forces of Nature

Force name	Relative Strength	Responsible for
Strong	1	Stability of the atomic nucleus.
Electromagnetic	10^{-2}	Atomic, molecular bonding.
Weak	10^{-6}	Radioactive decay processes.
Gravity	10^{-43}	Stability of space objects.

Starting with the great creationist physicist "Michael Faraday" in 1849, physicists have searched continually for a hidden relationship between gravity and the electromagnetic force. There is an ongoing effort to unify all four fundamental forces into a single equation or 'theory of everything', with no success thus far, Figure 14.19. Gravity remains the least understood force.

Gravity cannot be shielded in any way. Intervening objects, whatever their make-up, have no effect whatsoever on the attraction between two separated objects. This means that no antigravity chamber can be built in the laboratory. Neither does gravity depend on the chemical composition of objects, but only on their mass, which we perceive as weight (the force of gravity on something is its weight — the greater the mass, the greater the force or weight.) Blocks composed of glass, lead, ice or even Styrofoam, if they all have equal mass, will experience (and exert) identical gravitational forces. These are experimental findings, with no underlying theoretical explanation.

Fig.14.19: Gravity in Space

What really is gravity? How is this force able to act across the vastness of empty space?

And why does it exist in the first place? Science has never been very successful in answering these most basic questions about nature. Gravity cannot somehow slowly arise by mutation or natural selection. It was present from the very beginning of the universe. Along with every other physical law, gravity is surely a testimony to a planned creation.

Attempts to explain gravity have included invisible particles, called gravitons, that travel between objects. Cosmic strings and gravity waves have also been suggested, but none have been confirmed. We simply do not know how objects physically interact with each other over vast distances.

GRAVITY AND SCRIPTURE

Two Bible references are helpful in considering the nature of gravity and physical science in general. First, Colossians 1:17 explains that ***Christ is before all things, and by Him all things consist***. The Greek verb for consist (συνιστάω *sunistaō*) means to cohere, preserve, or hold together. Extrabiblical Greek use of this word pictures a container holding water within itself. The word is used in Colossians in the perfect tense, which normally implies a present continuing state arising from a completed past action. One physical mechanism used is obviously gravity, established by the Creator and still maintained without flaw today. Consider the alternative; if gravity ceased for one moment, instant chaos surely would result. All heavenly objects, including the earth, moon and stars, would no longer hold together. Everything would immediately disintegrate into small fragments, Figure 14.20.

A second reference, Hebrews 1:3, declares that ***Christ upholds all things by the word of His power***. Uphold (φέρω *pherō*) again describes the sustaining or maintaining of all things, including gravity. The word uphold in this verse means much more than simply supporting a weight. It includes control of all the ongoing motions and changes within the universe. This infinite task is managed by the Lord's almighty Word, whereby the universe itself was first called into being. Gravity, the 'mystery force', which is poorly understood after nearly four centuries of research, is one of the manifestations of this awesome divine upholding.

Fig.14.20: In Him All Things Hold Together

Table 14.2: Some Values of the Gravitational Force of Attraction between Various Objects

Pairs of Objects	Gravity force (kg at sea level)
You and a magazine you are reading	4.5×10^{-10} (10^{-9} lbs)
You and the moon	.00045
Two adjacent locomotives	.0022
You and the earth (at sea level)	your weight
Moon and Earth	1.8×10^{19}
Earth and Sun	3.6×10^{21}
NB: kg are actually units of mass (the units of force are N = Newtons, where the force on a 1 kg weight at sea level = 9.8 N). The weight (or force) exerted by a given mass depends on how close it is to the earth's gravity, so here sea level is assumed, Table 14.2.	

Einstein Gravity

Einstein said it is not a force in the way others are, but an effect of the curvature of space-time. According to general relativity, the effects of gravity are not instantaneous, but transmitted at the speed of light.

SPACE-TIME WARPS AND BLACK HOLES

Einstein' theory of general relativity regards gravity not as a force, but a curvature of space itself near a massive object. Even light, which traditionally follows straight lines, was predicted to bend while travelling through curved space. This was first shown when the astronomer Sir Arthur Eddington detected a change of a star's apparent position during a total eclipse in 1919, consistent with the light rays' being bent by the sun's gravity, Figure 14.21.

Fig.14.21: Gravity Warps Time/Space

Few scientists simulated view of a black hole in front of the Large Magellanic Cloud. The gravitational lensing effect, produces two enlarged but highly distorted views of the Cloud. Across the top, the Milky Way disk appears distorted into an arc.

General relativity also predicts that if a body were dense enough, its gravity would curve space so strongly that light could not escape at all. Such a body would absorb light and anything else caught by its intense gravity, and so is called a black hole. Such a body could be detected only by its gravitational effects on other objects, strong bending of light around it, and by intense radiation emitted by matter falling in.

BLACK HOLE SINGULARITY

All the matter inside a black hole is compressed into a singularity of infinite density. So instead, its 'size' is defined by its event horizon, a boundary surrounding the singularity such that nothing, not even light, can escape from within the boundary. Its radius is called the Schwarzschild radius, after the German astronomer "Karl Schwarzschild" (1873–1916), and is given by the formula $R_S = 2GM/c^2$, where c is the speed of light in a vacuum. If the sun were to collapse into a black hole, its Schwarzschild radius would be only 3 km (2 miles), Figurde 14.22.

Fig.14.22: The Center of a Black Hole: Massive Singularity – Curtsy - mondolithic.com

There is good evidence that a very massive star, after most of its nuclear fuel runs out, would have nothing to counteract a collapse under its own huge weight into a black hole. Black holes with the mass of a billion suns are thought to exist in the centers of galaxies, including our own, the Milky Way. Many scientists believe that the super-bright and extremely distant objects called quasars are powered by the energy released as matter falls into a black hole.

General relativity predicts that gravity also distorts time. This has also been confirmed by extremely accurate atomic clocks ticking a few microseconds per year slower at sea level than at high altitudes where Earth's gravity is slightly weaker. Near an event horizon, the effect is far more marked. If we monitored the watch of an astronaut approaching an event horizon, we would observe the watch tick ever more slowly. At the event horizon, it would be stopped, but we would never see that happen. Conversely, the astronaut wouldn't notice anything different about his own watch, but he would observe our clocks ticking faster and faster.

ASTRONAUT SUBJECTED TO BLACKHOLE TIDAL WAVES

The main danger to an astronaut near a black hole is tidal forces, caused by the fact that gravity is stronger on the parts of the body closer to the black hole than on the parts away from it. The tidal forces near a black hole with a mass of a star are much stronger than a hurricane, intense enough to stretch an onlooker into tiny pieces. However, while gravitational attraction decreases with the square of the distance ($1/r^2$), the tidal effect decreases with the cube of the distance ($1/r^3$), Figure 14.23.

Fig.14.23: Black Hole Tidal Disruption Flares Write their Signature in Cosmic Dust – Curtsy Image courtesy of NASA/JPL-Caltech

So, contrary to popular imaginings, large black holes have weaker gravitational (including tidal) forces at their event horizons than small black holes. Accordingly, the tidal forces at the event horizon of a black hole the mass of the observable cosmos, for example, would be less noticeable than the gentlest breeze.

Gravitational time dilation at and near an event horizon is the basis of the new cosmological model of the creationist physicist Dr "Russell Humphreys," in his book Starlight and Time. This model seems to provide one possible solution to the problem of seeing distant starlight in a young universe, and is currently a scientific alternative to the unbiblical 'big bang' theory which relies on philosophical assumptions outside science; apart from Stephen Hawking's theoretical discovery in 1974 that black holes would eventually evaporate via *Hawking radiation* due to a quantum mechanical effect.

Gravity and Motion - Isaac Newton

Isaac Newton published his discoveries about gravity and motion in 1687, in his masterpiece Principia, Figure 14.24. Some readers quickly concluded that Newton's universe left no room for God, since everything now could be explained with equations. But this was not Newton's view, as he clarified in the Principia's second edition:

'Our most beautiful system of the sun, planets, and comets could only proceed from the counsel and dominion of an intelligent and powerful being.'

Isaac Newton was not only a scientist but also a life-long biblical scholar. His favorite Bible books were Daniel and Revelation, in which God's plans for the future are described. Newton actually wrote more about theology than science.

Fig.14.24: *Isaac Newton (1642–1727) - Curtsy Wikipedia.org*

Newton gave credit to others such as Galileo. Newton actually was born the same year that Galileo died, in 1642. Newton wrote in a letter, 'If I have seen further, it is by standing on the shoulders of giants.' Shortly before his death, perhaps thinking of the mystery of gravity, Newton humbly wrote, 'I do not know what I may seem to the world, but as to myself, I seem to have been only like a boy playing on the seashore … now and then finding a smoother pebble or a prettier shell than ordinary, while the great ocean of truth lay all undiscovered before me.'

Newton is buried in Westminster Abbey. His Latin epitaph ends with the sentence 'Let mortals rejoice that there existed such and so great an ornament to the human race.'

DESIGNER GRAVITY

The force F between two masses m_1 and m_2, when separated by a distance r, can be

written as: $$F_9 = \frac{Gm_1m_2}{r^2}$$

where G is the gravitational constant, first measured by Henry Cavendish in 1798.

This equation shows that gravity decreases as the separation distance, r, between two objects becomes large but never quite reaches zero.

The inverse-square nature of this equation is intriguing. After all, there is no essential reason why gravity should behave in this way. In a chance, evolving universe, some random exponent like $r^{1.97}$ or $r^{2.3}$ would seem much more likely. However, precise measurements have shown an exact exponent out to at least 5 decimal places, 2.00000. As one researcher put it, this result seems 'just a little too neat.' We may conclude that the gravity force shows precise, created design. Actually, if the exponent deviated just slightly from exactly 2, planet orbits and the entire universe would become unstable.

GRAVITY COLLAPSING STAR REMNANT

Sir John Herschel named this large and beautiful circular feature, known as the Keyhole Nebula, within the Carina Nebula, in the 19th century. Its distance is about 8,000 light-years from Earth. This system contains some of the most massive and hottest stars known, with masses approximately 100 times that of our sun, and 10 times as hot, Figure 14.25. Nebulae such as these may contain within them 'black holes', resulting from the collapse of a star remnant under its own gravity. Such regions would have gravity so powerful that even light rays could not escape.

Fig.14.25: *Supernova remnant from a star collapse - Curtsy - NASA / SKYWORKS DIGITAL / SCIENCE PHOTO LIBRARY*

TIME AND SPACE

Fig.14.26: God Inhabits Eternity

Modern science has no difficulty with the idea that the familiar three dimensions (length, breadth, height) are not all there is to reality. Comprehending this may make our concept of the Creator easier to understand.

The Bible tells us that God inhabits eternity (Isaiah 57:15), Figure 14.26, that He dwells in Heaven (e.g., *Deuteronomy 26:15*), and also in light (*1 Timothy 6:16*). For convenience we could sum this up by saying that God lives in the heavenly dimension or the dimensions of eternity. By contrast, we live in a world of three dimensions of space (length, breadth, depth) and one of time.

Although God is omnipresent (present everywhere) in both the dimensions of eternity and in our universe, He also appeared in our world in specific ways at specific times. For example, the Spirit of God moved upon the face of the waters at Creation (*Genesis 1:2*), Jehovah God appeared to Moses in the burning bush (*Exodus 3:4–14*), and God the Son left the eternal dimensions (*John 1:1–2; Philippians 2:5–7*) and entered our world in a special way at the Incarnation (*Matthew 1:18; Luke 1:35; John 1:14*). After the Resurrection, Jesus appeared to His disciples on several occasions, including twice in a locked room (*John 20:19–31*), and at the Ascension Jesus left our world and returned to the heavenly dimension (*John 20:17; Acts 1:9*).

IMPARTING CREATION INFORMATION AND EVENT HORIZON

The information necessary for creation to occur existed in the mind of God, the Logos (*John 1:1* Greek *logos*=Word), outside space and time, and before these were created. That is, all creation existed in the mind of God, 'before the foundation of the world' (*Ephesians 1:4; 1 Peter 1:20*), and thus in the dimensions of eternity. God passed the information (and power) needed for the creation to occur, across the **Event Horizon** that separates eternity from space-time, in separate quanta or amounts, throughout the six days recorded in Genesis 1.

When Adam was formed and until he sinned, he was a hybrid (dimensionally speaking). In the Garden of Eden, he walked with God, who inhabits eternity, and he thus experienced something of the multi-dimensions of eternity, but he also existed in space-time. When he disobeyed God, God expelled him from Eden to prevent him from eating from the tree of life and living forever (*Genesis 3:22, 24*) in a sinful condition.

However, man was made for eternity. Therefore, God has provided a way whereby we can be reunited with Himself. Whereas Adam's sin closed the **Event Horizon** to the heavenly dimension for mankind, the forgiveness of sin opens this event horizon. This forgiveness comes through the death and resurrection of the Lord Jesus Christ, who paid the penalty for sin; it is accessed only through individual repentance and faith. Those who access God's forgiveness in this way experience fellowship with the eternal God in this life (*John 1:12; 17:23*), in a much more limited way than Adam did originally, and they will be united with God in the next life.

JESUS PASSED THROUGH SPACE TIME UNIVERSE

When the Lord Jesus Christ rose from the dead, His resurrection body left this space-time universe and entered the dimensions of eternity, passing through the *grave clothes* without disturbing them. It was the sight of the napkin which had been wound around the head of Christ 'wrapped together in a place by itself' (i.e., *still wound, and stiffened by the sticky mixture of myrrh and aloes used in embalming*), which caused the Apostle John to believe in the resurrection (*John 19:39–40; 20:7–8*). Although John might not have been aware of dimension theory, he realized, in effect, that Christ had transcended the dimensions of time and space.

As with the footprint in Flatland, Christ could appear in the locked room, or appear to and disappear from the disciples (e.g., on the road to Emmaus), unaffected by the event horizon that totally prohibits us from performing the same actions.

Jesus Crossed the Event Horizon

At the Ascension (*Acts 1:9*), Jesus crossed the event horizon between space-time and Heaven. Now, when believers pray, they access the risen Lord Jesus, through the Holy Spirit, who dwells in believers by faith (*John 14:16–17*).

HEAVEN, HELL, AND JUDGMENT

When a person dies, his/her soul/spirit exits this space-time world by crossing the **Event Horizon** of death. This is the boundary at which space and time end and timelessness begins. In the Bible, God tells us that beyond death there are Heaven, Hell, and judgment to come ('it is appointed unto man once to die, but after this the judgment'—*Hebrews 9:27*). The latter decides which of the two former we end up in, based on our faith (in this life) in Christ's atoning death and resurrection.

Heaven, Hell, and the final judgment all exist or occur within the dimensions of eternity. They can therefore never be examined by science, because they are all hidden behind an **Event Horizon**. This, however, does not mean that they are not real. The concept is perfectly scientific, and they are in fact the ultimate reality. The materialistic credo propounded for so long by atheists, that the 'here and now' of our space-time continuum is the only reality, is thus seen to be scientifically unsound. Skeptics, atheists, and others ignore this at their peril.

BLACK HOLES AND EVENT HORIZONS

When astronauts went to the moon, their spaceship had to attain a speed of more than 11.2 km/sec (25,000 mph), called the escape velocity, so they could break loose from the pull of Earth's gravity. Now imagine a sufficiently massive star, undergoing gravitational collapse, until its gravity was so strong that the escape velocity was 300,000 km/sec (186,000 miles/sec), or the speed of light.

Einstein's theory of general relativity predicts that within such a region, space would be so curved that no light, or anything else, could escape. Therefore, the star would be invisible (and so detectable only by its gravitational effects on nearby matter). The result would be a black hole, defined by Stephen Hawking as: 'a region of space-time from which nothing, not even light, can escape, because gravity is so strong.'

If a ray of light is directed near a black hole from outside, it will be deflected by the gravitational pull of the black hole and, if sufficiently close, will go into orbit around the black hole. (If it is directed at the black hole, it will be absorbed.)

The boundary of a black hole, where light from outside goes into orbit, or light rays from inside the black hole just fail to escape, is called an **Event Horizon**. It represents the point of no return around a black hole. The reason for this name is that no event can ever get out of this area, as there is no means of communicating the occurrence of any events. Thus, whatever happens within a black hole cannot be known by an observer from outside and hence cannot be known by science. This phenomenon has been called '*cosmic censorship*.'

CREATION - CURVED SPACE-TIME

A decade ago, the COBE (Cosmic Background Explorer) satellite showed slight temperature fluctuations in the background radiation of space. The new satellite WMAP (Wilkinson Microwave Anisotropy Probe) has now revealed greater detail about these 'cosmic ripples.' One conclusion is that the overall geometry of space is flat. To better understand this description of space, this section reviews the terms space-time, warped space, and space curvature. These concepts were first introduced in 1915 by Albert Einstein's general theory of relativity in an attempt to explain gravity. He wondered how the attractive force of gravity could act between objects widely separated in space. Isaac Newton had also struggled with this question two centuries earlier. Today, efforts continue to understand the gravity force and Einstein's explanation in terms of space curvature.

SPACE-TIME

Space itself refers to the familiar three dimensions of length, height, and width. Higher, unseen dimensions of space also may exist according to current thinking. Einstein's relativity theory fuses space with the passage of time. In this way time itself becomes a fourth dimension of the space-time continuum.

Time (t) can be given the units of distance by multiplying by light speed c, ct. This space-time connection is made because the actual length of a time interval depends on one's location in space.

If positioned near a massive object, for example, time will pass more slowly than it does in empty space. The effect is called gravitational dilation or the stretching of time. This time alteration was measured in 1962 using atomic clocks. Identical clocks were placed at the top and bottom of a water tower. The clock at the bottom of the tower, slightly closer to the massive Earth, ran nanoseconds slower than the elevated clock. Another way to express the space-time connection is that local time is defined in terms of the distribution of nearby matter. Clocks also run slower when they are moving at very high speeds, a separate aspect of relativity theory.

CURVED SPACE

Remember how a curved mirror distorts your image in a funhouse? Einstein's gravity theory predicts that matter or energy likewise distorts the fabric of nearby space. It is as if the familiar straight lines of length, height, and width can be bowed or twisted. As often happens, the mathematics were already in place when needed later to describe nature. The equations for such non-Euclidean geometry were first published by Bernhard Riemann in 1854. Hermann Minkowski also prepared the way for Einstein by developing space-time ideas in 1907. This ongoing development of intricate mathematical relations, in anticipation of future applications, is clear evidence of intelligent design and mathematical precision in nature.

A two-dimensional analogy is helpful in considering the meaning of curved space. Think of the flat surface of a waterbed. Place a bowling ball on the bed and it will sink into the resulting depression. The entire mattress surface becomes curved or warped downward, most noticeably near the ball. If a marble is now dropped unto the bed it will roll toward the bowling ball. The marble also may be drawn into a spiral motion around the bowling ball, somewhat like a planet orbiting the sun.

The waterbed comparison with curved space is helpful but is deficient in at least three ways. First, the two-dimensional mattress surface is deformed into the third familiar dimension of depth. However, the curvature of space is into a fourth dimension and cannot be visualized. If space curvature indeed occurs, we do not observe it because we are embedded in space. A second problem, the bowling ball sinks downward into the bed due to the Earth's gravity pull. In space, however, the gravitation of matter distorts its space surroundings whether or not any other objects are present. Einstein reasoned that this warping of space was itself an expression of gravity. In the solar system, planets are said to then move along orbit-shaped geodesic slopes or depressions in space caused by the sun. Matter tells space how to curve, and the resulting space curvature in turn tells objects how to move. A third shortcoming of the waterbed analogy is that massive objects also affect the passage of time in their vicinity.

EVIDENCE FOR CURVED SPACE

The initial evidence, which catapulted Einstein to fame, came from a 1919 eclipse of the sun. Astronomer "Arthur Eddington" made a special trip to the coast of West Africa for observations. During the eclipse, starlight was found to be deflected slightly by the sun, just 0.0005 degrees, consistent with Einstein's prediction. The simplest interpretation is that space is slightly disturbed in the sun's vicinity. In recent years there have been additional examples of gravitational deflection or lensing of distant starlight.

In some cases, the distortions of space appear to separate light source into multiple images. A twisting or dragging of space has also been reported in the vicinity of dense, spinning neutron stars. The very fabric of nearby space appears to become tangled and 'wound up', similar to batter being twirled by an eggbeater. Not all scientists accept the concept of curved space but it seems a plausible explanation of the data.

RELATED IDEAS

The term space curvature is also used to describe the overall geometry of the universe. If the universe is closed, its curvature then is said to be positive. Such a universe, if left to itself, would eventually stop expanding due to the gravity of all the matter in it, and collapse inward. Also, in a closed universe, parallel lines will eventually meet. If you travel outward in a straight line in such a universe, you would eventually return to your starting point. On the other hand, if the universe is open, its curvature is said to be negative and such a universe, left to itself, will expand forever. Parallel lines will diverge at great distances, and in straight-line travel you would never return to your origin. Thus far, measurements from the WMAP satellite suggests that the universe lies directly between the closed and open extremes, a geometry called flat, Figure 14.27.

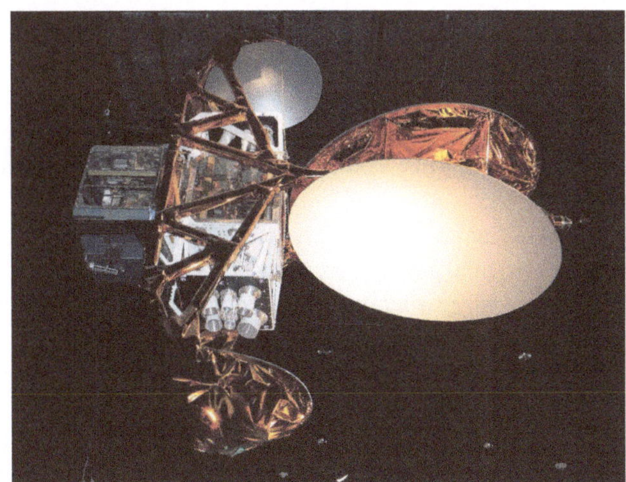

Fig.14.27: WMAP Satellite – Curtsy NASA

The discussion of the nature of space also includes exotic objects such as gravity waves (ripples in space/time), cosmic superstrings, white holes, worm holes and black holes. None of these items have been detected with certainty, including black holes. If black holes actually exist, they are locations of extreme space curvature where matter and light have become trapped. All these strange features may exist in deep space, along with other unknown objects not yet thought of. The universe surely contains many unknowns and surprises.

CREATION IMPLICATIONS

Some scientists have suggested models where the gravitational disturbance of space-time may help us understand the literal Creation Week. In such models, while 24-hour days passed in Earth's reference frame, billions of years of history actually transpired in deep space. The assumption is that there were greatly different time scales depending on one's location in space.

Such models also raise two interesting issues. First, just how far should we try to extend the current physical laws of the universe's operation to explain its origin during the Creation Week? We need to be aware that applying today's science to the initial events of creation may not be valid since supernatural activity took place on a grand scale during Creation Week. With regard to space-time, God may have added to the natural laws of operation by supernaturally stretching space.

The second issue concerns the extent to which time may be stretched by gravity. Accounting for deep time in space by gravitational time stretching, 10–15 billion years of history, is an extrapolation that is 1028 times greater than that observed so far with atomic clocks. Of course, we haven't observed such changes on Earth today, because gravity is so weak. But general relativity specialists agree that there is no limit to the time dilation—for example at the event horizon of a black hole, time stops completely. Therefore, an appropriate creationist cosmology can still make use of the principle. Humphreys' cosmology, for example, posits that during Creation Week Earth was inside such an event horizon, except of a 'white hole'—a black hole running in reverse. Some may wonder if it would be possible in future to manipulate clocks by compressing or stretching time scales. Could a person, in this way control his own destiny? However, Psalm 31:15 declares that 'My times are in his [God's] hand.' If the warping of space and time do indeed occur, it must be by God's direction. All relativistic time changes measured thus far are very small, only a microsecond or less, though they are real changes. This is somewhat similar to quantum mechanical effects which become significant only on the microscopic level.

For astronomers who are uncomfortable with a beginning for the universe, even a big bang beginning 10–15 billion years ago, the latest WMAP conclusion that the universe is flat may be something of a disappointment. They would probably prefer an eternal universe which continually oscillates inward and outward.

Three centuries ago, Isaac Newton wondered about the cause of gravity. More recently, Einstein proposed that the measured gravity force is actually caused by matter distorting space-time. However, the basic question still remains why matter distorts space in the first place. Gravity, the 'glue' which holds the universe together, remains a profound mystery.

BIBLICAL ASTRONOMY

Earth and Universe Center

It is popular today to deny any special recognition for the earth. Secular scientists tell us that we live on a speck of dust, circling a humdrum star in a far corner of an obscure galaxy! While this is all true, the earth remains of central importance. Scripture gives a refreshing contrast to secular thinking by declaring that earth and mankind are not an insignificant result of accidental evolution. The Book of Genesis states that our planet was created three days before the sun, moon, and stars. The purposes of the stars relate directly to the earth:

to provide a calendar system (*Gen. 1:14*) and to declare God's glory to men (*Ps. 19:1*). The earth is also a universal reference point in that Christ came here to walk among men, and will one day return. An unseen spiritual battle goes on for the souls of men, focusing on this earth and extending to high places (*Eph. 6:12*).

The earth is truly a spiritual center of the universe. It was once thought that the earth was physically located at the exact center of the universe, and furthermore that it did not move. This "geocentric" view is still held today by a few people, although Scripture does not require it and observation clearly shows the earth's movement. The earth revolves around the sun once a year. It also rotates on its axis every 24 hours. These motions add together in a dizzy combination. The earth's spin results in a surface speed of 1,000 miles per hour at the equator. The speed of the earth due to orbital motion around the sun is 66 times greater still. This is 30 times faster than a rifle bullet. During an average human's lifetime (70 trips around the sun), 41 billion miles are traveled. While you read this page, the earth has already traveled more than 1,000 miles!

Fortunately, we do not directly notice this motion since the earth's faithful gravity force ensures that both its atmosphere and inhabitants remain firmly in place. However, earth's motion is clearly shown by the westward movement of the sun, moon, and stars through the sky (see table 14.3).

We really don't know where the physical center of the universe is. If God's heavens are infinite in extent, then no center actually exists. But the question of the earth's physical position is less important than the spiritual reality of God's love for his people.

Table 14.3: Earth's Major Motions

Motion	Speed
Rotation on axis	1,000 miles/hour (at equator)
Revolution around the sun	66,600 miles/hour
Solar-system travel around the galaxy	500,000 miles/hour
Overall motion of the galaxy	1.1 million miles/hour

The "Big Bang" is usually defined as a random, chance event. As noted in the previous question, some instability supposedly developed in an original "kernel" of mass energy, and the universe ballooned outward. However, Scripture clearly rules out such an accidental origin. A modified version of the Big Bang theory says that when the explosive event happened, it was directed by God. This is the theistic-evolution approach, an attempt to compromise the Bible with evolutionary theories, and it, too, must be rejected because of its many conflicts with the order of events in Genesis. Table 14.4 contrasts some of the chronological discrepancies between the Bible's creation account and the "Big Bang" hypothesis.

Creation scientists maintain that what really happened at the time of creation is that God spoke and the earth appeared— he commanded and the heavens stood firm (*Ps. 33:9*)! All the many stars appeared suddenly and supernaturally in space. Scripture does not imply an explosion, although the universe must have experienced a sudden "explosive" input of ordered energy. Perhaps some of the astronomical data that seems supportive of the "Big Bang" theory, such as redshift and background radiation, needs to be looked at instead as evidence of a rapid creation. One secular variation of the "Big Bang" theory refers to an "inflationary" Big Bang, the suggestion being that the universe developed and matured very quickly in its first moments. In this particular theory, secular science seems to have taken one step in the creationists' direction. Further developments should be of interest in this area of theory and research.

Table 14.4: Chronological Discrepancies Between Scripture and Big Bang

Scripture	Big Bang
All elements made together	Elements beyond hydrogen and helium formed after millions of years
Earth formed before stars	Earth formed long after stars
Plants formed before the sun	Plants evolved after the sun
Sun formed on the fourth day	Sun formed before the earth
Sun, moon, and stars formed	Sun formed from older stars together

The Big Bang as it is understood today is an inadequate theory since there are many fundamental problems that are seldom mentioned in the pertinent literature. The following are some "missing links" in the theory:

1. **Missing Origin.** The Big Bang theory assumes an original concentration of energy. Where did this energy come from? Astronomers sometimes speak of origin from a «quantum mechanical fluctuation within a vacuum.» However, an energy source is still needed. Actually, there is no secular origin theory, since every idea is based on preexisting matter or energy.
2. **Missing Fuse.** What ignited the Big Bang? The mass concentration proposed in this theory would remain forever as a universal black hole. Gravity would prevent it from expanding outward.
3. **Missing Star Formation.** No natural way has been found to explain the formation of planets, stars, and galaxies. An explosion should produce, at best, an outward spray of gas and radiation. This gas should continue expanding, not form intricate planets, stars, and entire galaxies.
4. **Missing Antimatter.** Some versions of the Big Bang theory require an equal production of matter and antimatter. However, only small traces of antimatter (positrons, antiprotons) are found in space.
5. **Missing Time.** Some experiments indicate that the universe may be young, on the order of 10,000 years old. If true, then there is not sufficient time for the consequences of the Big Bang to unfold. A short time span would not allow for the gradual evolution of the earth, heavens, and mankind.
6. **Missing Mass.** Many scientists assume that the universe will eventually stop expanding and begin to collapse inward. Then it will again explode, and repeat its oscillating type of perpetual motion. This idea is an effort to avoid an origin and destiny for the universe. For oscillation to occur, the universe must have a certain density or distribution of mass. So far, measurements of the mass density are a hundred times smaller than expected. The universe does not appear to be oscillating. The necessary mass is "missing".
7. **Missing Life.** In an evolving universe, life should have developed everywhere. Space should be filled with radio signals from intelligent life forms. Where is everybody?

RED-SHIFT EXPLAINED PHENOMINA

The red-shift of starlight is a decrease in the energy of the light. This energy decrease results in a lengthening of the wavelength of the light, measured with an instrument called a spectrometer. Red is the rainbow color with the longest wavelength, hence the name "red-shift." Stars do not actually become red in appearance since the wavelength change is usually slight. Almost every star and galaxy is found to be red-shifted. The following list summarizes some of the alternative explanations for the origin of this stellar red-shift.

a. **Stellar Motion.** If a star moves outward from the earth, its light energy will be reduced and its wavelength stretched or red-shifted. Stars and entire galaxies show varying amounts of red-shift, therefore implying a variety of speeds for these objects. Police actually use this same effect with radar to measure the speed of cars. Stellar motion is often taken as evidence in support of the Big Bang theory. Stars are assumed to be speeding outward as a result of the explosion. This is not the only explanation of red-shift, however.

b. **Gravitation.** As light leaves a star, the star's gravity may slightly lengthen the wavelength of the light. A gravitational red-shift could also result from starlight passing near a massive object in space, such as a galaxy. As the light escapes from a strong gravity field, it loses energy, similar to what happens to a person struggling to the top of a mountain.

c. **Second-Order Doppler Effect.** A light source moving at right angles (tangentially) to an observer will always be red-shifted. This can be observed in the laboratory by using a high-speed turntable. A detector is placed in the center and a gamma radiation source is placed on the outside edge. The gamma energy is seen to decrease, or "red-shift," as the turntable speed increases. This is an intriguing explanation for stellar red-shift. When applied to stars, it implies that the universe may be in circular motion instead of radial expansion.

d. **Photon Interaction.** It is possible that light waves exchange energy during their movement across space and lose some energy in the process. A loss of light energy is equivalent to a "reddening" of its light. A theoretical understanding of this proposed "tired light" process has not yet been developed.

Any of these four explanations, alone or in combination, may be responsible for red-shift. We do not know enough about space to be certain of the source of stellar red-shift.

LIGHT PROPERTIES

Simple questions in science often have complicated answers, and light is an excellent example. Light has both particle-like properties and wave properties. That is, light sometimes behaves like invisible particles called photons.

These photons can collide with other particles such as electrons, and be deflected like microscopic marbles. At the same time, light also displays wavelengths which can act similar to water waves or sound waves. Scientists accept this unusual dual nature of light without completely understanding it.

We are familiar with visible light, of course. The sun produces the dramatic colors that brighten our day. Rainbows, blue skies, red sunsets all result from the separation of sunlight into its spectral colors. However, this visible light is only a small part of the total picture. The sun and other stars also emit many kinds of light that our eyes cannot see. You have heard of some of these forms of light: radio waves and microwaves, ultraviolet and infrared, x-rays and gamma rays. All of these strange varieties of light flood our sky continually. If we could see them, the heavens would appear to be bright with energy. Although this sounds dangerous, it should be noted that the microwaves from space are just a whisper, much weaker than those produced inside a microwave oven.

Also, most of the ultraviolet and x-rays are safely absorbed by the earth's atmosphere. In recent years, instruments have been designed to detect and learn from these invisible kinds of light. For example, infrared telescopes show us many new details of stars and galaxies, and receivers for the radio waves are built in the form of huge dishes. There is obviously much lighter in space than "meets the eye," and each variety of light, visible or invisible, has its own story to tell about the heavens.

Light was part of the initial creation, the opposite of darkness (*Gen. 1:3*). God is called "light" (*1 John 1:5*) and "Father of the heavenly lights" (*James 1:17*). This is a fitting title because light is pure, beautiful, and beyond human understanding. Christians are expected to be part of this image, since we are told to let our light "shine before men" (*Matt. 5:16*).

LIFE IN SPACE

All efforts to detect life beyond earth have failed so far. The search began with the moon, where astronauts walked during six lunar landings from 1969 through 1972. After it was concluded that the moon was a sterile, lifeless place, the search moved to the other planets and their moons.

Viking probes to Mars in 1976 performed experiments designed to detect life, including microscopic organisms, with negative results. Two unmanned Voyager craft whose destinations include Jupiter, Saturn, and Uranus—have taken thousands of pictures of the outer solar system. They reveal harsh, non-livable conditions everywhere. Searches of deepest space have been carried out by radio telescopes, instruments that are able to beam messages of greeting toward any planets that might be circling distant stars.

Radio telescopes also "listen" for any space messages that may be coming in earth's direction. During the past few decades, scientists have searched dozens of nearby stars for intelligible radio signals. The results are once again completely negative. At this point, it appears that life as we know it is unique to planet earth. This conclusion has been very upsetting to evolutionists, who believe that life began spontaneously on earth and that the same thing probably happened elsewhere in the universe.

FAVORABLE TELESCOPE

This is a practical question! Many beginning stargazers become discouraged because they are using the wrong equipment. Just as there are different levels of reference books for the study of Scripture, so there are levels of optical tools for studying the heavens. In both cases, the best advice is to start simply and build upward from that point.

One should make quality investments with lasting value. For the beginning astronomer, an initial aid might be binoculars, which should have wide-range optics and a power of about 7. (A label of 7 x 50 means that the magnification is 7, and the aperture or diameter of the front lens is 50 millimeters.) In many ways binoculars are superior to a small telescope. They are easy to use, give excellent views of the moon, planets, and star clusters, and they also work well for indirect viewing of solar eclipses and sunspots. Warning: The sun's image through binoculars should be projected onto a flat surface so it can be safely studied.

After one is familiar with a binocular view of space, a refracting or reflecting telescope can be considered. Seek some informed advice and, if possible, try several out before making a purchase. The initial experiences will help the stargazer decide which telescope is best for his or her own interests.

ANTHROPIC PRINCIPLE

This currently popular term in astronomy comes from the Greek word for man, *anthropos*. For any principle of science to be acceptable, there must be experimental results with general validity.

The Anthropic Principle which states that the universe is especially suited for the well-being of mankind, is one such assumption.

As just one of hundreds of examples; Consider the tides that the moon causes on earth. If the moon was closer to the earth, tides would be greatly increased. Ocean waves could sweep across the continents. The seas themselves might heat to the boiling point from the resulting friction. On the other hand, a more distant moon would reduce the tides. Marine life would be endangered by the resulting preponderance of stagnant water!

Mankind would also be in trouble because the oxygen in the air we breathe is replenished by marine plants. We can conclude that the moon is in the "correct" position for man's well-being. Even such details as the mass of protons and the strength of gravity have values that give stability to the universe and thus reinforce the Anthropic Principle.

The Anthropic Principle is a powerful argument that the universe was designed. Of course, whether it is an intricate watch or a beautiful planet, any design plan requires a designer! Evolution theory believes it has an answer to "design" in biological systems by hypothesizing ongoing processes of mutation and natural selection. Living things are said to change very slowly and improve with time.

There are many fundamental problems with evolution theory, not the least of which is that-in the case of the Anthropic Principle-the theory provides no answer at all. Whether describing tides, proton mass, or the earth's position in the solar system, is not a grand design present from the very beginning? These phenomena don't mutate or change with time. The negative response of secular science to new evidence of design is interesting in that it shows the extremes to which man will go to maintain a belief in the random origin of all things. It has even been proposed that there really is an infinite number of universes, each with a completely different set of physical properties. According to such thinking, our particular universe just happens to have conditions suitable for human life, and that is why we are here to enjoy it! Of course, there is no way to detect any "other" universes or comprehend their underlying principles. Scripture describes the creation of just one universe. It contains all things, including the clear marks of the supremely intelligent design of our creator God.

GOD BEFORE CREATION

The Bible teaches us that God existed before the universe, and that He is outside the universe and beyond the physical laws that govern it. In *Psalm 90:2*, Moses says, "Before the mountains were brought forth, or ever you had formed the earth and the world, from everlasting to everlasting you are God" Figure 14.28. When we think of God's pre-existence, we tend to think of God existing for billions of years before creation, but this is flawed. Instead, God is 'timeless'—God created time itself—yet He is also able to enter time and space to intervene in events at any point.

Fig.14.28: *God Before Time*

It is natural for us to ask: what was God doing in eternity past before He created the universe? Augustine cited one person's answer: "He was preparing hell for those who pry too deep." However, Augustine rejected such a flippant response. While we obviously cannot fully comprehend what God was doing 'before' creation, God did reveal some details about His pre-creation activity in Scripture. This helps us to understand His nature, as well as His relationship with and His plans for creation.

FELLOWSHIP OF TRINITY

Before the universe existed, there was only God—but He was not alone. God is a Trinity—the Persons of the Father, Son, and Holy Spirit all share in the same divine nature, and they are one God, Figure 14.29. This aspect of God's nature was not fully revealed in the Old Testament, but there are some intriguing hints of plurality in the Godhead in the Old Testament. For instance, while contemplating the incomprehensibility of God, Agur, the author of *Proverbs 30*, says:

Fig.14.29: Diagram describing the doctrine of the Trinity

Who has ascended to heaven and come down? Who has gathered the wind in his fists? Who has wrapped up the waters in a garment? Who has established all the ends of the earth? What is his name, and *what is his son's name?* Surely you know! (*Proverbs 30:4*, emphasis added).

Agur laments his ignorance (30:1 ff). The knowledge of eternal truths and power over the creation implied in the first questions is only possessed by God. So, when Agur asks "***What is his name***?" every good Jew would answer, "Yahweh." But then he goes on to ask, "and what is his son's name?" This would confound the Jews who lived before the revelation of the Trinity, because it explicitly indicates that God has a Son. But Christians living in light of the revelation of Jesus can see a clear reference to Jesus, the Son of God, in this proverb, long before God explicitly revealed the doctrine of the Trinity in the New Testament.

Because God is a Trinity, He enjoyed perfect companionship within Himself before He created the world. The Persons of the Father, Son, and Holy Spirit loved each other, shared in the divine glory, and had a perfect relationship to each other. A few places in Scripture give us hints of what this was like. In Jesus' prayer, often called the 'High Priestly Prayer' because of His intercession for believers, Jesus asks, "And now, Father, glorify me in your own presence with the glory that I had with you before the world existed" (*John 17:5*).

When God the Son became incarnate in the Person of Jesus Christ, He took on a human body and nature. During His earthly ministry, He set aside His divine glory and independent use of His divine powers which were rightfully His (*Philippians 2:7*). But in this prayer, Jesus anticipated the completion of His work on Earth and the return to His exalted place and the fellowship He longed for with the Father. This is also a passage which clearly teaches Jesus' pre-existence and deity. Jesus' whole ministry glorified the Father by doing His will, and in this prayer, Jesus anticipated when the Father would glorify Him in turn, just like they had done for eternity before the creation of the universe.

PLANNING - REDEMPTION AND GLORIFICATION

Jesus not only wanted to be glorified again and to be with the Father, but for His disciples to see His glorification as well. His expression of this desire shows the close relationship between Jesus and the Father. In *John 17:24*, Jesus says, "Father, I desire that they also, whom you have given me, may be with me where I am, to see my glory that you have given me because you loved me before the foundation of the world."

The whole reason for Jesus' incarnation, ministry, and death was to save a people for Himself. He wants His people to be with Him, and that must be the Father's will too, because it is the Father who gave them to

Him (*John 10:27–30*). And just as the pre-creation fellowship was characterized by love, love grounds the Son's request here. "The ultimate hope of Jesus' followers thus turns on the love of the Father for the Son."

It is important to understand the intra-Trinitarian fellowship because it guards against false impressions that God created people because He needed companionship or that He was lacking something. God already had perfect, complete fellowship within Himself. Also, the intra-Trinitarian love of God is a uniquely Christian doctrine, and makes it possible for love to be an *essential* trait of God's personality (*1 John 4:8*)—God does not need anyone outside Himself to love, because the members of the Trinity love each other.

Further, *Ephesians 1:4* teaches that God "chose us in him [Christ] before the foundation of the world, that we should be holy and blameless before him." This shows that God was not taken by surprise by the Fall, and had planned for it even before He created. He had already determined that Jesus would die for our sins and rise again. Accordingly, this passage sweeps through all of salvation history, from the work of God in eternity past, to the believer's present salvation and destiny in eternity future.

PLANNING HIS KINGDOM

A good architect does not start building before he considers the overall layout of the building. Therefore, it should not surprise us that God was planning His kingdom before He created the universe. God did not 'light the fuse' of any sort of 'big bang' and hope for the best—He created intentionally with a specific end in mind, and this end was pre-determined beforehand.

For example, *Matthew 25:34* tells us, "Then the King will say to those on his right, 'Come, you who are blessed by my Father, inherit the kingdom prepared for you from the foundation of the world.'" This reveals God had planned for His kingdom before creation, and that He always meant for believers to be co-heirs with Christ.

It is important to note how closely love is connected with God's plan for His kingdom. God didn't need a kingdom—He already had everything He needed within Himself before creation. Rather, He created the Kingdom for His Son, but also for His people, whom He wanted to bless. The ultimate purpose of salvation is to glorify God—but it also results in our glorification when we trust in Jesus.

CHRISTIANS FELLOWSHIP WITH THE TRINITY

In John's gospel, the believer's destiny is seen to be close fellowship with the Trinity. Unlike some false religions, the Bible does not teach that we will actually *become* gods; rather, in the Resurrection we will reflect the image of God as He intended us to. The sort of fellowship we will have with God is shown most clearly in Jesus' high priestly prayer for all those who believe in Him:

The glory that you have given me I have given to them, that they may be one even as we are one, I in them and you in me, that they may become perfectly one, so that the world may know that you sent me and loved them even as you loved me (*John 17:22–23*).

In His prayer for believers in all ages, Jesus is speaking about an intimate level of fellowship with the Trinity, including sharing in love and even glory. Therefore, in our resurrection bodies, the Christian will participate in the fellowship of the Trinity in the closest relationship possible for a mere creature. Accordingly, the foundation for the doctrine of Creation is in the good God who loves, and who exists in fellowship within Himself and with His creation, and who created out of the abundance of His love.

God did not need to create the universe, and He owes nothing to any of His creation. Yet He is shown again and again as a gracious Provider and Sustainer who blesses us even though we do not deserve it. Uniquely, Genesis presents God as fundamentally relational, and this affects our understanding of everything He does.

God Speaks!

A striking aspect of God's creating is that He *speaks* things into existence. In Genesis Chapter 1, regarding how everything came to be, we read "God said …" eight times. The first time (*Genesis 1:3*), we read "And God said, 'Let there be light,' and there was light." God spoke and it happened.

Only God can speak things into existence. The New Testament (NT) confirms this means of creating: "By faith we understand that the universe was created by the word of God, so that what is seen was not made out of things that are visible." God spoke everything into existence from *nothing*. The secular big bang idea has everything coming from nothing, but with no sufficient cause for this to happen (nothing begets nothing!).

Throughout Scripture we see the power of God's Word emphasized. Indeed, this is the reason for the authority of the Bible; it is God's Word—as it claims to be some 2,800 times.

The timeframe given in *Genesis 1*—six days—underlines the instantaneous nature of God creating things by His word on each day.

Furthermore, it is this immediate effect of God speaking in Genesis that shows us the divine nature of Jesus in the NT. When Jesus speaks, we see instant effects that are only possible because He is God the Son.

In the healing of the paralyzed man (*Matthew 9; Mark 2*), Jesus spoke, and the man walked straightaway—he did not need months of therapy!

With the storm on the Sea of Galilee, Jesus spoke, and the storm stopped *immediately*. Not only did the wind stop, but the waves too. Anyone familiar with the sea knows that after strong winds the waves continue for some time; they do not stop immediately the wind stops. The disciples knew that they had witnessed something that only God could do. They said, "What sort of man is this that even the winds and sea obey him?" (*Matthew 8:27*). The sailors on Jonah's ship also recognized the hand of God when the storm suddenly stopped.

In the feeding of the two crowds, Jesus created lots of extra bread and fish from *nothing* (*Matthew 14, 15; Mark 6, 8; Luke 9, John 6*). Only God can do this.

Saying that creation occurred over long ages of slow and gradual processes undoes these clear connections of Genesis 1 with the divinity of Christ. In contrast, the evidence for the biblical timeframe is getting stronger all the time.

We see also that creation thinking makes sense of things such as mass extinctions and nature conservation (the red panda). And we see lots of evidence that God did create living things, and 'after their kind.' Living things show that a super-intelligence created them, and our remarkable sun also reveals the hand of God, including His power/

FAITH REVEALS KNOWLEDGE

The 'hall of faith' in *Hebrews 11* is one of the passages that Christians are most familiar with. It is inspiring to look at the recounting of the biblical heroes, and spurs us on to greater faith. But the first example of faith that the author recounts is one practiced by all believers. *Hebrews 11:3* tells us:

"By faith we understand that the universe was created by the word of God, so that what is seen was not made out of things that are visible."

There is a lot of content packed in one small verse which is worth looking at in more detail.

By Faith

There are many truths in Scripture that we can only arrive at through special revelation. Nothing in nature would cause someone to arrive at the doctrine of the Trinity; we believe it because it is clearly taught in Scripture.

Likewise, nothing in nature would lead us to the absolute truth that God created in an orderly process over six 24-hour days. Our own human reason cannot lead us to this truth, either. We must believe it *by faith*. This does not mean that science has nothing to say about the matter, only that it cannot reach back to the first days of history and tell us what happened. Only an eyewitness can do that, and God was the only eyewitness to creation.

There is a great deal of scientific evidence *consistent with* six-day creation—but such evidence can never *prove* either creation or evolution, due to science's built-in limitations. It is not even enough to have 'an eyewitness' when the truths one is proclaiming are so important for our lives that we are literally staking our eternal existence on their trustworthiness. It requires an infallible, divinely-inspired record, which is what we believe the Bible is.

We Understand

But the faith to which we are called as Christians is not 'blind faith.' There are good reasons to believe the Bible's record, and we engage our minds as we think about this doctrine of creation. We apprehend what it teaches us about God, His nature, and our relationship to Him, as well as our relationship to the rest of nature. When we understand the doctrine of creation, we will apply it to all areas of our thinking. For instance, because human beings are created in the image of God, ***human life is sacred***, so ***we will be pro-life and anti-slavery***.

Because we are the stewards of creation, we will take reasonable steps to preserve the environment, without crossing the line into 'geolatry.' We could multiply examples; in short, it will affect our thinking in every aspect of our life and worldview.

It is also noteworthy that the 'we' in *Hebrews 11:3* is talking about believers. This is a truth that is not accessible to people who are not in the right relationship with God, because they have rejected the revelation that would have taught it to them. The idea that faith is preliminary to right knowledge is not new to the author of Hebrews. Much earlier, Scripture stated that "The fear of the Lord is the beginning of wisdom" (*Psalm 111:10; Proverbs 9:10*). This means that "the fear of Yahweh is where we must begin, if we are to have any hope of gaining wisdom."

The only way we can understand the true history of the world is by *faith*; by trusting in the true God who has revealed Himself to us through Scripture.

CREATION THROUGH THE WORD OF GOD

Because He is the Creator, God is the source of everything that exists. Unlike in pagan creation myths, however, God did not create the world as the outcome of a war with other gods; He created the world *by **His Word***. He spoke, and what He commanded to come into being, began to exist. Scripture uses God's identity as Creator to contrast God, who created humanity, with false gods, who are created by humans.

The New Testament, of course, teaches us that Jesus is the Word of God incarnate (the *logos*, *John 1:1*). Jesus claimed to be with the Father before creation, and He is the agent of that creation. In light of this fact, it is even more important that we believe what He says about creation, and Jesus Himself referenced God creating the world (*Mark 13:19*).

VISIBLE OUT OF INVISIBLE

If the creation was not made out of visible things, it had to be made from something not visible—the Word of God. And Scripture teaches that just as it was created by that Word, it continues to be dependent on the Word for its continued existence: "For by him [Jesus] all things were created, in heaven and on earth, visible and invisible, whether thrones or dominions or rulers or authorities—all things were created through him and for him. ***And he is before all things, and in him all things hold together***" (*Colossians 1:16–17*).

FAITH UNSEEN

The author of Hebrews goes on in the rest of chapter 11 to recount examples of Scriptural heroes who had faith in the unseen. Noah had never seen a global Flood; yet he believed God and built an ark which saved him, his family, and the animals. Likewise, Christians today believe in Jesus, whom we have not seen, but we trust Him to save us from judgment. Abraham went out to a land that he had never seen. Yet the author of Hebrews says that Abraham was looking forward to an even better country—the New Jerusalem (*11:10, 16*).

Materialist scientists pride themselves on only trusting what they can see and test in a lab. But they are either lying or deluded.

- They believe in things like the "big bang," past climate scenarios, evolution and many other things, real or otherwise, they have not seen.
- Second, they are only personally capable of 'doing science' in their own field. A biologist must *assume* that the chemists, physicists and other scientists on whose work his own research depends have done their own jobs correctly. And given that less than half of peer-reviewed published scientific research can be replicated, it seems that many put far too much trust in 'science'.

FAITH IN DUBIOUS CULTURE

It can be difficult to 'go against the current' of the culture, which is thoroughly revolutionized. Creation scientists are considered so wrong in many areas as to be worthy of only mockery, not serious intellectual engagement. There is thus a real temptation for many Christians to compromise on God's revealed Word so as to be accepted in an revolutionized world.

However, the Bible's record reveals that God rewards those who trust His Word, *especially* against a hostile, disbelieving culture. This does not mean having blind faith, but rather trusting God's promises based on what we know to be true of His nature. Christians trust Jesus with our eternal destiny based on the testimony of Scripture; shouldn't we also trust the same book when it tells us why we need to be saved in the first place?

TRIUNE GOD

Worshipping the true God requires at least minimal knowledge about who He is. And while none of us can understand God completely, He has revealed some truths about Himself in Scripture in a way we can understand. As Christians, we should want to understand God's revelation of His own character as clearly as possible. The Trinity is at the heart of God's self-revelation.

However, the Trinity is one of the most easily misunderstood doctrines; even many Christians are uncertain of what the Trinity means. Many unwittingly hold to doctrines that have been condemned as either heresy or serious error throughout Church history. Others are aware of the heterodox nature of their beliefs, but insist it was the Church, not the heretics, who were mistaken. In an era where theological teaching is often underemphasized in the Church, it is not surprising that there are fewer today than in the past who can confidently say what the Bible teaches about the Trinity.

TRINITY IN SCRIPTURE

Anti-Trinitarians will point out that the word 'Trinity' is not in the Bible anywhere—and they are correct. It is a technical term that arose later to communicate the entirety of the Bible's teaching about the triune nature of God without having to spell it all out every time. But each facet of the Bible's teaching on the Trinity *is* present in Scripture. These facets are:

1. There is one God.
2. The Father is God; The Son is God; the Holy Spirit is God.
3. The three Persons are distinct, and each is equally God.

In the Bible, the Persons of the Trinity are differentiated both in their relation to each other and to the Creation, but they are all called 'God'.

THERE IS ONE GOD

The Bible is clear: there is one God, and we should worship and serve Him only. *Deuteronomy 6:4*, states, "Hear, O Israel, Yahweh our God, Yahweh is One." In *Isaiah 45:5–6*, God says:

I am Yahweh, there is no other,
Besides me there is no God;
I gird you, though you do not know me,
that men may know, from the rising of the sun,
and from the west, that there is none besides me,
I am Yahweh, and there is no other.

And in *Isaiah 45:21–22 He says:*

There is no other god besides me,
a righteous God and a Savior;
there is none besides me."
"Turn to me and be saved,
all the ends of the earth!
For I am God, and there is no other.

The apostle Paul affirms that "God is one" (*Romans 3:30*). He says, "There is one God, the Father, from whom are all things and for whom we exist" (*1 Corinthians 8:6*). He writes elsewhere, "For there is one God, and there is one mediator between God and men, the man Jesus Christ" (*1 Timothy 2:5*). James, the half-brother of Jesus, also acknowledges "You believe that God is one; you do well. Even the demons believe and shudder!" (*James 2:19*). But even though James is speaking about the importance of having more than just doctrinal correctness, the people to whom he is writing "do well" to believe that God is one.

God is the only appropriate object of worship. Angels are not to be worshipped (Colossians 2:18), and the *holy* angels refuse worship (*Revelation 19:10; 22:8–9*). Neither are the sun, moon, or stars to be worshipped (*Deuteronomy 17:3*). Idols are not to be worshipped (*Leviticus 19:4*). People are not to be worshipped either; the apostles never permitted themselves to be worshipped (*Acts 10:25–26; 14:13–18*). Herod is struck down for the sin of accepting worship (*Acts 12:20–24*).

However, as we will see below, it is not just God the Father that deserves to be worshipped. God the Son (Jesus) and God the Holy Spirit also receive praise and worship in Scripture. Therefore, if the Bible presents the Father, the Son, and the Holy Spirit as appropriate objects of worship, they all must be God.

PLURALITY OF GODHEAD OLD TESTAMENT

Even in Genesis, we see an indication of the plurality of the Godhead—meaning that God is made up of more than one Person. In *Genesis 1:26* He says, "***Let us*** make man in our image, after our likeness." He is using a first-person plural pronoun ('us'), but 'image' and 'likenesses' are in the singular, suggesting a plurality in the Godhead, but also of absolutely the same nature. Likewise, in *Genesis 3:22* He says, "Behold, the man has become ***like one of us*** in knowing good and evil."

In *Isaiah 6:8*, God uses the first person singular and plural in the same sentence: "Whom ***shall I*** send, and who will go ***for us***?"

Other places in the Bible distinguish one person called "God" or "the Lord" or "Yahweh" from another called "God." For example, in *Psalm 110*, a Davidic Psalm, says, "Yahweh says to my Lord: 'Sit at my right hand until I make your enemies a footstool for your feet.'" Incidentally, this Psalm is the single-most quoted Scripture in the New Testament, and Jesus is always said to be the One addressed by God in those references.

TRINITY IN THE NEW TESTAMENT

The New Testament has even clearer references to the Trinity. At Jesus' baptism, the Spirit descended on Jesus in the form of a dove, and the Father's voice spoke from Heaven—so all three of the members of the Trinity were clearly present (*Matthew 3; Mark 1*). Jesus commands believers to be baptized "in the name of the Father and of the Son and of the Holy Spirit" (*Matthew 28:19*).

Paul's letters are full of Trinitarian formulas. For instance, *Romans 8:3–4* (emphases added)

For **God** has done what the law, weakened by the flesh, could not do. By sending his own **Son** in the likeness of sinful flesh and for sin, he condemned sin in the flesh, in order that the righteous requirement of the law might be fulfilled in us, who walk not according to the flesh but according to the **Spirit**.

This makes sense in light of the background in which it is given. Paul is not advocating three separate gods, but one God in three persons. God the Father sent God the son to take away sin, and God the Spirit helps us to live according to the forgiveness granted to us by God the Father, but only after the work of the Son was accepted. Thus, contrary to certain "Oneness" groups, the three Persons are *distinct*, not merely different modes or manifestations of one Person.

And verse 16:

The **Spirit** himself bears witness with our spirit that we are children of **God**, and if children, then heirs—heirs of God and fellow heirs with **Christ**, provided we suffer with him in order that we may also be glorified with him.

These Trinitarian formulas are even more apparent when we realize that the Father is often designated with the Greek *theos* (God), and the Son with *kyrios* (Lord), as in *1 Corinthians 12:4–6*:

Now there are varieties of gifts, but the same **Spirit**; and there are varieties of service, but the same **Lord**; and there are varieties of activities, but it is the same **God** who empowers them all in everyone.

Please observe the use of the word "Lord" (*kyrios*) instead of "Son" (*huios*) or Christ, as in the passages above.

These formulas often appear in the benedictions to the letters. For instance, in *2 Corinthians 13:14*.

The grace of the **Lord Jesus Christ** and the love of **God** and the fellowship of the **Holy Spirit** be with you all.

And this isn't unique to Paul. The author of Hebrews states (*2:3–4*):

How shall we escape if we neglect such a great salvation? It was declared at first by **the Lord** (*kyrios*, an obvious reference to Jesus), and it was attested to us by those who heard, while **God** also bore witness by signs and wonders and various miracles and by gifts of the **Holy Spirit** distributed according to his will.

And (*9:13–14*):

For if the blood of goats and bulls, and the sprinkling of defiled persons with the ashes of a heifer, sanctify for the purification of the flesh, how much more will the blood of **Christ**, who through the eternal **Spirit** offered himself without blemish to **God**, purify our conscience from dead works to serve the living God.

Peter uses this formula (*1 Peter 1:2*)

According to the foreknowledge of **God the Father**, in the sanctification of the **Spirit**, for obedience to **Jesus Christ** and for sprinkling with his blood.

John says (*1 John 4:13–14*)

By this we know that we abide in him and he in us, because he has given us of his **Spirit**. And we have seen and testify that the **Father** has sent his **Son** to be the Savior of the world.

Jude exhorts (1:20):

But you, beloved, building yourselves up in the most holy faith and praying in the **Holy Spirit**, keep yourselves in the love of **God**, waiting for the mercy of our **Lord Jesus Christ** that leads to eternal life.

Even though the *word* 'Trinity' is never used in the New Testament, the *teaching* is clearly there, so much so that a non-Trinitarian doctrine would substantially alter the message of the Bible itself.

Thus, "God is one" and "God is a Trinity." But some people are confused about how the Persons of the Godhead relate to each other. The Bible teaches that each Person is fully God and shares all the attributes of deity.

THE FATHER IS GOD

This is perhaps the least contested point—all the historical heresies affirmed that the Father is God, but err in how they saw the relationship between the Persons of the Godhead, or in the identity of the other Persons. The Father is the one who speaks things into being in *Genesis 1*. He sent the Son in the Incarnation (*John 8:42*). And the Father sends the Spirit (*John 14:26*). The Father is clearly an appropriate object of worship (*John 4:21–23*).

THE SON IS GOD

Hebrews 1:3 says that Jesus "is the radiance of God's glory and the exact representation of his being." Humans are created in the 'image and likenesses' of God, meaning that we are like God in some ways, but far more than that is attributed to the Son. The Greek translated "exact imprint of his nature" is χαρακτὴρ τῆς ὑποστάσεως αὐτοῦ, (*charactēr tēs hypostaseōs autou*), and means basically that Jesus is exactly identical to the Father—there is no attribute of the Father that the Son does not have in equal measure. There is no way in which Jesus does not resemble the Father. Jesus teaches the same thing when He said, "Whoever has seen me has seen the Father" (*John 14:9*), and Paul says, "In him the whole fullness of deity dwells bodily" (*Colossians 2:9*). The writer of Hebrews reinforces this a few verses later when he quotes God himself, "But of the Son he says, 'Your throne, O God, is forever and ever,'" (*Hebrews 1:8*), showing that God addresses the Son also as God.

JESUS EXISTED BEFORE HIS BIRTH

Jesus, unlike mere human beings, existed before His birth. Speaking of Jesus the Son, the Gospel of John starts out with, "In the beginning was the Word, and the Word was with God, and the Word was God" (*John 1:1*). He is called "the one and only (μονογενῆς, *monogenēs*) God who is in the bosom of the Father" (*John* 1:18).

Paul says, "He is the image of the invisible God, the firstborn of all creation. For by him all things were created, in heaven and on earth, visible and invisible, whether thrones or dominions or rulers or authorities—all things were created through Him and for Him" (*Colossians 1:15–16*). Anti-Trinitarians point to this verse to claim that Jesus was only a created being, even if He was exalted. But this same verse says "by him all things were created," meaning that Jesus Himself could not have been created, or else He would come under 'all things'. 'Firstborn' in this instance simply means that Jesus has the *privilege* of the firstborn, something that was very meaningful in a time when the firstborn expected to receive a double portion of the inheritance. So, in this case 'firstborn' (Greek *prototokos*) does not mean 'first created' (Greek *protoktisis*), but simply denotes His superior position.

Hebrews 13:8 says that "Jesus Christ is the same yesterday and today and forever." In *John 5:26*, Jesus claims, "For as the Father has life in himself, so he has granted the Son to have life in Himself." But only God is self-existent.

Jesus is also viewed as a proper recipient of worship in the New Testament. After the Resurrection, Thomas calls Jesus, "My Lord and my God!" (*John 20:28*). A righteous person who received praise due only to God would deflect it immediately (see how Paul and Barnabas reacted in *Acts 14:8* ff.)—but Jesus didn't, indicating that he thought it was proper. He even says, in effect, "You *finally* believe in me!" *Titus 2:13* calls Jesus "our great God and Savior", as does *2 Peter 1:1*. Paul refers to "Christ, who is God over all" (*Romans 9:5*). Every time Jesus is worshipped in Scripture, it is cited with approval. Indeed, He demands equal honor with the Father:

> … that all may honor the Son, just as they honor the Father. Whoever does not honor the Son does not honor the Father who sent him. (*John 5:23*).

THE HOLY SPIRIT IS GOD

Some think of the Holy Spirit as a sort of impersonal, nebulous 'force'—and many people think of spirits as ghostly ethereal beings. But the Holy Spirit is clearly a Person in Scripture (so a 'Him', not an 'it').

When Ananias and Sapphira lied about the price of the field they sold, Peter said, "Ananias, why has Satan filled your heart to lie to the Holy Spirit and to keep back for yourself part of the proceeds of the land? … You have not lied to man but to God" (*Acts 5:4*). So, the Holy Spirit is equated with God. Later in the book, it's even clearer, because the Holy Spirit uses two first person pronouns—thus there can be no doubt that He is a Person:

> ... the Holy Spirit said, "Set apart for me Barnabas and Saul for the work to which I have called them." (*Acts 13:2*)

David attributes omnipresence to God's Spirit when he says, "Where shall I go from your Spirit? Or where shall I flee from your presence? If I ascend to heaven, you are there! If I make my bed in Sheol, you are there!" (*Psalm 139:7–8*). Paul attributes omniscience to Him: "For the Spirit searches everything, even the depths of God" (*1 Corinthians 2:10*).

The Bible is clear that only God can give spiritual life (*1 John 3:9*), but Jesus said, "unless one is born of water and the Spirit, he cannot enter the kingdom of God. That which is born of the flesh is flesh, and that which is born of the Spirit is spirit" (*John 3:5–6*). If only God can give spiritual life, and the Spirit gives spiritual life, then the Spirit must be God.

Furthermore, when we understand the Father and the Son to be fully God, that the Spirit is equally divine follows from the Trinitarian verses cited below. As "Wayne Grudem" explains:

Once we understand God the Father and God the Son to be fully God, then the Trinitarian expressions in verses like *Matthew 28:19* ("baptizing them in the name of the Father and of the Son and of the Holy Spirit") assume significance for the doctrine of the Holy Spirit, because they show that the Holy Spirit is classified on an equal level with the Father and the Son. This can be seen if we recognize how unthinkable it would have been for Jesus to say something like, "baptizing them in the name of the Father and of the Son and of the archangel Michael"—this would give to a created being a status entirely inappropriate even to an archangel.

Some believe that since there isn't a verse which straightforwardly says, "Worship the Spirit" that the Spirit is not a valid object of worship. But if God is worthy of worship, and the Spirit is God, then the Spirit is worthy of worship.

THE CREATOR IS TRINITY

At creation, the Father spoke the commands that caused things to come into existence. Jesus was the agent of that Creation (the *Logos* who John talks about in *John 1*; also *Hebrews 1:2*). Speaking of Jesus (*cf. Colossians 1:13*), *Colossians 1:16–17* says:

For by him all things were created, in heaven and on earth, visible and invisible, whether thrones or dominions or rulers or authorities—all things were created through him and for him. And he is before all things, and in him all things hold together.

Genesis also teaches that the Spirit of God was present and active in creation, hovering over the face of the waters (*Genesis 1:2*). *Ecclesiastes 12:1* uses the plural "***Creators***" although this is often masked in translation — it's interpreted as a 'plural of majesty' by people who don't see the Trinity in the Old Testament, but there are no other instances of 'plurals of majesty' other than places where the Trinity 'has to be' explained away.

DOCTRINE OF TRINITY

The various facets of the doctrine of the Trinity are taught clearly throughout Scripture, and the true believer will accept the doctrine of the Trinity. Throughout Church history, practically every other way of understanding the relationship between the Persons of the Godhead has been rejected as heresy. The doctrine is so interwoven with how salvation works that to reject the Trinity is tantamount to a rejection of the Gospel, because the alternatives violate the nature of God and/or the status of Jesus as God. So, historically, those who do not believe in the Trinity have been rejected as not Christian. This is especially important to remember today, as various Trinity-rejecting sects are asking to be recognized as Christians.

Finally, it is important to understand why Christian scientists make such a strong stand on this issue—it is because ultimately their mission is not just about 'design' or about 'a creator', or attacking evolution. It is all about and for Jesus Christ, His glorious Gospel and the expansion of His Kingdom. And this issue of who God is, and particularly who Christ is, is inescapably vital and foundational to the Good News of salvation. Jesus said in *John 8:24*: "I told you that you would die in your sins, for unless you believe that I am He you will die in your sins." The 'he' is missing from the Greek, so literally Jesus is saying (emphasis added): "Unless you believe that ***I am*** you will die in your sins." This is a clear reference to the deity of Jesus, for he is directly equating himself to God the Father (*Ex 3:14*). If we willfully reject Jesus' claim to deity, we in effect nullify His saving grace. It could hardly be more serious.

APPENDIX

i. The escape velocity is the velocity needed for an unpowered object to escape the gravitational field of a planet or star. It is given by the formula $v_e = \sqrt{2GM/R}$, where G is the universal gravitational constant, M is the mass of the planet or star, and R is its radius.

ii. 'Black holes are more than just theoretical concepts. They are, first, direct predictions of general relativity, which is backed by a great deal of experimental evidence.' Russell Humphreys, *Starlight and Time,* Master Books, CO 80936, USA, 1994, p. 22.

iii. Stephen Hawking, *A Brief History of Time*, Bantam Books, London, p. 194, 1988.

iv. The radius of an event horizon around a black hole is called the Schwarzschild radius, and is given by the formula $R_S = 2GM/c^2$, where c is the speed of light in a vacuum. It is named after the German astronomer Karl Schwarzschild (1873–1916), who in 1916 predicted the existence of collapsed stellar bodies that emit no radiation. If the sun were to collapse into a black hole, the radius of its event horizon would be only 3 km (2 miles).

v. He is therefore beyond or transcendent to it—i.e., not subject to its limitations (*Genesis 1:1*).

vi. Note that black hole event horizons are one-way, while the Flatland event horizon is two-way.

vii. God, on the other hand, is omnipresent and intimately aware of all that we say and do (cf. the 3-D being *vis-a-vis* the Flatlanders). He is 'nearer than breathing, closer than hands and feet.' See Psalm 139:7–14.

REFERENCE

1. For the technically minded, $G = 6.672 \times 10^{-11}$ Nm^2kg^{-2}
2. Thompsen, D., 'Gravity very precisely', *Science News* 118(1):13, 1980.
3. MacRobert, A., Turning a corner on the new cosmology, Sky and Telescope 105(5):16–17, 2003.
4. Hawking, S., A Brief History of Time, Bantam Books, New York, 1988. Cowen, R., Neutron stars twist Einstein's theory, Science News 158(10):150, 2000.
5. Humphreys, R., Starlight and Time, Master Books, Green Forest, 1994.
6. DeYoung, D., Creation and quantum mechanics, Impact 305: i–iv, 1998
7. Cosner, L., Our Triune God, 18 October 2012; creation.com/triune-god.
8. Phillips, D.J., *God's Wisdom In Proverbs*, p. 84, Kress Biblical Resources, The Woodlands, TX, 2011.
9. Cosner, L., The incarnate Word, 25 December 2014; creation.com/incarnate-word.
10. Cosner, L., *Did Jesus believe Genesis?* 11 July 2015; creation.com/Jesus-genesis.
11. Ioannidis, J.P.A., Why most published research findings are false, *PLoS Med* 2(8):e124 | doi: 10.1371/journal.pmed.0020124.
12. Augustine, *Confessions* 11:12:14, AD 400.
13. Carson, D.A., *The Gospel According to John*, Pillar New Testament Commentary, Eerdmans: Grand Rapids, MI, p. 570, 1991.

14. Here and following, the divine name יהוה is purposely left in its transliterated state; it is also rendered YHVH, YHWH, and the Lord.
15. See Grigg, R., Who really is the god of Genesis? *Creation* 27(3):37–39, 2005.
16. See Sarfati, J., Islam, testimony, and the Trinity, 27 May 2012. Return to text.
17. One might say that Paul meant that all things were created through Christ, except Christ Himself. But if that were the case, we might expect Paul to clarify that, as he clarified in 1 Corinthians 15:27 that the Father is not part of the 'all things' that are put in Christ's subjection.
18. Grudem, W. *Systematic Theology* (Grand Rapids: Zondervan, 1994), p. 237.
19. Isaiah 42:5; 2 Corinthians 4:6, remembering that *theos* (God) is usually used for the Father.
20. But see *Young Literal Translation*'s rendering, "Remember also thy Creators in days of thy youth … "
21. Of course, this is in the case of willful rejection of the doctrine, not a case of being in error because of ignorance.

UNDERSTANDING TIME DILATION

WHAT IS TIME

Time dilation is the slowing of time as perceived by one observer compared with another, depending on their relative motion or positions in a gravitational field.

Time dilation is the lengthening of the time interval between two events for an observer in an inertial frame that is moving with respect to the rest frame of the events (in which the events occur at the same location), Figure 15.1.

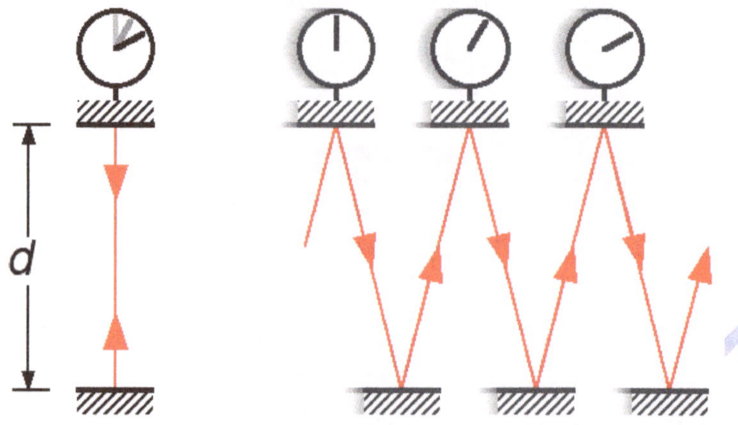

*Fig.15.1: Time Dilation with respect to the Observer – **d** is The Distance*

Einstein realized that time is relative and passes at different rates for different people.

Fig.15.2: Time Dilation Refers to the Idea that Time is Relative and Passes at Different Rates for Different Observers, Depending on Certain Factors. Curtsy - (Image: Shutterstock)

Time dilation refers to the seemingly odd fact that time passes at different rates for different observers, depending on their relative motion or positions in a gravitational field.

Here's how that works. Time is relative. As counterintuitive as those sounds, it's a consequence of Einstein's theory of relativity.

In everyday life, we're used to speed being relative — so, for example, a car traveling at 60 mph (97 km/h) relative to a stationary observer would be seen as moving at 120 mph (193 km/h) by a driver going in the opposite direction at the same speed, Figure 15.2.

This same phenomenon also impacts time. Depending on an observer's relative motion or their position within a gravitational field, that observer would experience time passing at a different rate than that of another observer. This effect, known as time dilation, becomes detectable only under certain conditions, although at a low level, we're subject to it all the time. Let's take a closer look at the theory of time dilation and some of its consequences, including GPS errors and the famous twin paradox.

Time dilation is the slowing of time as perceived by one observer compared with another, depending on their relative motion or positions in a gravitational field. It›s a consequence of Einsteinian relativity, in which time is not as absolute as it might appear; the rate at which it passes is different for observers in different frames of reference.

Einstein's starting point was the fact that light always has the same measured speed regardless of the observer's own motion, according to the late Michigan State University physics professor. This seemingly innocuous assumption inevitably leads to the conclusion that "moving clocks run slowly." This phrase is often used as a concise description of time dilation, but it's somewhat misleading because of the emphasis it places on clocks, which are only relevant insofar as we use them to measure time. But we really ought to think of time dilation as "an unexpected truth about space and time, rather than as a property of the clock?

Fig.15.3: Space Shuttle Discovery lifts off from Kennedy Space Center as onlookers watch July 26, 2005, in Titusville, Florida. Curtsy - (Image: Mario Tama/Getty Images)

The shuttle crewmembers would have experienced time dilation and thus would have perceives the trip as taking less time than Earthlings on the ground, Figure 15.3.

The theory of relativity has two parts — special relativity and general relativity — and time dilation features in both. The principle that the speed of light is the same for all observers plays a key role in special relativity. One of its consequences, according to Boston University physicist, is that two observers moving at a constant speed relative to each other measure different times between the same events. But the effect becomes noticeable only at velocities approaching the speed of light, commonly symbolized by c.

Imagine a spaceship traveling at 95% of the speed of light to a planet 9.5 light-years away. A stationary observer on Earth would measure the journey time as distance divided by speed, or 10 = 0.95/9.5 years. The spaceship crewmembers, on the other hand, experience time dilation and thus perceive the trip as taking only 3.12 years. (The math here is a little more complicated, but we›ll get to it later.) In other words, between leaving Earth and reaching their destination, the crewmembers age a little over three years, while 10 years have passed for people back on Earth.

Although really striking situations like this call for enormously high speeds, time dilation occurs on a more modest scale for any kind of relative motion. For example, a regular flier who crosses the Atlantic every week would have experienced about a thousandth of a second less time than a non-traveler after 40 years, according to "How to Build a Time Machine" (St. Martin's Griffin, 2013). The book also explains how the kind of speeds needed for more impressive feats of time dilation can occur in the real world, at least in the case of short-lived elementary particles called *muons*. These are created when cosmic rays hit Earth's upper atmosphere, and they can travel at nearly the speed of light. The *muons* are so unstable that they shouldn't last long enough to reach Earth's surface, yet many of them do. That's because time dilation can extend their lifetimes by a factor of five.

TIME DILATION AND GRAVITY

Gravity is slightly weaker on the top floor of a high building than at ground level, so the time dilation effect is also weaker higher up. Time goes faster the farther away you are from Earth›s surface. Even though the effect is too small to detect with human senses, the time difference between different altitudes can be measured using extremely accurate clocks, as West Texas A&M University physics professor Christopher Baird describes on his website.

To see a more dramatic example of gravitational time dilation, we need to find somewhere with much stronger gravity than Earth, such as the neighborhood around a black hole. NASA has considered what would happen if a clock were put in orbit 6 miles (10 kilometers) from a black hole having the same mass as the sun. It turns out that when viewed through a telescope from a safe distance, the clock would take around an hour and 10 minutes to show a difference of 1 hour.

TIME DILATION EQUATION

Einstein's original time dilation equation is based on special relativity. As daunting as the equation looks at first glance, it's not that difficult if we have a scientific calculator and work through the formula step by step, Figure 15.4. First, take the

- speed v of the moving object and
- divide it by c, the speed of light,
- and square the result.
- This should give you a number
- somewhere between 0 and 1.
- Subtract this from 1, and
- take the square root;
- then invert the result.

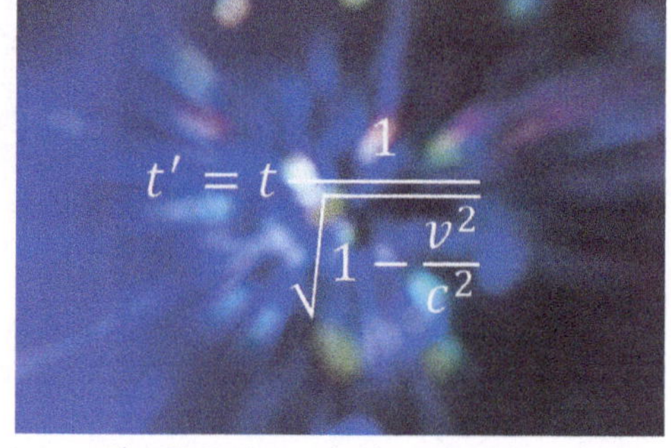

Fig.15.4: *Time Dilation Equation – Curtsy - Shutterstock*

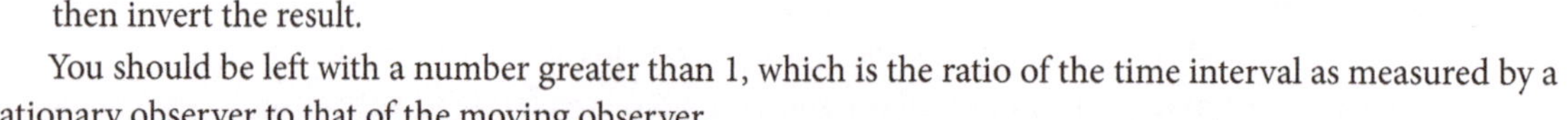

You should be left with a number greater than 1, which is the ratio of the time interval as measured by a stationary observer to that of the moving observer.

If that sounds like too much work, you can use an online calculator provided by Georgia State University. Just

- type in the speed, v,
- as a fraction of c, and
- the corresponding time ratio will appear automatically.

The same website also has the analogous formula relating to gravitational time dilation.

THE DILATION IN SPACE

Fig.15.5: *Taking time dilation and gravity into account, the Voyager 1 spacecraft, launched in 1977, turns out to be 1.2 seconds younger than Earthlings. Curtsy - NASA/JPL*

Time dilation is of double relevance to spacecraft, due both to their high speeds and the changing gravitational fields they experience. In 2020, a group of students at the University of Leicester in the U.K. computed the time dilation effects on NASA's Voyager 1 probe in the 43 years following its launch in 1977, Figure 15.5. Special relativity predicted that Voyager has aged 2.2 seconds less than we have on Earth. But general relativity partially counterbalances this. We experience stronger gravity than the spacecraft, so in this sense, the probe has aged around 1 second more than we have. Combining the two effects, Voyager still turns out to be younger than Earthlings, but by only about 1.2 seconds.

GPS SATELLITES PRECISION

Calculations like these may seem frivolous, but they can be very important in situations in which precise timing is critical. In the case of the GPS satellites used for navigation, for example, timing errors of just a few nanoseconds (billionths of a second) can lead to a positioning error of hundreds of meters, which is clearly unacceptable if you›re trying to pinpoint a specific address. To achieve the desired accuracy, the GPS system has to account for time dilation, which can amount to 38 microseconds (millionths of a second) per day, according to Richard W. Pogge, a distinguished professor of astronomy at The Ohio State University. As in the Voyager example, both special and general relativity contribute to this figure, with 45 microseconds coming from gravitational time dilation and minus 7 microseconds from the speed-related effect.

TWIN PARADOX

One of the most mind-bending consequences of time dilation is the so-called *twin paradox*. In this thought experiment, one identical twin lives on Earth while their twin takes a round trip to a distant star at velocities approaching the speed of light. When they meet up again, the traveling twin — thanks to time dilation of the special relativistic kind — has aged far less than the one who stayed at home. The apparent "paradox" comes from the mistaken belief that the situation is symmetrical — in other words, that you could also say the traveling twin is stationary relative to the Earthbound twin, meaning that the Earthling would have aged less than the star-voyaging twin.

But that's not the case, because the situation isn't symmetrical. When special relativity talks about relative motion, it's referring to motion at *constant speed in a straight line*. That's not the case here. Because the twins are together at the start and end of the journey, the traveler has to accelerate from a standstill to top speed and then, at some point, turn around and head back in the opposite direction, before eventually decelerating to a stop again. These phases of acceleration and deceleration bring in general relativity, because they have similar effects to a gravitational field, according to "Paradox: The Nine Greatest Enigmas in Physics "(Crown, 2012). When the math is worked out to account for this acceleration, it turns out that, in something akin to time travel, the spacefaring twin does indeed age more slowly than the Earthbound one.

WHY CLOCKS IN MOTION SLOW DOWN ACCORDING TO RELATIVITY THEORY

Fig.15.6: Time Dilation According to Einstein – *Curtsy: Pete Linforth from Pixabay*

On September 26, 1905, Albert Einstein's revolutionary paper *Zur Elektrodynamik bewegter Körper* (*On the Electrodynamics of Moving Bodies)* was published in the scientific journal *Annalen der Physik* (*Annals of Physics*), Figure 15.6. In it, Einstein describes his theory of special relativity, which provides fundamentally new definitions of the concepts of space and time.

Special relativity reconciles electromagnetic theory (more specifically, Maxwell's equations) with mechanics by introducing significant changes to Newtonian mechanics for speeds approaching the speed of light, Figure 15.6.

891

3. *Zur Elektrodynamik bewegter Körper;*
von A. Einstein.

Daß die Elektrodynamik Maxwells — wie dieselbe gegenwärtig aufgefaßt zu werden pflegt — in ihrer Anwendung auf bewegte Körper zu Asymmetrien führt, welche den Phänomenen nicht anzuhaften scheinen, ist bekannt. Man denke z. B. an die elektrodynamische Wechselwirkung zwischen einem Magneten und einem Leiter. Das beobachtbare Phänomen hängt hier nur ab von der Relativbewegung von Leiter und Magnet, während nach der üblichen Auffassung die beiden Fälle, daß der eine oder der andere dieser Körper der bewegte sei, streng voneinander zu trennen sind. Bewegt sich nämlich der Magnet und ruht der Leiter, so entsteht in der Umgebung des Magneten ein elektrisches Feld von gewissem Energiewerte, welches an den Orten, wo sich Teile des Leiters befinden, einen Strom erzeugt. Ruht aber der Magnet und bewegt sich der Leiter, so entsteht in der Umgebung des Magneten kein elektrisches Feld, dagegen im Leiter eine elektromotorische Kraft, welcher an sich keine Energie entspricht, die aber — Gleichheit der Relativbewegung bei den beiden ins Auge gefaßten Fällen vorausgesetzt — zu elektrischen Strömen von derselben Größe und demselben Verlaufe Veranlassung gibt, wie im ersten Falle die elektrischen Kräfte.

Beispiele ähnlicher Art, sowie die mißlungenen Versuche, eine Bewegung der Erde relativ zum „Lichtmedium" zu konstatieren, führen zu der Vermutung, daß dem Begriffe der absoluten Ruhe nicht nur in der Mechanik, sondern auch in der Elektrodynamik keine Eigenschaften der Erscheinungen entsprechen, sondern daß vielmehr für alle Koordinatensysteme, für welche die mechanischen Gleichungen gelten, auch die gleichen elektrodynamischen und optischen Gesetze gelten, wie dies für die Größen erster Ordnung bereits erwiesen ist. Wir wollen diese Vermutung (deren Inhalt im folgenden „Prinzip der Relativität" genannt werden wird) zur Voraussetzung erheben und außerdem die mit ihm nur scheinbar unverträgliche

Fig.15.7: *Albert Einstein as a young man and the front page of his special relativity paper "Zur Elektrodynamik bewegter Körper" ("On the Electrodynamics of Moving Bodies").*

THE TWO POSTULATES OF SPECIAL RELATIVITY

Special relativity is based on two basic postulates:

- *Principle of relativity*: The same laws of physics are valid in all inertial reference systems (a reference system not undergoing acceleration).
- *Universality of the speed of light*: The speed of light in the vacuum, usually denoted by c, is the same for all inertial observers, and it does not depend on the motion of the emitting source, Figure 15.7.

The first postulate, the principle of relativity, was first enunciated in 1632 by the Italian astronomer, physicist, and engineer, Galileo Galilei (1564–1642) in his book *Dialogue Concerning the Two Chief World Systems*. The second postulate, the universality of the speed of light, "was Einstein's great and radical contribution to relativity" Figure 15.8.

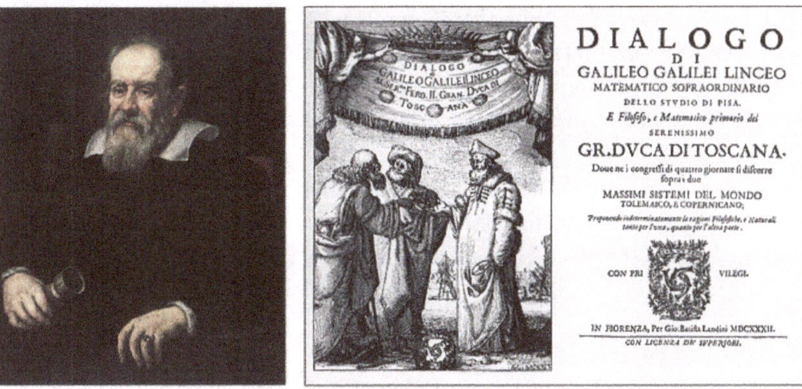
DIALOGO
DI
GALILEO GALILEI LINCEO
MATEMATICO SOPRAORDINARIO
DELLO STVDIO DI PISA.
E Filosofo, e Matematico primario del
SERENISSIMO
GR.DVCA DI TOSCANA.
MASSIMI SISTEMI DEL MONDO
TOLEMAICO, E COPERNICANO;
CON PRI VILEGI.
IN FIORENZA, Per Gio:Batista Landini MDCXXXII.

Fig.15.8: Galileo Galilei and his book *Dialogue Concerning the Two Chief World Systems where he* enunciated the principle of relativity.

RELATIVITY OF SIMULTANEITY

Suppose we have two events happening in two different locations. The relativity of simultaneity means that, if these events occur simultaneously in a given reference frame, they may not be simultaneous in a different reference frame.

Consider a railroad car, moving at constant speed along a straight track. In the center of the car, a flash of light is given off, and the light spreads out isotopically at speed c. Since the distances between the light source and the two ends of the car are equal, an observer inside it will detect the light reaching the front and back ends of the car simultaneously. However, to an observer outside the car, the two events (the arrival of the light at the front and back ends of the car) are not simultaneous, Figure 15.9.

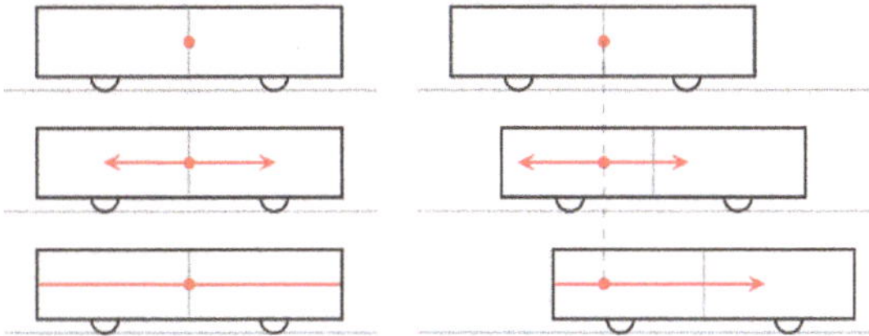

Fig.15.9: *An illustration of the relativity of simultaneity.*

This occurs because, as the light spreads out, the railroad car moves forward, and therefore, from the point of view of the observer outside, the light traveling to the back end of the car has a shorter distance to travel. Since for both observers, the speed of light is equal to *c* (according to the second relativity postulate), the outside observer will perceive the first event occurring before the second (note that the observers must correct for the time the signal takes to reach them), Figure 15.9.

STARLIGHT TIME

We attempt to illustrate more firmly than ever how light from the most distant stars would have reached Earth in a very short time.

Many stars are millions, even billions of light-years away, a light-year is the distance light travels in one year, Figure 15.10. Therefore, it would seem obvious that it takes light millions of years to reach us from those stars. Nonetheless, how can this be in a universe that, the Bible makes plain, is only thousands of years old?

Fig.15.10: Observatory of Galaxies Light Travel Time to Reach Earth Problem

For many, this is a huge stumbling block to accepting the straightforward creation history in the Bible. But Genesis history is crucial to the logic of the gospel, a good world, ruined by sin, to be restored in the future, so this is a *major* issue.

UNACCEPTABLE ANSWER

A common response from believers is to say that God could have created the light 'on its way'. However, light from distant stars also carries information—galaxies rotating, stars exploding. Accordingly, if that information never left the star, but was created 'en-route', this means that in a 6,000-year-old universe, anything we see beyond 6,000 light-years away would be a phony light show.

This is because the information in the light beams recording those events would have necessarily been created 'en-route', thus the events would never have taken place in the actual stars themselves. In other words, it would involve a God of truth creating a completely unnecessary false history. It's really not far removed from the embarrassing suggestion around Darwin's time that God created the fossils in the rocks.

Fig.15.11: Vast and Young Universe

THE RIGHT APPROACH

When asked about this conundrum, we have often, prior to suggesting possible answers, talked about facing God to give an account. We would not like to have to say, 'Lord, I did not believe the straightforward teaching of your Word on recent creation, all because we, with my very limited knowledge, could not figure out how you managed to do the trick of making a universe that was at once immensely large and relatively young, Figure 15.11.'

To say it is 'impossible' in the light of modern physics is really very naïve. Intuition says that it is 'impossible' for time to change, or the length of objects to change depending on such things as the speed of travel. Or for observations on one object to instantly affect another one on the other side of the universe. Yet experiments on Einstein's relativity and quantum mechanics have demonstrated that such effects are real, Figure 15.12.

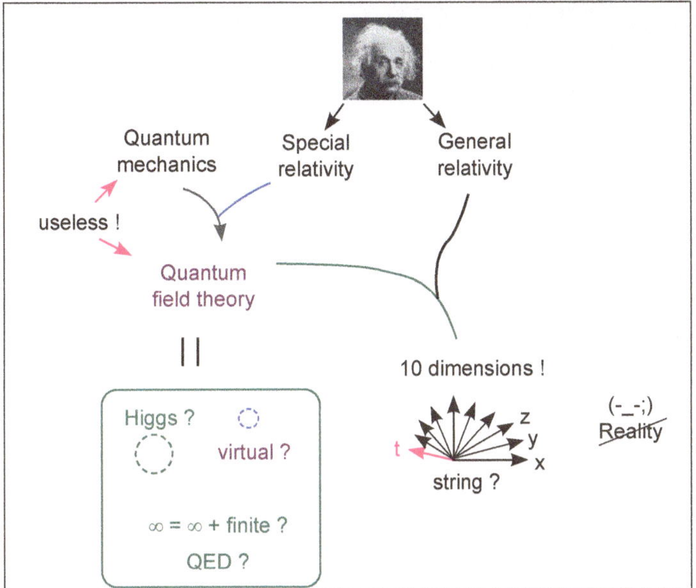

Fig.15.12: Quantum mechanics + Special relativity = Quantum Field Theory

LONG-AGERS HAVE MUCH THE SAME PROBLEM

Creation scientists' speakers also like to point out that the darling of long-agers, the big bang idea, has its own light-travel-time problem. There are apparently points in the distant universe which are today all at the same temperature, yet they are so far apart that there has not been anywhere near enough time for energy travelling at the speed of light, the speed limit of the universe, to cross that distance to equilibrate the temperature. So even though the big bangers' model allows them billions of years, they need billions of years *more* time than that, Figure 15.13.

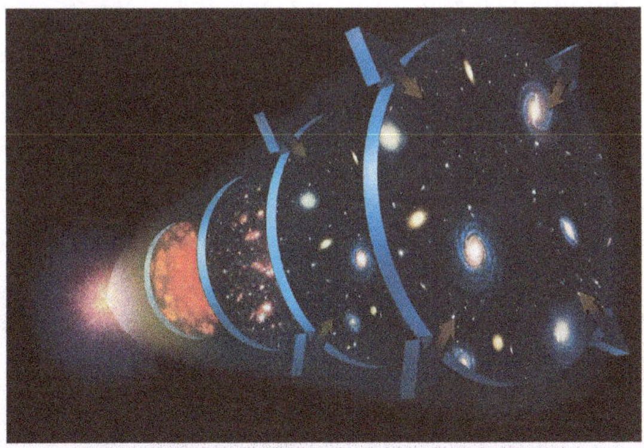

Fig.15.13: How Old is The Universe?

Big-bangers have suggested all sorts of exotic solutions to their puzzle, including that the laws of physics must have changed, or that there was expansion faster than the speed of light. So, they can hardly point the finger at creationists who propose similarly esoteric-sounding solutions to essentially the same problem.

Fig.15.14: Two of the large telescopes at the summit of the extinct volcano Mauna Kea, Hawaii, used to image the heavens. Left telescope is the James Clerk Maxwell 15 m sub-millimeter telescope. Right telescope is the Caltech 10.4 m sub-millimeter telescope. Curtsy Photo by John Hartnett

EARLIER SUGGESTIONS

Years ago, some scientists reported enthusiastically about the evidence for the notion that the speed of light (c) had changed historically. This is not completely 'dead in the water', but most creation physicists we know of, though they might wish 'c decay' to be true, find insurmountable physical hurdles with this idea.

Changing c, speed of light 186,282 miles/s causes huge numbers of other changes in physics, some of which should show up in the light from distant stars. These problems have not been overcome to date, Figure 15.14.

A STRETCH IN TIME

A breakthrough came in 1994 with creationist physicist Dr Russ Humphreys' pioneering work *Starlight and Time*, still a mind-popping and important read. Humphreys reminded readers that we have long known from Einstein's General Relativity, and the experimental support for it, that *time* does not flow at a constant rate everywhere, Figure 15.15.

Fig.15.15: This gold-plated record was sent into space on Voyager 1 and 2 spacecraft carrying languages and samples of man's existence into outer space. Curtsy Image by NASA

Clocks in gravitational fields of different strengths, for instance, will run at different speeds. And since we know that God ***'stretched out the heavens'*** at creation, this opens the door to all sorts of gravitational (and thus time) effects. The whole universe could have been created in six earth-rotation days, 6,000 years ago, as measured by Earth clocks, while there was simultaneously enough 'time' (by galactic 'clocks') for light to reach us from billions of light-years away.

Humphreys said that it was unlikely that his was going to be the 'last word' on the subject, and he encouraged others to develop these ideas further. Now Dr John Hartnett, a well-known creation scientist, have done just that, in *Starlight, Time and the New Physics*. For some time now, with the support of the University of Western Australia where he is an Associate Professor, he has been engaged in research on what he calls the 'fudge factors' in big bang thinking. These are mysterious unknown forms of matter and energy (called 'dark matter' and 'dark energy'). Their existence is not directly observed, which is why they are called 'dark'. They are postulated in order to try to explain certain observations about the rotation of galaxies and the expansion of the universe, for instance, and to solve some of the problems of the big bang. Additionally, the author is providing a distinctive understanding of DarkMatter/Dark Energy in chapter 28.

However, in Hartnett's book, he reminds us of other instances in the history of physics where weird unknown things were proposed (a hidden planet lurking behind the sun, Figure 15.16, for instance) to try to 'explain away' uncomfortable observations that did not fit known physics. This required new physics, for example Einstein's General Relativity, which added to, without discrediting, Newton's laws of gravity.

The new physics in the book's title is the expanded form of relativity of the late secular Israeli cosmologist Moshe Carmeli. The book is designed for the intelligent layman, with all the supporting data in technical appendices suitable for your local neighborhood cosmologist (or village sceptic) demanding equations. Hartnett shows how exquisitely the *observational data* fits Carmelian cosmology, with no need for any dark matter or dark energy at all, seriously undermining big bang-ism.

Fig.15.16: *A Planet Lurking Behind the Sun – Curtsy Getty*

THE UNIVERSE HAS A CENTER!

Fig.15.17: *Is There A Center of The Universe? Curtsy - ELI SHAYOTOVICH*

But this is only the beginning of the exciting implications of this revolutionary finding by the author of this book in addition to the book by Hartnett. The Biblical accounts and the recent observational data that overwhelmingly leads to the conclusion that the universe must have a center, with our galaxy somewhere near it. By:

- plugging this 'galactocentric' universe into Carmeli's equations, and

- then adding the biblical 'stretching out of the heavens' by God at creation (big bangers can scarcely disagree, as they call something like that 'inflation'), what 'falls out' is an astonishing by-product, Figure 15.17. Namely, that there are built-in gravitational time-dilating effects such that:
  1. Adam would have seen light from most of the same stars we see today when he looked up on the sixth day and
  2. light from even the most distant quasar, billions of light-years away from Earth, would reach us now within the six or so thousand years since creation.

Please observe that this is not the product of wishful thinking, or mathematics designed to give that outcome—just a straightforward chain of scientific reasoning:

a. Carmelian cosmology fits the *observational* data overwhelmingly, data which has been so puzzling that mysterious (and now unnecessary) fudge factors have been invented.

b. *Observations* also indicate that we are in a galactocentric universe.

c. Combining a) and b) mathematically in addition to the biblical account shows that it is *inevitable* that an initial rapid ***'stretching out'*** completely eliminates the so-called light-travel time problem for biblical creation.

As the author of this book and John Hartnett would agree, only the Bible, not science, gives us absolute certainty. Models are there to be overthrown, or developed further—in time, new physics may be needed again for future conundrums and discrepancies. But at the least, John's book shows that it is totally misguided ever to simplistically assume that the Bible's account has been discredited by science.

These are indeed exciting times for creation cosmologists, and we hope and pray that the eyes of many thousands now racked by doubt due to this once seemingly intractable issue.

TIME TRAVEL PROBLEM

Solutions proposed for the starlight-travel-time problem by several creation scientists, cosmologists, and the author fall within one of five categories. Probably only two of the categories hold any hope of a solution, Figure 15.18. However, any solution must be self-consistent and the type of solution adopted affects which astronomical arguments can be used as valid evidence for a young universe. A new cosmological model, of the same class as Humphreys' white-hole cosmology, is presented, which fits the observational evidence from the cosmos.

Fig.15.18: Universe Travel Time - Curtsy - FlashMovie/Shutterstock

GALAXIES LIGHT-TRAVEL-TIME TO EARTH

As has been often repeated in creation scientists' literature, the starlight-travel-time problem is particularly important to solve. The problem is simply that in the time available since creation (about 6,000 years) there has not been enough time for light to get to Earth from even the nearest neighbor galaxies (1.5 to 3 million years travel time at constant speed of light *c*) let alone the most distant galaxies (billions of years travel time at constant *c*). How then do we see them and how did Adam see them?

Fig.15.19: *Faster than the Speed of Light? Curtsy - Christopher S. Baird*

One common solution that has been presented, and continues to appear, is that the speed of light was enormously faster around Creation Week and has slowed down since (*c*-decay). A good example of this may be found in a book by Burgess, which has recently been reviewed. The review describes a rapid aging process for stars and a faster speed of light, Figure 15.19.

The universe was accelerated like fast-forwarding a videotape, and after all the light information reached the Earth the rates were reduced to what we now measure.

The problem with this model is that the stars would disappear from view as the light slowed down, subsequently taking millions and billions of years to get to Earth. Also, such light arriving at the Earth would show enormous observable blueshifts. It doesn't. A more ingenious mechanism is needed to overcome such obvious objections.

R.E. Kofahl describes an appealing scenario of the ***heavens being stretched*** out and the speed of light being up to 600 billion times the present value. Again, this presents the same problem: once the speed of light slowed down, how do we now see the stars?

The stars provide us with information in the starlight that we see. If the speed of light had been enormously faster in the past, we should be able to detect that in the starlight. Unless a plausible mechanism can be demonstrated, that doesn't lead to absurd physical implications, these types of scenarios will always fail.

NO VALIDITY TO LONG AGES UNIVERSE

As an argument against the validity of long ages in the universe and for recent creation, it is not uncommon for creation scientists to point out some astrophysical feature (e.g., the high dispersion velocities of stars in galaxies) that is inconsistent with the assumed long ages in big bang cosmology. The claimers then use this as evidence for short ages (i.e., 6,000 years) in the cosmos, consistent with a creationist view. But surely that type of argument is only valid in the framework of the creation scientists model adopted.

You can't have a Humphreys' type model, with time running faster in the cosmos than on Earth and as a result billions of years pass, *and* use the short age argument together. Within the framework of the adopted model, for example, there may still be insufficient time for the observed spirals to wind up. In the big bang conjecture, all galaxies in the universe formed at the same epoch only a billion years after the big bang, which is alleged to have occurred 12–18 billion years ago. So, the question may still be asked, 'why are there still spirals?' Why haven't they all wound up? This would still be a cogent creationist argument. Self-consistency is essential or we have no argument, Figure 15.20.

Fig.15.20: How Do We Know the Age of The Universe? Curtsy - Constantine Johnny via Getty Images

The whole underlying problem may be a reluctance by creation cosmologists to break with the idea that *time is absolute* and that it has always flowed at a constant rate all throughout the universe. Humphreys' white-hole model made such a break and has generally been well received by creation scientists. Probably this is because his model involves accelerated time increments happening in the cosmos *during* 24 hour periods on Earth. It needs to be made very clear that in the cosmos billions of years of ordinary Earth time may have passed, while only 6 x 24-hour days passed on Earth. But a valid mechanism describing how this happened has yet to be discovered, Figure 15.21.

Fig.15.21: Six-Day Creation – Curtsy Answer in Genesis

'ECONOMY' OF MIRACLES

The Humphreys' model uses an 'economy' of miracles and as a result relies heavily on a particular solution of Einstein's field equations from general relativity to explain the mechanics of the cosmos. In terms of apologetic value, this approach is very appealing but observationally there are difficulties. Also, it is important to remember that God was not bound to any laws of physics until the end of the Creation Week. After it ended, the Word says 'He rested'. Maybe the solution to the starlight travel time problem is in this fact that the conservation laws we observe today were not yet all operating.

WHERE ARE THE SOLUTIONS

There are five possible areas of explanation, all consistent with the text of Genesis, that still maintain the 6 x 24-hour literal days. They are:

1. That the language of Genesis is phenomenological language (describing appearance). In this case, stars were made millions and billions of years before Day 4, but in such a manner that the light from all stars, no matter how far away, all arrived at the Earth on Day 4 and so would have been seen first at that moment. This is then a reference frame time-stamping events from that moment they are seen on Earth. Newton's time convention describes this idea. The long-term survival of this model, lies with scriptural interpretation, for example, whether the phenomenological view is consistent with *Ex. 20:9,11*, which reads, '**Six days you shall labor, and do *all* your work: For in six days the LORD made heaven and earth, the sea, and all that in them is, and rested the seventh day** ... '. The emphasized word 'all' seems to restrict the work being done before to the Creation Week period where 6 days pass on Earth. The phenomenological interpretation puts the actual physical creation of the stars before the six days begin and is 'seen' as happening on Day 4 on Earth. Note that Newton's physical interpretation is questionable.
2. That clocks in the cosmos in the past have run at much higher rates than clocks on Earth. Especially during Creation Week, clocks of the exact same type on the edge of the universe ran something like 10^{13} times faster than clocks on Earth and therefore light from such regions had plenty of time to get to Earth in a matter of days, not millions or billions of years. The Burgess model is of this type. This hypothesis is not as simple as it first seems and the light coming from the cosmos carries information that makes the model testable. We can compare clock rates on Earth today with clock rates in sources on galaxies in the cosmos and we should still see a difference. However, I contend that there are no observations that support this hypothesis. In fact, observational evidence suggests the contrary. Light from those sources that have faster clock rates should be blue shifted compared to Earth clocks. It is not.
3. That clocks on Earth in the past have run at much slower rates than clocks in the cosmos. Especially during Creation Week clocks of the exact same type on Earth ran about 10^{13} times slower than clocks at the edge of the universe and therefore light from the edge of the universe had plenty of time to get to Earth in a matter of days as recorded by Earth clocks, not millions or billions of years. Humphreys' model is of this type. The perception of time to someone on the Earth looking at astronomical clocks, during this period, would be that they are running very fast. The hypothesis is simpler than number 2 and not equivalent. It is important to realize that this description requires that the universe have a preferred frame of reference. There is evidence that this is the case and it appears the Earth is actually near the center of the universe. The language of Genesis puts the Earth in a reference frame that is special, in the center of God's will and plan. A new model of this type is suggested below.
4. That the speed of light was enormously faster in the past, of the order 10^{11}c to 10^{12} c. This may have been the case during Creation Week and then the light slowed enormously to the present value. Again, this model is testable, especially with astronomical observations, such as measurements of the fine structure constant. This hypothesis has been advanced in the past by creationists Setterfield and Norman, who placed considerable weight on the precision of a few historical astronomical determinations of the speed of light. The idea is currently in vogue in the secular community, but they are not dealing with timescales on Earth of only 6,000 years. The observational evidence available to us today clearly precludes this model. It is absolutely not viable, unless there is and has been a complicated balance of changes in many 'so-called' constants over observable history. But "Occam's razor" would tell us that this is not the case. Another model in this category is the Harris model. It starts with an infinite speed of light at creation. Then, after the Fall, it changes to the current value as a function of time and linear distance from Earth.

Like an expanding bubble spreading out through the universe, the speed of light drops from an infinite value to the current value at the surface of the bubble. One problem with this model may be the massive blueshifts resulting from a change of infinite to finite speed of light. Also, the fine structure of the atomic spectra must change from a stage of no fine structure to the current state as the bubble passes. This would be observable in starlight. It isn't.

5. **Mystery and miracles**! This last option I have to include because the Creator God revealed in the Bible *is* a **God of miracles**. It is probably true that if we were looking a miracle in the face, we might try to reason a naturalistic mechanism for it. ***God does intervene in the physical world*** and during those times the laws of physics are obviously 'put on hold' (or rather, added to). However, ***I don't believe God commits fraud***. Creating a beam of light from source to observer so that the observer appears to see current information must also mean there is a whole stream of information in the beam that is false. But the question may be asked whether God created the light from the stars just outside the solar system that carries current and accurate information from those stars? Yes, He could have, ***but when it is a miracle, it is usually understood*** and/or revealed. For example, when supernova 1987A exploded in the Large Magellanic Cloud, did it explode 200,000 years ago or in 1987? God could have miraculously translated the light across 200,000 light-years distance of space instantly (as if the photons passed through a wormhole) and then just outside the solar system let it move at the speed *c*. This hypothesis is untestable and seems implausible.

HUMPHREYS' WHITE-HOLE COSMOLOGY

Humphreys' White-Hole Cosmology (HWC) model is an excellent attempt to address this important question in creationist cosmology. However, it seems to suffer from a few deficiencies. In this model, all the matter of the universe expanded out through a 'white hole' during Creation Week to form the cosmos. At the same time space expanded with the matter, moving by virtue of that expansion. Due to gravitational time dilation, clocks on Earth near the center of this spherically-symmetric, bounded and finite distribution of matter ran slower than clocks throughout the cosmos. The farther out one looks the faster clocks would appear to run compared to Earth clocks. But because Earth clocks are, at least initially, deep in a gravitational well, they are running slow and the clocks in the cosmos are less affected by gravity and run fast. Let's say for clocks free from gravity that they run at a *normal* rate, the same as most clocks run today on or near the Earth. (Let's not concern ourselves with small corrections due to relative motion or gravitational potential near Earth).

If this picture was still the state of the universe that we see today, then starlight would be blue-shifted (a gravitational effect) and that blueshift would be greater at greater distances from the Earth. This is not what is observed. We, in fact, see redshifts that are small in magnitude compared to the required magnitudes for the needed blueshifts. The HWC model however also involves a 'timeless' Euclidean zone where the time coordinate in the general relativity spacetime metric becomes spacelike during the expansion stage. This timeless region collapses as material expands out through the 'white hole' and eventually it disappears as it reaches Earth.

As addressed in another paper, this too has its problems both in its mathematical description and conceptually, as there is insufficient time-dilation locally (between nearby galaxies at least). As a result, there remains a difficulty in explaining how light from nearby galaxies would get to Earth in 6,000 years or less.

A NEW MODEL

We propose a new model of type 3. During Creation Week, all clocks on Earth, at least up to Day 4, ran at about 10^{-13} times the rate of *astronomical* clocks. Actually, the rate is a parameter of the model. All *astronomical* clocks in the cosmos run at the same rate that we would measure any *normal* clock today. They have always done so except under special circumstances where they might have been affected by gravity. During this time the rotation speed of the newly created Earth was about 10^{-13} times the current rotation speed as measured by astronomical clocks, but normal by Earth clocks. By the close of Day 4 the clock rates on Earth rapidly speeded up to the same rate as the *astronomical* clocks. All of this was maintained under God's creative power before He allowed the laws of physics to operate 'on their own' at the end of Creation Week.

An 'observer' on Earth at this time looking at the heavens would have seen apparently accelerated motions. Conversely, an 'observer' outside our solar system would observe apparently very slow advance of time on Earth clocks. In fact, only in an extra-solar system frame of reference would Earth clocks appear to be running slow. This effect would allow millions and billions of years to pass in the cosmos, while only a few 24-hour days pass on Earth. Hence the light from the most distant stars traveling at the normal speed, *c*, would have plenty of time to get to Earth. Of course, I am not suggesting there were any such observers, except the Creator, but He doesn't live within time.

The question might be raised as to the spatial region of this special frame around the Earth where clocks run slower up to or during Day 4 of creation. To be consistent with Scripture it doesn't necessarily need to include the whole solar system. However, it may have, because light from anywhere in the solar system can reach Earth within about 8 hours. If the special frame was confined to the solar system, we could call it 'young'. If the special frame was confined to the Earth only, we could call the solar system 'old'. The difference would make the model testable.

However, to be self-consistent with other evidence that makes the solar system appear 'young', we would place the boundary of the special frame at least outside the solar system. So, then this is consistent with the ***Young Solar System (YSS)*** model. Further investigation is required though to see if this is consistent with other age estimators within our region of space.

Of course, the stars were made on Day 4. In order for Adam to see light from the nearest stars (other than the sun), on Day 6, it is necessary that the edge of the special Earth frame not extend much beyond Pluto.

Therefore, due to the massive time dilation effect, during Creation Week, Adam would have been able to see starlight on Earth coming from the visible stars of at least our own galaxy. The light coming from supernova 1987A travelled most of its journey through a portion of Day 4 of Creation Week, when the Earth clock rates were very slow. It arrived at the Earth in 1987, some 200,000 *astronomical years* after it departed.

This model is simple in design and makes no unusual predictions about past events. It is similar to Humphreys' model with some important differences. Time after the end of Day 4 is linear in the whole universe and may be understood in the normal commonsense way.

Time during Creation Week up to Day 4 is highly non-linear but only on Earth (and possibly the surrounding solar system), and nowhere else throughout the cosmos. (Please observe: the HWC model employs different rates of clocks and different passage of time in the cosmos in a highly non-linear fashion, which should be detectable from Earth today.)

In our model, the general matter distribution of the stars and galaxies in the universe is the universal frame of all reference clocks. Generally, these *astronomical* clocks have ticked at the same rate. Clocks on Earth since Day 4 also have ticked at the same rate as these universal clocks. Only clocks on Earth up to the close of Day 4 ticked much slower compared to the universal reference clocks. The model does not employ any general relativistic effects as does HWC but it doesn't impose any implausible conditions either. *The Creation Week period, by definition, is not expected to be a period where natural law explanations apply.*

There are a few points about this model that should be stated here:

1. It has low apologetic value, because in terms of extra-solar system observations it makes no unusual predictions.
2. In terms of locally elapsed time since creation, this model does imply that objects within the solar system are much younger than objects outside it. Therefore, even though further investigation needs to be undertaken, there is some evidence for a young sun but it may also be argued that God created the sun mature as it was especially important for life on Earth.
3. There is the question of where and what type of boundary should be postulated that once enclosed the 'slow' zone. Was it a sharp or gradual transition to 'astronomical' clock rates, and what observational consequences might be expected?

CALCULATIONS

Let's do a few simple calculations. Let us suppose that the relative rate of clocks on Earth compared to *astronomical* clocks during Creation Week was

$$\frac{\partial t_0}{\partial t} = 10^{-13} \qquad \text{...(1)}$$

where t_0 represents time on Earth and t represents time in the cosmos (same for all clocks everywhere except on Earth). By integrating over the 24 hours of Day 4 (assuming = 0.003 years approximately), we can calculate the time available in the cosmos for a photon to travel to Earth. It follows from (1),

$$\int_0^t dt = \int_0^{0.003} \left(\frac{\partial t}{\partial t_0}\right) dt_0 = 1013\ (0.003) = 30 \text{ billion years} \qquad \text{...(2)}$$

There is more than sufficient time during Creation Week. And since light now arriving on Earth left the stars sometime during Creation Week, it had plenty of *astronomical years* to nearly get to Earth. The rest of the journey has been made in the 6,000 years since creation. No accelerated speeds have been assumed, just the constant speed of light that has been repeatably measured for the past 300 years. It is not necessary to suppose that light from all stars in the universe arrived by the close of Creation Week, but at a minimum from our own Milky Way galaxy and maybe farther out to the Virgo Cluster of the order of 70 million light years. The specific dilation rate in (1) is an adjustable parameter of the model, which would determine the extent to how far starlight travelled during Day 4.

EXPANSION OF THE COSMOS

The issue of whether or not the universe rapidly expanded during the Creation Week is not crucial to this model; however, it seems the scriptures demand it. Verses like Job 9:8, 37:18; Psalm 104:2; Isaiah 40:22, 42:5, 44:24, etc., may have their fulfilment in an expansion scenario. Since the model provides plenty of *astronomical time* during Days 1 to 4 on Earth, ***God could have stretched the heavens out to the billion light-years scales in this period of time, while forming the stars and galaxies on Day 4***. And light travelling at constant *c* still would have gotten to Earth in little time as measured by Earth clocks. A mature creation that is seen as an expanding universe may also be part of the description.

The amount and passage of time in the cosmos is pertinent to the creation scientists because we need to interpret the evidence within a self-consistent framework of the model we adopt.

Therefore in a model of type 1 or type 3, which incorporate *astronomical* time, explanations of the rotation curves in galaxies, the Tully-Fisher law or the apparent excess of mass inferred from the dynamics of equilibrium clusters of galaxies become an issue to creation cosmologists.

A new model, of a type similar to Humphreys', has been described that allows billions of years to pass in the cosmos but only 24 hours on Earth during Day 4. In this model, the laws of physics are suspended while creation is in progress and enormous time dilation occurs between Earth clocks and *astronomical* clocks. This solves the light-travel-time problem faced by creation cosmologists and makes all astronomical evidence fit the Genesis account. No non-physical requirements are placed on the model.

The author of this book does not conform to this model of creation. The major issue is the concept that God Suspends the laws of physis on Day 4 is inconsistent with God's creation of time. The laws of physics were created simultaneously with the creation of time as they are inseparable. Suspending one from the other creates substantial contradictions to the consistency of creation. I strongly believe that stretching is not only one event to allow light to reach Earth on Day 4 but allowing the universe to maintain stability through continuous stretching clearly observed today as well today experimentally proven accurate.

Admittedly, the life of plants, birds, and marine life would not be affected by this model.

STRETCHING OF THE HEAVENS

One or more leading creation cosmologists have become committed to the view that scriptural references to the '***stretching of the heavens***' cannot possibly refer to cosmological expansion, labelling such a position eisegesis. But have they, in fact, moved in the direction of eisegesis themselves in some of their interpretation of these verses?

There is no strong scriptural reason for creation scientists to be wedded to the view that these 'stretching' verses cannot possibly have any present-day context. If expansion is still occurring, God may intend this to be a present-day witness to mankind. Accepting that some form of expansion might be occurring today does not pre-suppose acceptance of big bang cosmology. On the contrary, big bang presuppositions have led secular cosmology today into a severe crisis, Figure 15.22.

Fig.15.22: God Stretched Out Heavens, Like Tent Curtains – Curtsy

Creation scientists should take advantage of this situation in our apologetics, while also explaining that possible present-day expansion of the heavens, while not required, is also not inconsistent with Scripture. Ultimately, we need balance in how we approach this issue.

THE SCRIPTURAL ARGUMENT

What does the phrase, the '***stretching of the heavens***' (***natah ha-shamayim***), refer to in the Scriptures?

The phrase appears in the books of Job, 2 Samuel, several of the Psalms, Isaiah, Jeremiah, and Zechariah, but it does not appear in Genesis nor anywhere in the Pentateuch. Did God create the heavens (the stars and galaxies) and then stretch these within a fixed empty space, or did He stretch space as a whole and the galaxies' positions within it?

Did all this occur during Creation Week and then end, or are 'the heavens' still being stretched today? Alan Pace recently referred to Dr Russell Humphreys, "now famous 17 verses on the stretching out or spreading out of the heavens" in the latter's 1994 book, Starlight and Time. The implication of this analysis, also citing Dr John G. Hartnett and Dr Charles Taylor, is that the Hebrew word ***natah*** should not be interpreted as relating to expansion. I respectfully disagree. I believe that an expansionist interpretation of these verses, that is, the view that a 'stretching of the heavens' may still be occurring today, remains perfectly reasonable within their scriptural context. Whether this is actually occurring today, however, remains an open question.

THE HARTNETT/HUMPHREYS VIEW

Formerly, Hartnett thought these verses could describe a cosmological expansion of space,4 but he no longer believes that. He asserts that the very "idea that the biblical text could at all allude to expansion of space … now seems quite preposterous." He published this revised view in 2011. Thereupon Humphreys, who previously held an expansionist perspective on the 'stretching' verses as well, also came to change his point of view, agreeing with Hartnett. Hartnett has recently reaffirmed his view that the Hebrew verbs cited in these passages "cannot be used for [describing] cosmological expansion". Those who might hold the latter view he accuses of eisegesis. The expansionist view is more and more under assault within the some creation scientists community—it is apparently seen as merely an extension of presuppositions related to "Big ang" cosmology (inflation, dark energy, etc.). Whether true or false scientifically, the idea that possible present-day expansion in the context of the 'stretching of the heavens' verses should be rejected on scriptural grounds is, in my opinion, unwarranted.

The 'Hartnett/Humphreys view', as I shall refer to it here, constrains how we should interpret many of these verses by limiting them to the idea of the heavens and/or of space being able to be so stretched only so far 'as a tent'. Humphreys asks: "why would God compare the material being stretched to such materials as tent curtains, which can extend their dimensions by only a few percent before tearing?" It may be a good question to ask with respect to the model that Humphreys is developing, but perhaps it is the wrong question to ask with respect to what I believe to be the chief intention of Scripture in these passages.

FABRIC OF TENT/CURTAIN ANALOGY

In my assessment, the Hartnett/Humphreys view places an unwarranted focus on the question of fabric, both in presenting a particular creationist model and in critiquing "the rubber-sheet analogy of modern big bang cosmology." While the fabric referred to in these verses is not irrelevant in discussing this topic, it may be secondary to the actual role of the tent in nomadic life and to what the biblical writers may have been intending to convey.

A desert-based Bedouin tent is the closest thing we have today to understanding how the nomadic ancient Israelites might have viewed and responded to the scriptural descriptions of the heavens 'as a tent,' Figure 15.23. When one thinks of an ordinary tent for a poor Bedouin family living in Israel or Jordan today, it is not usually very large. However, it is home to that family. It is also a place of refuge from the dangers of the desert and all that is outside the tent. Meanwhile, the head of the family is the master of everything that takes place inside the tent. It is his domain. The tent sets a boundary, so to speak, of the master's absolute authority.

Fig.15.23: Bedouin Tent in the Desert

Within the scriptural context, God is the Master of the heavens—they are His domain; they are His tent. His authority and power extend throughout. These are some of the broad concepts that the writers of Scripture sought to convey, I believe, when, under the guidance of the Holy Spirit, they used the phrase 'as a tent' or 'as a curtain.' I don't think that they were focusing on the fabric of the tent or curtain nor trying to describe a specific kind of cosmological model. Others may have a different view.

There is safety and security in the tent. All of the ancient Near East nomadic customs of protection are conferred upon the guest who has come to visit and is under the roof of the tent. One can picture a visitor lying down on a carpet inside the tent, and, perhaps while eating a meal, looking up through an opening in the tent at the great expanse of the night sky overhead. In a similar way, God's tent—the heavens—cover all of us; we are under His power, authority and protection.

In Bedouin culture today, the greater or grander the power and authority of the Bedouin sheik or personage, the larger the tent will likely be to house his family, relatives and guests. A pastor-friend of mine in Israel and his wife have been guests among many Bedouins in Israel and Jordan over the years and know Bedouin culture very well. They were once feted by a powerful sheik in a large Bedouin tent. They agree with the general proposition that 'the greater the personage, the greater the tent', and also with the view that a tent might be enlarged as needed to entertain a greater number of guests. Thus, there is no reason to assume that the analogy in Scripture to the heavens being stretched 'as a tent' must necessarily be constrained or limited by a single type of fabric. The tent might simply be expanded as needed, irrespective of the fabric used.

Psalm 104:2, Isaiah 40:22, Isaiah 54:2

Psalm 104:2 is one of the key verses that refers to "stretching out heaven like a [tent] curtain." In that same verse we also see God covering Himself "with light as with a cloak". He is "clothed with splendor and majesty." These descriptions depict God's greatness as being far beyond anything man can comprehend. 'Stretching the heavens' is expressed in the same vein.

'God's tent' represents the vastness of Creation. This is poetic language describing the greatness of God and His power, not an exact physical description of a universe that can only stretch as far as some fabric in a tent. This view is echoed in Isaiah 40:22, where the scripture describes God stretching the heavens "like a curtain ... like a tent to dwell in". Note that it is God (not man) who would be doing the dwelling if He sought to do so within the vastness of space! The incredible vastness of the universe was a notion beyond the wildest imaginings of the ancients. Majesty and greatness are the attributes conveyed in these passages, which seem much more aligned with the concept of vastness than with the image of a constrained universe where the fabric of space or of the heavens themselves might tear if stretched too far.

Humphreys also seeks to enlist Isaiah 54:2 on behalf of his position: "Enlarge the place of your tent; Stretch out the curtains of your dwellings, spare not ...". He writes:

"It is likely that the outer coverings of the tabernacle in the wilderness were stretched taut ... to prevent them from flapping in the wind. This is an example from Old Testament times of applying tension to a fabric without having much extension of its length or width."

However, this passage actually implies the exact opposite meaning in my view—it is not discussing tension, tautness, or constraint, but rather expansion, growth, and increase. God was preparing His people for something way beyond what they had been used to. Linking Isaiah 54:2 to the Tabernacle and to the question of 'stretching the heavens' seems, if the reader will pardon the expression, quite a stretch in itself—we need to dispense with that image entirely.

Our focus should instead be on why the Israelites are being told by God to "enlarge the place of [their] tent". The reason He is telling them this is because they "will spread abroad to the right and to the left ... your descendants will possess nations. And they will resettle the desolate cities" (v. 3). There is no sense here of limitation or constraint. Instead, the 'tent' of Israel is to be greatly enlarged to accommodate all of the new territory and nations that will come under the Israelites' purview. The Lord is saying to Israel, in effect, 'Get ready, I am going to expand you beyond anything you can imagine'.

STRETCHING EVENTS

According to Genesis, in the beginning God made the expanse [raw-kee-ah] and called it 'heaven' (Genesis 1:8). Neither placing the lights in the expanse of the heavens, nor the creation of the stars (Genesis 1:14–16) necessarily refer to any stretching or spreading. Later, when the Flood takes place and the floodgates of heaven are opened (Genesis 7:11; 8:2), there is also no reference to stretching. The idea that the 'stretching of the heavens' had to occur during Creation Week is not demanded by Scripture. It might have occurred then, or it might have occurred at a later time, or there may have been a combination of these events.

Hartnett and Humphreys, however, assume that most or all of the 'stretching of the heavens,' whatever it consisted of, occurred during Creation Week. Pace agrees. Humphreys claims: "Many of the seventeen verses connect the stretching with events of the Creation Week." He concludes that "the stretching (an increasing of tension) occurred during the first six days of Creation, and was completed (stopping the increase of tension) during that period." While possibly true, this is unsubstantiated on scriptural grounds in terms of forcing us to accept that it had to occur during Creation Week. This is because there are no passages in Scripture that directly connect the 'stretching of the heavens' with the act of Creation. It is merely an assumption by some creationists.

Job 9:8

Job is believed to be one of the oldest books in the Bible. Job 9:8 says: "He alone stretches out the heavens and treads on the waves of the sea." The context seems to be in the present tense, not the past. Consider the last part of the verse: "[He] ... treads on the waves of the sea ...". Think of Jesus Himself walking upon the water (Matthew 14:25 and elsewhere). The image of God treading on the waves of the sea is a present description of His continuing power—it has nothing to do with Creation Week. Since the last part of Job 9:8 is clearly not tied to Creation Week, we need not assume that the first part of the verse must relate to Creation Week.

Irrelevant Verses

Looking at the Hebrew text, Humphreys asserts that several of the 'stretching' verses "are qal perfect, implying a past action", while two other verses (2 Samuel 22:9–10 and Psalm 18:8–9) "follow a qal perfect verb with a waw consecutive prefixing a qal imperfect verb, which implies past action." But these two particular passages (2 Samuel and Psalm 18) relate to God coming down and 'bowing the heavens' in a theophany. Hebrew scholar David Brewer states: "The imagery here is similar to what we see when the Lord descended to Mount Sinai (Exodus 19:16–19)." These verses are about God's judgment and rescue, not about Creation or Creation Week. These passages are irrelevant to the question of when the 'stretching of the heavens' occurred and should be dropped from the discussion.

PAST ACTION - CONTINUING EFFECTS

Humphreys also refers to Isaiah 45:12; 48:13 and Jeremiah 10:12; 51:15, stating that these are all 'qal perfect, implying past action.' Yes, these verses do imply past action. But does that mean that the action has been completed? Isaiah 40:22 may provide greater insight to our understanding. Here we see ***natah*** used as a ***qal*** active participle (He '… is stretching'), followed by a waw consecutive with the verb maw-thakh' as a qal imperfect, which might be translated as 'and He has spread them out like a tent …'. So, we may have past action in some of the relevant passages, but this does not necessarily signify completed action.

Apart from how we may interpret these verses, Humphreys asserts that, even with all of the stretching (an increase of tension) of the heavens occurring during Creation Week, the "results of the increase, such as a slow increasing of the gravitational potential of the cosmos, could still be occurring to this day."

In other words, we might have an action during Creation Week (the initial stretching), but there may be after-effects of that stretching up to the present day (in Humphreys' view, possibly "a slow increasing of the gravitational potential of the cosmos"). But in that context, if there are any after-effects of past stretching into the present day, regardless of what they are, then it is also plausible to infer that current expansion might be among those after effects in terms of how we interpret the text.

Viewed in that context, the debate over whether these verses must be interpreted as implying past completed action collapses. The after-effects of the stretching, whether they occurred during Creation Week or later, could be being described by the writers of Scripture as having *both a past and present component* to them. This doesn't mean that present-day expansion is occurring, but it does mean that we have no strong reason to rule it out on scriptural grounds.

It is also very clear that God has a present-day relationship with the stars and galaxies. The scripture says: He "leads forth their host by number; He calls them all by name". And because of the greatness of His power, "Not one of them is missing" (Psalm 40:26). The heavens today "are telling of the glory of God … their expanse is declaring the work of His hands" (Psalm 19:1).

Not one of the stars is lost or misplaced. God knows where each one is, and each has a specific place—what a wonderful analogy this was not only for ancient Israel but also for us today! Isaiah compares this incredible truth to God's relationship with the Jewish people, asking: "Why do you say, O Jacob, and assert, O Israel, 'My way is hidden from the Lord…?'" (Isaiah 40:27). God's 'stretching of the heavens' is one of several witnesses primarily to the Jewish people that He is a God who keeps His covenants.

A Witness to Mankind

Much of Isaiah 42 is devoted to describing God's majesty and His care for His people. Verse 5 refers to "God the Lord who created the heavens and stretched them out …" The 'creating' and the 'stretching' could be interpreted as two separate events. If so, Scripture places no obligation on the text that the latter event must have occurred during Creation Week. Whatever the case, both events are witnesses of God's power to feeble mankind.

Isaiah 48:13 has a similar passage. The Lord "founded the earth" and His "right hand spread out the heavens." Next, He says, "When I call to them, they stand together." God may have been stretching the heavens and then fixed them in place, or this passage may mean something else. We see God creating, stretching and calling out. The purpose of the passage is not to give us an exact chronological description but rather to give us a glimpse of God's majesty and power. This passage serves as a witness to us and especially to the people of Israel, to get their attention: "Listen to Me, O Jacob, even Israel whom I called; I am He, I am the first, I am also the last …" (Isaiah 48:12).

The notion of a limited, barely stretchable 'tent/curtain' view in these verses as the only view consistent with Scripture is unsupported. We all must be careful to avoid placing our own preconceived ideas onto Scripture, seeking the Holy Spirit's aid at every turn to interpret Scripture properly. The question of the 'stretching of the heavens' and what that really means remains unresolved. While one can agree with Hartnett "that it is not possible to categorically state that Scripture requires that the universe is expanding at all [emphasis in original]," the possibility that the heavens may currently be being stretched (expanded) is also not inconsistent with Scripture.

THE APOLOGETICAL ARGUMENT

The apologetical argument flows from our assessment of Scripture. Secular observers today certainly believe in a currently expanding universe. That, of course, does not make it true, but a simple belief in cosmological expansion today— stripped of its "big bang" presuppositions—is not unscriptural in the way that, for example, belief in macroevolution is. The idea that the 'stretching of the heavens' might still be occurring today should not automatically be equated with what Hartnett calls the 'dark science' regarding "the notion of expansion of the fabric of space." Perhaps there is another alternative.

The Purpose for 'Stretching the Heavens'

What was or is the real purpose of God's stretching the heavens?

We don't know—perhaps it has a utilitarian purpose so that the universe is not static. However, the broader purpose of stretching the heavens, as Scripture makes clear, was and is to be a witness to mankind of God's greatness and glory! This is extremely important. The numerous references to 'stretching the heavens' are mostly apologetical in tone and intent. They imply that mankind should be aware of this stretching, that knowledge of its existence or occurrence would be an argument for showing God's greatness and power. Here then, perhaps, is a question at least as important to ask as the question about the nature of the fabric of the tent (or space)—why would God tell us in His Word over and over again that He had stretched the heavens unless He also provided some evidence that He had stretched or is continuing to stretch the heavens?

God is speaking to us through His inerrant Word, telling us that He has indeed 'stretched the heavens' in the past and/or is still stretching the heavens in the present and that this is evidence of His glory. And who is supposed to be the primary audience for this evidence? —the people of Israel, the Jewish people.

In Zechariah 12:1, the 'stretching of the heavens' is cited by the Lord as one of three events defining His majesty and power and proclaiming what He is yet to do with the people of Israel. Those three events are:

1. stretching the heavens;
2. laying the foundation of the Earth; and
3. forming the spirit of man within him.

In this passage, God is citing His credentials as a witness and testimony that He is yet to do something that is extraordinary and beyond human understanding: He is going to redeem unrepentant Israel. ('Redeeming Israel' is also related to God's promises in the New Testament in Romans 11.)

When Zechariah penned his words under the guidance of the Holy Spirit, he could have had no understanding (unless God revealed it to him supernaturally) of the incredible vastness of space, of hundreds of billions of galaxies, or of the speed of light. But God Himself, speaking though the prophet, uses this example of stretching the heavens to show His power and authority.

The other two events—laying the foundation of the Earth and forming the spirit of man within him—were things that Zechariah would have understood. But 'stretching the heavens? As a prophet, he was only repeating what the Lord had told him. God used Zechariah as His mouthpiece to give "The word of the Lord concerning Israel." And that word had to do with the 'last days' and what God is yet going to do when all nations of the earth are gathered against Jerusalem (12:3). God will deliver the people of Israel, and then they shall look on Him whom "they have pierced" (the Messiah) (12:10).

The 'stretching of the heavens' in Zechariah thus stands as a testimony against the world by the living God. In Isaiah 51, the Lord rebukes the children of Israel, saying, you "have forgotten the Lord your Maker, [the One] who stretched forth the heavens and laid the foundations of the earth ..." (Isaiah 51:13). These scriptures are primarily about the Jewish people. They are intended both as a witness for salvation to those who will believe and a witness against those who will not, both in the prophets' days and in our own.

MISGUIDED COSMOLOGISTS

Many of the best-known cosmologists today are self-proclaimed atheists of Jewish background. They accept cosmological expansion as a given, based on big bang cosmology, including alleged inflation of the universe, and related factors. As a creation scientidst, I reject those presuppositions, as well as attempts to derive an age of the universe based on them. At the same time, believing that these passages might refer to present-day expansion of the universe does not thereby obligate me to accept the whole rotten edifice of big bang cosmology. Can these two things be divorced from each other? That is the question.

The biblical passages about the 'stretching of the heavens' are stark statements made centuries ago by the Hebrew prophets—perhaps, in part, as a witness to our own unbelieving generation. To me, this is really more an apologetical or missiological question at this point in time rather than primarily an academic or scientific one. We are in a struggle for men's souls. If pointing to the possible expansion of the heavens can challenge non-believers to look into Scripture and God's promises, I believe that we have a wonderful opportunity here to use this as part of our witness. Meanwhile, I am also concerned that we, as the creation scientists community, do not, as the saying goes, 'cut off our nose to spite our face' by unnecessarily dismissing one possible interpretation of Scripture at a moment in time when it can give us great advantage.

Their Cosmology may be Used Against Them

I have always deeply appreciated Dr Henry Morris's book, That Their Words May Be Used Against Them and the spirit behind it regarding evolutionists' quotes and worldview. (This year is also the twentieth anniversary of its publication). We can certainly use similar approaches in our apologetics when it comes to aspects of big bang cosmology without embracing the big bang itself, in the spirit that 'Their Cosmology May Be Used Against Them.' What does modern-day cosmology claim?

Three of the most important indicators that appear to support cosmological expansion include:

- redshift measurements of distant galaxies according to Hubble's Law;
- the predictions of general relativity; and,

- based on supernovae data, the change in the rate of expansion derived from measurements of the purported cosmological constant.

Hartnett has examined these and related factors in depth and their pros and cons with respect to expansion in two key articles that appeared in writing.

REDSHIFT AND HUBBLE'S LAW

As readily admitted by a leading cosmologist, while Hubble's Law may be "almost exactly true nearby, [it is] … not necessarily true over a large fraction of the observable universe." Beyond that, Hartnett has recently shown that 'the greater the redshift, the greater the distance rule' upon which big bang cosmology and the Standard Model of expansion are based may not hold for quasars and active galactic nuclei (AGNs). Thus, while redshift measurements may indicate an expanding universe (and most cosmologists believe that they do), Hartnett has noted that some of the data can fit "a static universe with a simple Euclidean non-expanding space just as well as … the standard concordance BB model, Figure 15.24"

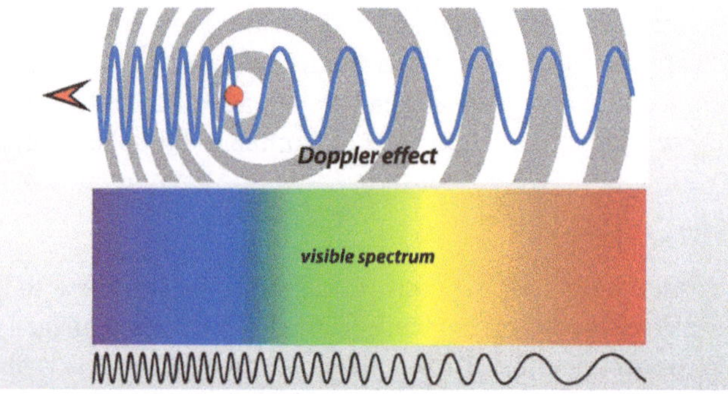

Fig.15.24: Red Shift Hubble's Law

GENERAL RELATIVITY

General relativity predicts that the universe is expanding. Whether this is interpreted as other galaxies moving away from us at tremendous speeds or whether space itself is expanding, general relativity asserts "that these two equivalent viewpoints … are equally valid," General relativity also allows for the possibility that space itself may be expanding faster than the speed of light: according to physicist Max Tegmark, "while nothing is allowed to move faster than light through space, … space itself is free to stretch however fast it wants to."

Supernovae Data and Cosmological Constant

In recent decades, cosmologists have considered type 1a supernovae explosions as a form of calibrated 'standard candles' used as yardsticks for measuring distances. Celebrated efforts by two research teams in 1998 resulted in measurements of dozens of supernovae in numerous galaxies. Figure 15.25, shows a photo of 1994 types 1a supernova. According to secular cosmologists, these measurements from type 1a supernovae (SN1a) indicate that the universe is expanding and that the expansion rate may be accelerating.

Fig.15.25: Brightest Super Nova Ever Seen – Type Ia Supernova, SN1994d

Hartnett himself stated in 2011: "The type 1a supernova (SN) measurements are the very best evidence for an expanding universe [emphasis in original]." At the same time, he also examines the many assumptions that are built into the interpretation of that evidence. For secular cosmologists, however, the supernovae measurements confirmed an extremely tiny but positive cosmological constant, known as Lambda (Λ), at a stunning measurement of 10–120 power smaller than what was estimated from theory—a knife-edge so fine as to defy comprehension.

Conflict between theory and observation concerning this value has produced a deep crisis within physics and cosmology today. It is difficult to over-estimate the depth of this crisis, which hit the physics and cosmology communities like a 'proverbial ton of bricks,' according to well-known Jewish atheist physicist Leonard Susskind. Susskind acknowledged: "No missing mathematical logic is going to explain that." Alluding to William Paley's famous watch analogy, self-proclaimed atheist and NASA scientist Carlos I. Calle has referred to this apparent extreme fine-tuning measurement of the cosmological constant as the "the biggest watch of all". Physics writer Brian Greene admitted that when he first heard of the supernovae measurements, his first reaction was: "It just can't be."

This crisis in cosmology has been so extensive that it has helped propel multiverse theory as the only way out for many secular cosmologists, since they cannot believe that a designer designed this universe with such apparent fine-tuning.

In other words, their own theory and observation have led them to this cosmological dead-end (from their perspective). This has then led to the ridiculous notion that there are near-endless trillions upon trillions of universes, or that 'all possible universes exist' and that we just happen to find ourselves in 'one that contains life' (the so-called anthropic principle). Dissenting Jewish atheist philosopher Thomas Nagel, whose critique of the standard neo-Darwinian materialist worldview in his book, Mind & Cosmos, has upset so many of his colleagues, rightly calls this notion of the multiverse a 'cop-out.'

Thus, the supernovae data present a huge problem for secular cosmologists. Further, these SN1as "can equally be telling us that the presumptuous assumption of the Cosmological Principle is not a certain doctrine upon which to build one's worldview". The Cosmological Principle assumes that there is 'no unique center, and no edge' in our universe. This is a separate issue from the question of expansion, but it shows how much impact the supernovae data have had.

Since the rest of the scientific world today assumes that the universe is expanding, we can and should use this conundrum of secular cosmology in our apologetics. We can do so without embracing big bang cosmology with all of its presuppositions. We can point out to non-believers that present-day 'science' has reached a complete dead-end on this issue and is thoroughly confused, making up 'fudge factors' to try to get out of this dilemma, while at the same time showing that, if the universe is indeed expanding at present, Scripture pointed to that fact long ago. This can serve as a powerful witness to our world today.

We know from the Word of God that the heavens (and/or space) were indeed stretched in the past, either during Creation Week or later. They may also still be being stretched or expanded today. There is no scriptural requirement to exclude this possibility. Thus, I believe that Hartnett has gone too far to conclude: "To suggest that these texts describe cosmological expansion of space ... is not justifiable and is pure eisegesis."

If believing that current cosmological expansion might be occurring (with respect to these texts) is eisegesis, then the same charge might be applied to the Hartnett/Humphreys view as well—to the extent that perhaps more is being inferred from the text with respect to the tent fabric analogy than the text supplies.

In 2011, after examining all the current scientific evidence for and against an expanding universe, Hartnett stated: "it is impossible to conclude either way whether the universe is expanding or static. The evidence is equivocal." Given that reality on the scientific front, what we need now is balance on the scriptural side as well, allowing both views to have their proverbial 'day in court' to see where true science leads within a biblical framework.

DISTANT STAR LIGHT AND BIBLICAL TIMEFRAME

When creation scientists do talks around the world, Figure 15.26, one of the most-asked questions people would ask, 'How can you believe in a straightforward biblical time-frame and explain distant starlight?'

The 'Problem' is formulated thus:

- The biblical time-frame is about 6,000 years since creation ("In the beginning God created the heavens and the earth", *Genesis 1:1*)
- There are stars that are millions and billions of light years distant that we can see. Travelling at the speed of light, the light from those stars would take millions or billions of years to travel to Earth, depending on the distance, so that we can see them.

Fig.15.26: *The Genesis account makes it clear that the creation of the heavenly bodies was just as miraculous as the creation of the first people.*

Skeptics of the Bible's history often frame this as a 'gotcha' question because they think that there is no satisfactory answer. Theistic evolutionists, who accept the grand evolutionary tale and 'add God,' often pose this question. Likewise with so-called 'old earth creationists' who wish to defer to the secular timeframe but retain other parts of the Bible's account. They think that such an unanswerable question destroys belief in the Bible's timeframe—and thus Genesis 'must' be poetic, 'just theology,' or 'the days are long periods of time,' etc.

The question does not trouble me for three reasons:

1. Creation Week Entailed a Series of Miracles.

Throughout the account in *Genesis 1*, the Bible says that God spoke things into existence—eight times, "And God said … ". And after He spoke it is often concluded with, "and it was so."

The New Testament tells us:

"By faith we understand that the universe was created by the word of God, so that what is seen was not made out of things that are visible." (*Hebrews 11:3*).

The word of God brought the universe into existence from something that is not visible, something unlike ordinary visible/tangible matter and energy. This is consistent with the scientific conclusion that the matter and energy that comprise the universe cannot be eternal. Thus, the cause of the universe must be *super*natural.

And therefore, the Bible describes the Word of God as ***powerful***:

"So shall my word be that goes out from my mouth; it shall not return to me empty, but it shall accomplish that which I purpose, and shall succeed in the thing for which I sent it." (*Isaiah 55:11*)

The New Testament tells us that this agency of God, the Word by which He created everything, was none other than the Lord Jesus Christ:

In the beginning was the Word, and the Word was with God, and the Word was God. He [Jesus Christ] was in the beginning with God. All things were made through him, and without him was not anything made that was made. (*John 1:1–3*) [Jesus Christ].

Genesis 2 tells us that God made the first man and woman. He took dust and made the man, Adam (*Genesis 2:7*), and took his rib and fashioned the woman, Eve, the mother of all.

I have never had anyone demand that I explain to them how God made a man from dust. And yet there is this demand that we explain how God could have created the stars such that we can see the light from distant stars.

The Genesis account makes it clear that the creation of the heavenly bodies was just as miraculous as the creation of the first people:

And God said, "Let there be lights … And God made the two great lights—the greater light to rule the day and the lesser light to rule the night—and the stars." (*Genesis 1:14–16*).

Was this any less a miracle than the creation of the man from dust? And yet there is a demand for a naturalistic explanation for how God did this! This seems to me to be quite inconsistent and unreasonable to demand such a thing as a condition of believing the Bible's account, especially the timeframe.

It's also interesting that the timeframe of six days with a seventh day of rest—the basis of our 7-day week (*Exodus 20:11*)—underlines the miraculous nature of God's actions. And that is part of the problem for those who don't or won't believe the timeframe, such as theistic evolutionists and long age creationists. When they refuse to believe the timeframe, they then tend to think of 'creation' in a naturalistic way, over billions of years. And then secular ideas of how things came to take precedence over the Bible's clear account. Thus, the miraculous nature of Creation Week takes a back seat and so we have this demand for a naturalistic explanation for how we can see distant starlight.

The major point: Creation Week involved a series of miracles, one after the other. Thus, these are things for which we can provide no natural explanation. We do not know how God could speak the stars into existence, and thus we cannot know how He created things in such a way that we can see the light from celestial objects millions and billions of light years distant.

In other words, the question tacitly denies the supernatural nature of the Creation Week events. In doing so it robs God of His omnipotence and limits Him to work only in ways that we can understand. This results in a very diminished view of God. In effect, those who do this are constructing a god compatible with their own limited understanding, which is a form of idolatry.

2. There are Potential Explanations Anyway.

Christian astrophysicists have proposed various explanations as to how God *might* have created things in such a way that even Adam and Eve would have been able to see distant starlight. This section may involve some technical explanations. However, we will try to make it as easy as possible to understand. The ideas are mind-stretching because they seem to conflict with our everyday experience of the world.

Time Dilation Models

As described in the previous chapter Einstein is famous for discovering that time is not constant but is affected by movement (speed) and gravitational forces. This is known as Special Relativity and General Relativity, respectively.

When an object moves very fast, time for that object slows down, or even stops at the speed of light. Also, when an object is in the presence of a massive object, which provides a strong gravitational attraction, time slows down. These effects are measurable and thus have been verified by experiment. They are so real that GPS satellites, which depend on precision clocks for global positioning calculations, must have their on-board clocks adjusted for the lesser gravity (being up in the sky) and for the speed of movement of the satellite.

Now we can try to imagine God creating the universe, "stretching out the heavens" (*Psalm 104:2*) on Day 4 of Creation Week. This would entail massive gravitational forces and enormous differences in speeds, both of which would change the time 'out there' compared to planet Earth. Thus, in one Earth Day (Day 4) an enormous amount of time could transpire 'out there' allowing ample time for the light to travel to Earth. Additionally, God has stretched the heavens for this purpose and still stretching them to maintain a perfect tuned order of the universe. Various models have been proposed based on these ideas.

Discovering Light/Time

Well, this is not so new; it goes back to Einstein's era. Einstein's findings of time changing with motion means that the speed of light (c) cannot be measured in one direction. It can only be measured for the round trip (two-way speed). This gives the average speed, but we cannot know if the speed is the same in both directions.

Einstein *assumed* that the speed of light in all directions was the same (called the synchrony convention) but he realized that it was an assumption without proof.

Physicists have been wracking their brains ever since Einstein to try to figure out how to measure the speed of light in one direction. For example, we might want to fire a laser beam at the moon and measure how long it takes to get there. However, we need to place a clock on the moon that is telling the same time as a clock on Earth. That is a problem, because as we fly the clock to the moon, the time changes on the clock! Putting it another way, the only way the observer on the moon can know when the light beam is sent from Earth is to be sent a message, which then travels at the speed of light. It all becomes totally circular!

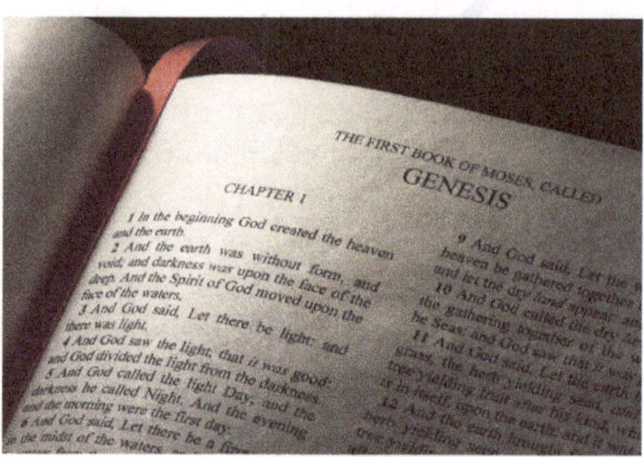

Fig.15.27: God Calls us to Humbly Submit to Him and His Word

There has been quite some discussion about this in recent times, including from secularists—that the speed of light might not be the same in all directions.

So, what if the speed of light towards us was infinite and the speed away from us was c/2? That would give an average speed of c, which we measure. There is no way that we can know that this is not so. And that would mean that light from distant stars would reach Earth the instant that they were created, Figure 15.27. No problem!

3. Everyone Believes in Miracles!

Those who demand a naturalistic explanation (no miracles allowed!) for distant starlight from Christians don't seem to realize that the standard 'big bang' secular view of origins entails miracles—but without a miracle worker!

The problem is that the distribution of the background radiation in the universe is fairly uniform, but there has not been enough time for radiation (at the speed of light) to disperse over such a large universe. This is called the '***horizon problem***'. It is really the big bang's very own 'light time-travel' problem. To 'explain' this, cosmologists invoked a period of super-fast expansion of the universe—much faster than the speed of light—for a brief time just after the 'bang.' This was dubbed 'inflation.' What started it, how it could proceed, and what stopped it are all mysteries. These are in effect naturalistic miracles, with no sufficient cause or explanation. They are used to prop up a theory that would not work without them.

Therefore, it is not that miracles are not allowed in explaining origins. Ironically, they are only disallowed when it comes to biblical creation, which the Bible says is miraculous!

Bible-believing Christians are 'streets ahead' of secularists here because we have an all-powerful God who is able to do things beyond our ken.

Great is our Lord, and abundant in power; his understanding is beyond measure. The LORD lifts up the humble; he casts the wicked to the ground. (*Psalm 147:5–6*)

GOD CALLS US TO HUMBLY SUBMIT TO HIM AND HIS WORD

I was encouraged recently by this testimony from someone who changed to believing God's Word:

"My world view of creation by evolution changed to biblical six-day creation through reading creation material about 30 years ago. It changed my unsure belief into a sure faith, trusting God at his Word.

Thanks for standing on the side of God, we endeavor to call people back to the authority of God's Word.

Light-Travel Time - Big Bang

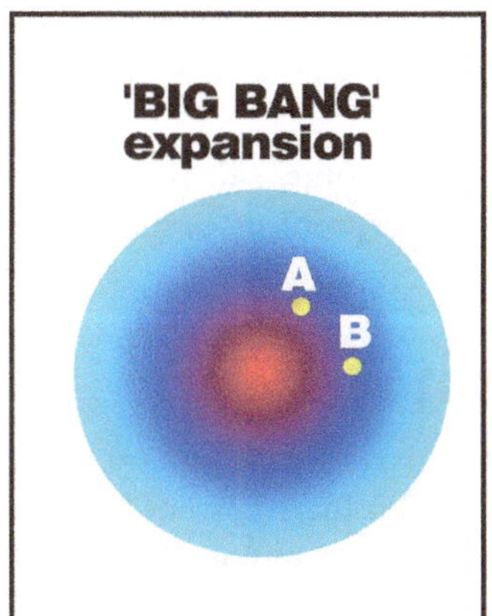

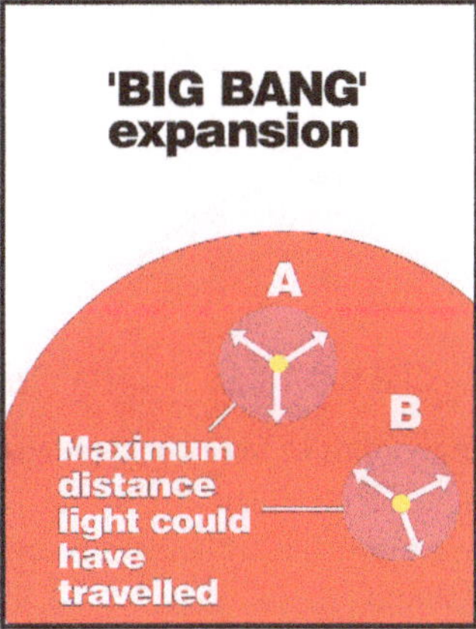

Fig.17.28: (1) Early in the alleged big bang, points A and B start out with different temperatures. (2) Today, points A and B have the same temperature, yet there has not been enough time for them to exchange light.

The 'distant starlight problem' is sometimes used as an argument against biblical creation. People who believe in billions of years often claim that light from the most distant galaxies could not possibly reach earth in only 6,000 years. However, the light-travel–time argument cannot be used to reject the Bible in favor of the big bang, with its billions of years. This is because the big bang model *also* has a light-travel–time problem, Figure 15.28.

Prelude

In 1964/5, Penzias and Wilson discovered that the earth was bathed in a faint microwave radiation, apparently coming from the most distant observable regions of the universe, and this earned them the Nobel Prize for Physics in 1978. This Cosmic Microwave Background (CMB) comes from all directions in space and has a characteristic temperature. While the discovery of the CMB has been called a successful prediction of the big bang model, it is actually a *problem* for the big bang. This is because the precisely uniform temperature of the CMB creates a light-travel–time problem for big bang models of the origin of the universe.

The Problem

The temperature of the CMB is essentially the same everywhere—in all directions (to a precision of 1 part in 100,000). However (according to big bang theorists), in the early universe, the temperature of the CMB *would have been very different at different places in space due to the random nature of the initial conditions.* These different regions could come to the same temperature if they were in close contact. More distant regions would come to equilibrium by exchanging radiation (i.e., light). The radiation would carry energy from warmer regions to cooler ones until they had the same temperature.

The problem is this: even assuming the big bang timescale, there has not been enough time for light to travel between widely separated regions of space. So, how can the different regions of the current CMB have such precisely uniform temperatures if they have never communicated with each other? *This is a light-travel–time problem.*

The big bang model assumes that the universe is many billions of years old. While this timescale is sufficient for light to travel from distant galaxies to earth, it does not provide enough time for light to travel from one side of the visible universe to the other. At the time the light was emitted, supposedly 300,000 years after the big bang, space already had a uniform temperature over a range at least ten times larger than the distance that light could have travelled (called the 'horizon'). Therefore, how can these regions look the same, i.e., have the same temperature?

How can one side of the visible universe 'know' about the other side if there has not been enough time for the information to be exchanged? This is called the 'horizon problem,' Secular astronomers have proposed many possible solutions to it, but no satisfactory one has emerged to date.

IGNORING GOD CREATED - DISCREDIT WISDOM

The big bang requires that opposite regions of the visible universe must have exchanged energy by radiation, since these regions of space look the same in CMB maps. But there has not been enough time for light to travel this distance. While many Biblical creation scientists understand that 'God Created' the universe, from one end to the other vast end simultaneously," there are others still attempting to hypothesize how was the universe created.

However, some creation scientists and big bang supporters have proposed a variety of possible solutions to light-travel–time difficulties in their respective models. So big-bangers should not criticize creationists for hypothesizing potential solutions, since they do the same thing with their own model.

The horizon problem remains a serious difficulty for big bang supporters, as evidenced by their many competing conjectures that attempt to solve it. Therefore, it is inconsistent for supporters of the big bang model to use light-travel time as an argument against creation scientists, since their own notion has an equivalent problem.

Attempts Overcome Big Bang's Light-Travel–Time Problem

Currently, the most popular idea is called '***inflation***'—a conjecture invented by Alan Guth in 1981. In this scenario, the expansion rate of the universe (i.e., space itself) was vastly accelerated in an '***inflation phase***' early in the big bang. The different regions of the universe were in very close contact before this inflation took place. Thus, they were able to come to the same temperature by exchanging radiation *before* they were rapidly (**faster than the speed of light**[13]) pushed apart.

According to inflation, even though distant regions of the universe are not in contact today, they were in contact before the inflation phase when the universe was small.

However, the inflation scenario is far from certain. There are many different inflation models, each with its set of difficulties. Moreover, there is no consensus on which (if any) inflation model is correct. A physical mechanism that could cause the inflation is not known, though there are many speculations.

There are also difficulties on how to turn off the inflation once it starts—the 'graceful exit' problem. Many inflation models are known to be wrong—making predictions that are not consistent with observations, such as Guth's original model. Also, many aspects of inflation models are currently unable to be tested.

Some astronomers do not accept inflationary models and have proposed other possible solutions to the horizon problem. These include: scenarios in which the gravitational constant varies with time, the 'ekpyrotic model' which involves:

- a cyclic universe, scenarios in which light takes 'shortcuts' through extra (hypothetical) dimensions,
- 'null-singularity' models, and
- models in which the speed of light was much greater in the past
- (Creationists have also pointed out that a changing speed of light may solve light-travel–time difficulties for biblical creation.)

In light of this disagreement, it is safe to say that the horizon problem has not been decisively solved. However, the author believes that creation took place simultaneously in a peaceful, and calm manner enabling the CMB to be distributed equally throughout the entire universe from one end to another end.

MODERN SCIENCE IN THE MIND OF CREATION SCIENTISTS

Fig.15.29: Galaxy Triplet ARP 274. NASA, ESA, and M. Livio and the Hubble Heritage Curtsy - Team (STScI/AURA)

As six-day creation scientists, can we know what God did when he created this vast universe?

If we agree that God created the universe, and it was created in a form that is essentially like we observe today—a mature creation—very large, tens of billions of light-years across, Figure 15.29—very old in appearance, in terms of processes we observe—then we have two possibilities within the creationist worldview:

1. God created everything 6,000 years ago and we cannot know how He did it in Creation Week but somehow we can see the whole visible universe, including " ... events which lie entirely beyond our limited understanding of nature;" or,
3. God created everything 6,000 years ago and we can (in principle) know how He did it in Creation Week as much as we are able to see the whole visible universe.

We can take the position that we cannot know how God did it because it was supernatural and beyond our understanding. However, we should not make untenable claims such as that supernovae (exploding stars) represent death and hence must have occurred after the Fall.

(A supernova is a light show resulting from exploding gas. It cannot be construed as death in the biblical sense.) Or even the claim that modern physics (that developed post-1905, starting with Einstein's three papers published in *Annalen der Physik*, Figure 15.30, which dealt with the photoelectric effect (quantum theory), special relativity and Brownian motion) is all wrong.

One idea that has developed is that modern quantum theory, modern special and general relativity and hence modern astrophysics and cosmology, which include both of the latter, are wrong. Some creationists even reject these modern ideas, preferring only classical physics, while others claim we cannot even know the physics of this universe, Figure 15.29.

OPERATIONAL OR HISTORICAL SCIENCE

But there is a big difference between repeatable, operational science and something that is extrapolated back into the unknown unobservable past, what has been called in creationist circles 'historical science'. The science done in the lab, which includes modern-clock tests of special relativity, hence of modern physics, yields reliable repeatable results that are consistent with that theory. It was because of the very notion that the Bible promoted a consistent reliable creation, hence consistent laws of nature, that modern science developed in the first place. It is because those laws are stationary that we can know anything at all about the universe by our own observations.

Fig.15.30: *While working in the Swiss patent office in Bern in 1905, Albert Einstein developed and published his ideas on special relativity, the photoelectric effect (for which he won the Nobel Prize in 1921) and Brownian motion.*

If we are to make the assumption that we cannot know, or that the laws of nature we test in the laboratory are not the same as those we observe elsewhere in the cosmos (excluding the idea that what we do know is incomplete), then we have no basis to test any hypothesis about the universe. Taking that idea further, since we cannot travel to the nearest star, why not suppose that the laws of nature and the structure of the universe are such that all stars lie within a four-light-year radius of Earth? That idea could never be disproved because it is always possible to say the laws and structure of the universe are consistent with this notion. And we would not have a light-travel-time problem.

Scientists offered many ideas, numerous technical papers and offered countless presentations revolving around the rejection of most of modern physics, e.g., trying to find a model for the simplest atomic species without quantum theory.

This does not seem to be appropriate as it ignores the last one hundred years of research. If second position stated above is taken, a straight-forward reading of Genesis as true history, we would not need to say that everything in the universe must be 6,000 years old, as measured by processes in their own frame of reference. That is not contradictory of the creation timeline.

However, those processes measured by Earth clocks must have taken less than 6,000 years to happen. God's creation is knowable and understandable (at least those aspects limited to the physics we know today) to us as humans. He made the universe in a way that is rational and reasonable, and the efforts since the development of modern science, say, over the last thousand years, have revealed a lot of truth. (Of course, along the way we have had to throw out a lot of error.)

MODERN SCIENCE IS RELIABLE

Modern physics, by and large, is reliable; we can test relativity with GPS satellites and even with Earth-bound modern atomic clocks. Every time we use a device with a laser, we are using something developed from quantum theory. Nevertheless, that is not the same thing as understanding what happened in the cosmos, in the past, billions of years ago, based on the assumption of the constant speed of light and the size of the universe, Figure 15.30. It is not so clear, and we cannot interact with the universe like we can with our experiments in the lab. The former is not really repeatable science and hence it is very weak in its predictive power.

However, the second position stated above is another way of saying that we can trust the Lord, we can trust modern science, where it is testable, and we can, in principle, know what God did. One of the reasons is that the laws of nature are God's Laws; He created them. Although the idea of many biblical miracles is that they involve highly unusual rates of change (changing water into wine, calming a storm, etc.), these miracles are the exception, not the rule.

Therefore, this is not about the rejection of 'millions or billions of years' *per se*; it is really about the truthfulness of God's Word. We really should not even use terms like 'young-earth creationism' (YEC) or 'old-earth creationism' (OEC); we could instead adopt 'biblical creationism,' or to be clearer 'straight-forward history in Genesis biblical creationism' (SHGBC). Because surely it is about bringing people back to a clear understanding of the veracity of the Word of God, from Genesis all the way through.

'Young earth,' meanwhile, implies its age is young compared to the supposed long geologic ages, which are contrary to the Genesis timeline. And so often arguments from the other side are a caricature of the creationist position, or a straw man argument, but some may have been once held by creationists. We should hold ourselves to a higher standard. We will never satisfy the sceptics who reject the idea that God's Word can be relied upon, but we must challenge those who 'put their heads in the sand' over the last hundred years of modern science.

STARLIGHT TRAVEL TIME DILEMMA

If we accept all observations about the universe, realizing they are tainted with certain assumptions, which may be wrong, then creationists have a starlight-travel-time problem. This is true if we believe only 6,000 years have passed since the creation of the most distant light sources, and that they were all created at that time, as measured by normal Earth clocks, and we hold to the convention that the timer was started when the star was created.

However, if the timer was started when the light first arrived on Earth, when someone first saw the event, then this is the Anisotropic Time Convention, and there is no light-travel-time problem. There is nothing to answer. Or if Earth clocks ran slow during Creation Week compared to all other clocks in the cosmos, there would be billions of years of process going on out there, and plenty of time for light to get here in the past 6,000 years.

This is a relativistic effect. In all cases the universe is large, and normal, testable physics applies. It also allows for a certain starting point we could call '***mature creation***', as the place where God started. The only difference is we do not know which is the correct model or time convention. Maybe none of them are, but we should keep looking within the realms of modern testable physics.

APPENDIX

i. This information is often quite detailed; for example, a burst of neutrinos (which pass through the matter that slows light) precedes a burst of light. This is quite different from the creation of a mature Adam, or mature fruit-bearing trees. Here the 'appearance of age' is a *necessary* consequence of creating a grown adult, e.g., there is no necessity to create a full-blown picture show of events that did not happen in order to have a mature universe. It would be all deception.

ii. A famous proponent of this was Philip Gosse in his book *Omphalos*. See the book *Green Eye of the Storm*, by T.J. Rendle-Short. Some claimed that God did this to fool us, or test our faith.

iii. All time references in the Bible are clearly invoking the frame of reference of Earth-based clocks, with the earth-rotation day as the basis. So the existence of the whole universe dates from 6,000 years ago, period.

iv. If the speed of light was much greater in the past, either the frequencies were higher due to higher excitation energies of the sources or the received wavelengths are shortened by the Doppler effect. In either case, referenced against standard sources on Earth, such light would appear blue-shifted.

v. It may be more accurately classified as a hybrid between my categories 2 and 4. But it does have a strong element of this type 2.

vi. Consider the clock rates at emission and reception. In category 2 at emission, clocks in the distant cosmos were running faster than Earth clocks now run at reception. In category 3 at emission, clocks in the distant cosmos were running at the same rate as Earth clocks now run at reception. Only during a few days of Creation Week were Earth clocks running slower on receiving the light.

vii. Occam, William of Occam (or Ockham) (1284–1347) was an English philosopher and theologian. His work on knowledge, logic and scientific inquiry played a major role in the transition from medieval to modern thought. He based scientific knowledge on experience and self-evident truths, and on logical propositions resulting from those two sources. In his writings, Occam stressed the Aristotelian principle that entities must not be multiplied beyond what is necessary. This principle became known as Occam's (or Ockham's) Razor or the law of parsimony. A problem should be stated in its basic and simplest terms. In science, the simplest theory that fits the facts of a problem is the one that should be selected. www.2think.org/occams_razor.shtml.

viii. Astronomical years measure time applicable to astronomical objects. The ordinary years we measure on Earth now are also identical and have been since the end of Day 4 of Creation Week.

ix. Astronomers no longer believe ellipticals wound up from earlier spiral forms because most have little angular motion. They are more like motionless blobs. However, in the time available to a spiral galaxy since the big bang it could have wound around about 500 times.

x. 'Young' means that the age by Earth clocks is < 6,000 years.

xi. 'Old' means the age by Earth clocks is of the order of millions or billions of years.

xii. A non-static universe seems to be an inevitable conclusion considering gravity to be only an attractive force.

xiii. Tully-Fisher law: observed luminosity of spiral galaxies varies as the fourth power of their rotational velocities.

xiv. The Hebrew verb natah is variously translated as 'to stretch', 'to stretch out', 'to extend', 'to spread out', 'to bend', and related meanings.

xv. Unless one assumes that God's creative act itself of placing the stars in the expanse of the heavens actually equates to the 'stretching of the heavens', which it might.

xvi. Mary Beth de Repentigny noted in 2011: "Creation scientists do not necessarily have a philosophical problem with the idea that the expansion rate of the universe might be accelerating, as the data from type 1a supernovae could be telling us." Repentigny, M., What are type 1a supernovae telling us?

xvii. There are notions that the universe must be no more than 6,000 light-years in radius, to the most distant galaxy, because the universe is only 6,000 years old, and even that all stars must not be more than a few light-days away, because Adam was created only two days after the stars were, and he could see stars. The problem with these ideas is that we have a repeatable method of measuring parallax of about a thousand stars from the earth as it traverses the sun, and we can measure those distances by that method, which you could say is a direct method, based on what we know of reliable science on Earth. The ESA satellite Hipparcos mapped a hundred thousand stars in the galaxy, all of which were more distant than two light-days. Hipparcos also confirmed Einstein's prediction of the effect of gravity on starlight.

REFERENCES

1. Pace, A., What does the Bible say about the fabric of space? J. Creation 30(1):51, 2016. See page 53, table 1, for a recent chart on the 17 verses identified by Humphreys. In the chart, Psalm 102:4 is an error and should be 104:2. Humphreys' book is Starlight and Time: Solving the puzzle of distant starlight in a young universe, Master Books, Green Forest, AR, pp. 66–68, 1994.
2. Taylor, C.V., Waters above or beyond? J. Creation 10(2):211–213, 1996; creation. com/images/pdfs/tj/j10-2/j10_2_211-213.pdf.
3. Hartnett, J.G., Starlight, Time and the New Physics, Creation Book Publishers, pp. 102–103, 2007.
4. Hartnett, J.G., Expansion of Space—A Dark Science, Answers Research J. 7: 457, 2014.
5. Hartnett, J.G., Does the Bible really describe expansion of the universe? J. Creation 25(2):125–127, August 2011; creation.com/images/pdfs/tj/j25_2/ j25_2_125-127.pdf.
6. Humphreys, D.H., New view of gravity explains cosmic microwave background radiation, J. Creation 28(3):106–114, 2014.
7. Hartnett, J.G., Our eternal universe, J. Creation 30(3):105, 2016.
8. Emails to and from the author with Hebrew scholar David Brewer, February- March 2015. We see a similar theme and imagery in Psalm 144:5.
9. Morris, H., That Their Words May Be Used Against Them, Institute for Creation Research, Green Forest, AR, 1997.
10. Hartnett, J.G., Does observational evidence indicate the universe is expanding? —part 1: the case for time dilation, J. Creation 25(3):109–114, December 2011; and, Does observational evidence indicate the universe is expanding—part 2: the case against expansion, J. Creation 25(3):115–120, December 2011.
11. Kirshner, R.P., The Extravagant Universe, Princeton University Press, Princeton, NJ, p. 158. 2002.
12. Hartnett, J.G., The Lyman-α forest and distances to quasars, J. Creation 30(3):9, 2016. Hartnett, J.G., ref. 20, part 2, p. 119.
13. Tegmark, M., Our Mathematical Universe: My Quest for the Ultimate Nature of Reality, Alfred A. Knopf, New York, pp. 47–48, 2014.
14. Panek, R., The 4% Universe: Dark Matter, Dark Energy, and the Race to Discover the Rest of Reality, ch. 8, Hello, Lambda, Houghton Mifflin Harcourt, New York, pp. 140–163, 2011; Riess, A. et al., Observational Evidence from Supernovae for an Accelerating Universe and a Cosmological Constant, arxiv. org/abs/astro-ph/9805201, 1998.
15. However, this is not necessarily true. As de Repentigny has pointed out: "an objective handling of the data does not necessitate this interpretation" (de Repentigny, ref. 9, p. 127). 27. Hartnett, J.G., ref. 20, part 1, p. 111.
16. Susskind, L., Cosmic Landscape: String Theory and the Illusion of Intelligent Design, Little, Brown and Company, New York, p. 185, 2006.
17. Calle, C.I., The Universe—Order without Design, Prometheus Books, Amherst, New York, p. 19, 2009.

18. Greene, B., The Hidden Reality: Parallel Universes and the Deep Laws of the Cosmos, Alfred A. Knopf, New York, p. 142, 2011.
19. I critique the current secular crisis in cosmology, multiverse theory and fine-tuning in Melnick, A.J., The Quantum World: A Fine-Tuned Multiversal Reality? J. Interdisciplinary Studies 27:45–60, 2015.
20. Nagel, T., Mind and Cosmos: Why the Materialist Neo-Darwinian Conception of Nature is Almost Certainly False, Oxford University Press, New York, p. 95, fn 8, 2012.
21. de Repentigny, ref. 9, p. 127. Jake Hebert and Jason Lisle have recently published two extensive articles in the Creation Research Society Quarterly regarding galactic luminosity and the question of special location in the universe. They describe the cosmological principle as "the assumption that on the largest distance and angular scales, there are no special places or directions in the cosmos". They state: "The biblical worldview may well allow for this principle, but it does not require it." Herbert, J. and Lisle, J., A Review of the Lynden-Bell/Choloniewski Method for Obtaining Galaxy Luminosity Functions. Part I, Creation Research Society Quarterly 52(3):177–178, 2016.
22. Hartnett, J.G., ref. 6, J. Creation 25(2):125. Elsewhere he has repeated that charge and added "and not good exegesis" (Hartnett, ref. 5, p. 457).
23. Lisle, J., Light-travel time: a problem for the big bang, *Creation* 25(4):48–49, 2003; creation.com/lighttravel.
24. Coles, P. and Lucchin, F., *Cosmology: The Origin and Evolution of Cosmic Structure*, John Wiley & Sons Ltd, Chichester, p. 91, 1996.
25. Peacock, J.A., *Cosmological Physics*, Cambridge University Press, p. 288, 1999. However, the existence of CMB was actually deduced before big bang cosmology from the spectra of certain molecules in outer space.
26. Peebles, P.J.E., *Principles of Physical Cosmology*, Princeton University Press, p. 404, 1993.
27. For convenience, the commonly understood term CMB will be used without implying that the radiation peaked at the same wavelength in all epochs of the model. Infrared radiation is part of the spectrum of light.
28. This is an internal inconsistency for the big bang model. It is not a problem for a creation model; God may have created the distant regions of the universe with the same temperature from the beginning.
29. Misner, C., Mixmaster Universe, *Phys. Rev. Lett.* 22(20):1071–1074, 1969. Ref. 1, p. 136.
30. Lightman, A., *Ancient Light*, Harvard University Press, London, p. 58, 1991. \
31. This notion does not violate relativity, which merely prevents objects travelling faster than c *through* space, whereas in the inflation proposal it is space *itself* that expands and carries the objects with it. Kraniotis, G.V., String cosmology, *Int. J. Mod. Phys. A* 15(12):1707–1756, 2000.
32. Wang, Y., Spergel, D. and Strauss, M., Cosmology in the next millennium: Combining microwave anisotropy probe and Sloan digital sky survey data to constrain inflationary models, *Astrophys. J.* **510**:20–31, 1999. .
33. Coles, P. and Lucchin, F., *Cosmology: The Origin and Evolution of Cosmic Structure*, John Wiley & Sons Ltd, Chichester, p. 151, 1996. Levin, J. and Freese, K., Possible solution to the horizon problem: Modified aging in massless scalar theories of gravity, *Phys. Rev. D (Particles, Fields, Gravitation, and Cosmology)* 47(10):4282–4291, 1993.
34. Steinhardt, P. and Turok, N., A cyclic model of the universe, *Science* 296(5572):1436–1439, 2002.
35. Chung, D. and Freese, K., Can geodesics in extra dimensions solve the cosmological horizon problem?, *Phys. Rev. D* 62(6):063513–7, 2000. .

36. Célérier, M. and Szekeres, P., Timelike and null focusing singularities in spherical symmetry: A solution to the cosmological horizon problem and a challenge to the cosmic censorship hypothesis, *Phys. Rev. D* **65**:123516–9, 2002.
37. Albrecht, A. and Magueijo, J., Time varying speed of light as a solution to cosmological puzzles, *Phys. Rev. D* **59**(4):043513–16, 1999. Clayton, M. and Moffat, J., Dynamical mechanism for varying light velocity as a solution to cosmological problems, *Phys. Lett. B* 460(3–4):263–270, 1999.
38. For a summary of the c-decay implications, see: Wieland, C., Speed of light slowing down after all? Famous physicist makes headlines, *J. Creation* 16(3):7–10, 2002. This is known as the 'horizon problem'. Lisle, J., Light-travel time: a problem for the big bang, *Creation* 25(4):48–49, 2003, <creation.com/lighttravel>.
39. Humphreys, D.R., *Starlight and Time,* Master Books, Colorado Springs, USA, 1994.
40. Hartnett. J., *Starlight, Time and the New Physics*, Creation Book Publishers, Georgia, USA, 2007.
41. Demick, D. and Wieland, C., In the middle of the action, *Creation* 28(1):52–55, 2005, <creation.com/quantized>.
42. Norman, T. and Setterfield, B., The atomic constants, light and time, *SRI International Invited Research Report,* Menlo Park, 1986.
43. Burgess, S., *He Made the Stars Also*, Day One Publications, Surrey, 2001. Recently reviewed in *CRSQ* **39**:39, 2002.
44. Kofahl, R.E., Letter to the Editor: Speculation concerning God's 'big bang', *CRSQ* **39**:64, 2002.
45. Bernitt, R., Fast stars challenge big bang origin for dwarf galaxies, *TJ* 14(3):5–7, 2000.
46. Humphreys, D. R., *Starlight and Time,* Master Books, Colorado Springs, 1994.
47. Hartnett, J. G., Look-back time in our galactic neighborhood leads to a new cosmogony, *TJ* 17(1):73–79, 2003.
48. Newton, R., Distant starlight and Genesis: conventions of time measurement, *TJ* 15(1):80–85, 2001.
49. Hartnett, J.G., Distant starlight and Genesis: is 'observed time' a physical reality? *Letters, TJ* 16(3):65–68, 2002.
50. Humphreys, D.R., Our galaxy is the center of the universe, 'quantized' red shifts show, *TJ* 16(2):95–104, 2002.
51. Cho, A., Light may have slowed down, *Newscientist.com,* www.newscientist.com/news/print.jsp?id=ns 99991158, 2001.
52. Hartnett, J.G., Is there any evidence for a change in c? Implications for creationist cosmology, *TJ* 16(3):89–94, 2002.
53. Harris, D.M., A solution to seeing stars, *CRSQ* 15(2):112–115, 1978.
54. Humphreys, D.R., New vistas of space and time, *TJ* 12(2):195–212, 1998.
55. Humphreys, D.R., More on vistas, *TJ* 13(1):55, 1999.
56. Worraker, W.J., Look-back time in Humphreys' cosmology, *TJ* 15(2):46–47, 2001.
57. Davies, K., Evidence for a young Sun, *ICR Impact* 26:1–4, 1996. Note that even though the question of the neutrino emission has been answered (see Newton, R., 'Missing' neutrinos found! No longer an 'age' indicator, *TJ* 16(3):123–125, 2002) the questions Davies discusses relating to the oscillation periods are still outstanding.
58. Faulkner, D., The young faint sun paradox and the age of the solar system, *ICR Impact* 300:1–3, 1998.
59. Worraker, B. J., MOND over dark matter? *TJ* 16(3):11–14, 2002.

60. Hartnett, J.G. and DeYoung, D.B., Mature creation and seeing distant starlight, *J. Creation* 25(1):46–47, 2011.
61. *Death only applies to nephesh chayyāh*, translated 'living soul' or 'living creature' with the breath of life; see Sarfati, J., The Fall: a cosmic catastrophe, 21 February 2005.
62. Hartnett, J., 'Cosmology is not even astrophysics', 3 December 2008.
63. But doesn't quite roll off the tongue like 'YEC' does.
64. Lisle, J.P., Anisotropic Synchrony Convention—A Solution to the Distant Starlight Problem, *Answers Research Journal* 2:191–207, 2010; answersingenesis.org/articles/arj/v3/n1/anisotropic-synchrony-convention.
65. Humphreys, D.R., *Starlight and Time*, Master Books, Colorado Springs, CO, 1994; Humphreys, D.R., New time dilation helps creation cosmology, *J. Creation* 22(3):84–92, 2008.
66. Hartnett, J.G., A new cosmology: solution to the starlight travel time problem, *J. Creation* 17(2):98–102, 2003; Hartnett, J.G., *Starlight, Time and the New Physics*, Creation Book Publishers, Brisbane, 2007; Hartnett, J.G., A 5D spherically symmetric expanding universe is young, *J. Creation* 21(1):69–74, 2006; I recommend the second edition of the book, where a misinterpretation of the type of time dilation the model involves has been corrected.

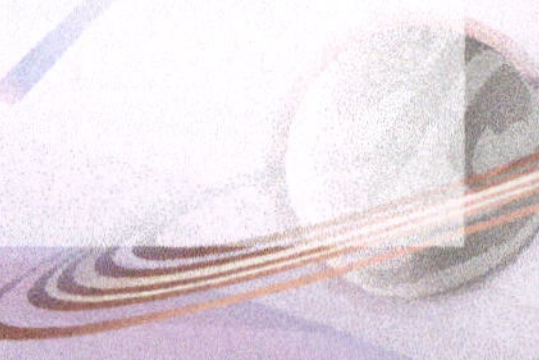

16

UNDERSTANDING THE UNIDENTIFIED FLYING OBJECTS UFO'S

THE UFO CHARACTER

On the shores of the Mediterranean Sea, Figure 16.1, since my childhood I used to spend a great deal of time contemplating about my simplistic comprehension of the flying saucers hovering above the sea in the wide blue sky. I have never seen any except in hazy pictures or in graphic illustration. I imagined them to be a different type of flying kites, however, without strings. Throughout the journey of my life being acquainted with different cultures, societies, and people from all walks of life world-wide. I then realized that the simple childlike toy of a flying saucer roaming the sky is much deeper than I perceived.

Fig.16.1: My Favorite Childhood Shores Alexandria, The King's Palace, at the Mediterranean Sea – Curtsy Mediterranean sunset by Stefano Politi Markovina

As I advanced in my academic career studying science and technology, and to my surprise I found that my deeper intellect was still resonating inside me wondering about my childhood Flying saucers. Let alone the far deeper claims of many people of different tongues and different topography are deeply concerned about the UFO Aliens, Figure 16.2. Aliens, who are flowing in the sky, sometimes in day light and sometimes lurking in the cover of darkness irrespective which territory in the face of the Earth. Deeper yet, they are sweeping through our minds spreading fears and deep anxieties.

Fig.16.2: U-2s, UFOs, and Operation Blue Book – Curtsy - US Gov

Years passed hoping such unknown UFO events may dissipate by time. How mistaken, indeed, I was. In particularly, an awesome responsibility was placed on my shoulders when I was assumed the headship of the "Advanced Technology and Science Department at Columbia University, NY. USA. I found myself facing a serious task to scientifically investigate such mysterious unknown Identifying Flying Objects!

SCIENTIFICALLY, WHAT ARE THE UFO?

Let us identify the problem scientifically, to discover logical solutions that may shed some light onto the mysterious nature of the UFO. My plan of investigations was as formalized follows:

a. Observe the problem carefully.
b. Document the frequency of the appearance Events.
c. Measure and compute physical characteristics of the UFO.
d. Analyze the accumulated data.
e. Formulate some preliminary findings and understanding of the UFO.
f. Discuss the findings with my academic colleagues.
g. Formulate a logical conclusion of the UFO.

I devotedly began to tackle the UFO matter step by step according to my stated plan.

OBSERVE THE PROBLEM CAREFULLY

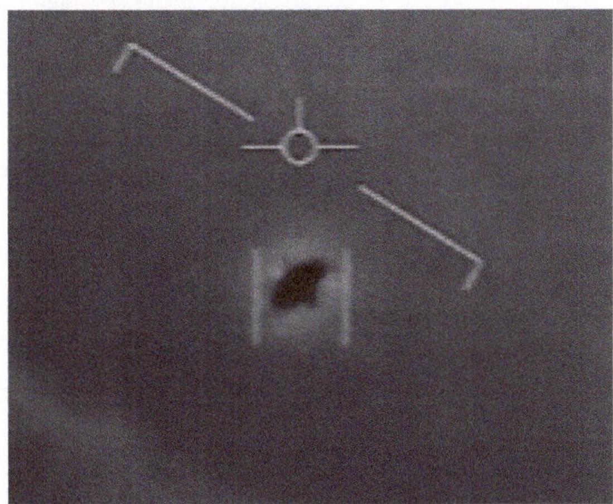

Fig.16.3: Pentagon formally releases Navy UFO videos - A Screengrab from a Navy Video Shows what the Military Calls "Unidentified Aerial Phenomenon." Curtsy - U.S. Navy

Figure 16.3 - Three Navy videos of strange flying objects that have circulated among UFO enthusiasts and investigators for years were officially released by the Pentagon, USA. The three videos appear to show small airborne craft flying and maneuvering at very high speeds. The Navy confirmed the *videos are real.*

The videos were shot by the Advance Targeting Forward Looking Infrared (ATFLIR) pods on Navy F/A-18s. One of them — called "FLIR1" — was videoed by an F/A-18 operating off the aircraft carrier Nimitz off the coast of San Diego on Nov. 14, 2004.

"After a thorough review, the department has determined that the authorized release of these unclassified videos does not reveal any sensitive capabilities or systems, and does not impinge on any subsequent investigations of military air space incursions by unidentified aerial phenomena," the Pentagon said in a released statement.

"(The Department of Defense) DOD is releasing the videos in order to clear up any misconceptions by the public on whether or not the footage that has been circulating was real, or whether or not there is more to the videos."

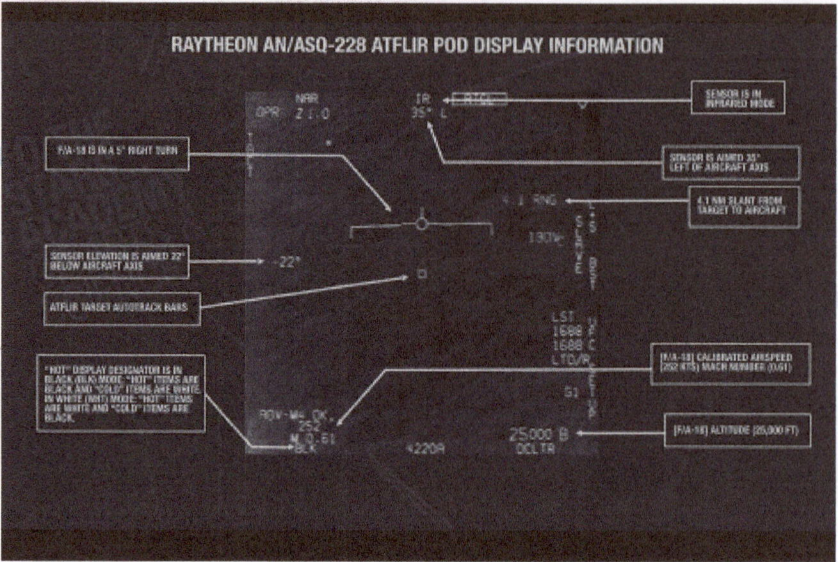

Fig.16.4: *The Pentagon acknowledged the videos have already been circulating in public but offered no explanation of the aerial phenomena seen in them.*

"The aerial phenomena observed in the videos, Figure 16.3, and Figure 16.4, remain characterized as "unidentified," the Pentagon said.

In September, the Navy acknowledged, for the first time, that the videos were real after they were published by former Blink-182 guitarist Tom DeLonge's to the Stars Academy of Arts & Science in Encinitas.

Fig.16.5: *UFO sighting sparked by mysterious darting lights in Wisconsin, says report*

In rural areas of Wisconsin in the US, people reported seeing unexplained lights in the night sky. These lights darted across the sky in a swift motion. Onlookers said they could be from Unidentified Flying Objects, according to reports, Figure 16.5.

The Daily Mail released footage that shows bright white lights swiftly traveling across the sky. The region in the video is the sky above farmland in Fredonia and West Bend in the Badger State. The metadata from the videos released suggests that they were taken from different locations in the region.

The light took three rounds in front of them, which cleared their doubt that it definitely wasn't an owl. They then saw several such white lights in the sky above. Both of them were hazed at the spectacle and started taking a video.

Five miles away from Kimberly, Ken saw similar things from his home in Fredonia. The video taken by Ken shows four lights moving toward the horizon, and then a fifth light joined them, and they all continued moving in the same direction

Fig.16.6: *A United States Coast Guard photographer, Shell R. Alpert, took a photograph that allegedly shows unidentified flying objects flying in a "V" formation at the Salem, Massachusetts, air station at 9:35 a.m. on 16 July 1952, through a window screen. (Official U.S. Coast Guard photograph: 5554. Library of Congress Control Number. 2007680837)*

High-altitude testing of the U-2 soon led to an unexpected side effect—a tremendous increase in reports of unidentified flying objects (UFOs) [known today as unidentified aerial phenomena (UAP)]. In the mid-1950s, most commercial airliners flew at altitudes between 10,000 and 20,000 feet and military aircraft like the B-47s operated at altitudes below 40,000 feet. Consequently, once U-2s started flying at altitudes above 60,000 feet, air traffic controllers began receiving increasing numbers of UFO reports, Figure 16.6.

Such reports were most prevalent in the early evening hours from pilots of airliners flying from east to west. When the sun dropped below the horizon of an airliner flying at 20,000 feet, the plane was in darkness. But, if a U-2 was airborne in the vicinity of the airliner at the same time, its horizon from an altitude of 60,000 feet was considerably more distant, and, being so high in the sky, its silver wings would catch and reflect the rays of the sun and appear to the airliner pilot, 40,000 feet below, to be fiery objects. Even during daylight hours, the silver bodies of the high- flying U-2s could catch the sun and cause reflections or glints that could be seen at lower altitudes and even on the ground. At this time, no one believed manned flight was possible above 60,000 feet, so no one expected to see an object so high in the sky.

Not only did the airline pilots report their sightings to air traffic controllers, but they and ground-based observers also wrote letters to the Air Force unit at Wright Air Development Command in Dayton charged with investigating such phenomena. This, in turn, led to the Air Force's Operation Blue Book. Based at Wright-Patterson, the operation collected all reports of UFO sightings. Air Force investigators then attempted to explain such sightings by linking them to natural phenomena. Blue Book investigators regularly called on the [Central Intelligence] Agency's Project staff in Washington to check reported UFO sightings against U-2 flight logs. This enabled the investigators to eliminate a majority of the UFO reports, although they could not reveal to the letter writers the true cause of the UFO sightings. U-2 and later OXCART flights accounted for more than one-half of all UFO reports during the late 1950s and most of the 1960s.

From Gregory W. Pendlow and Donald E. Welzenbach. The CIA and the U-2 Program, 1954-1974. Central Intelligence Agency, 1998, pp. 72-73.

While Some UFO sightings were genuine, there are however, some UFOs aren't necessarily alien spacecraft. And some purported UFOs aren't UFOs at all. Take the example from Apollo 16, Figure 16.7.

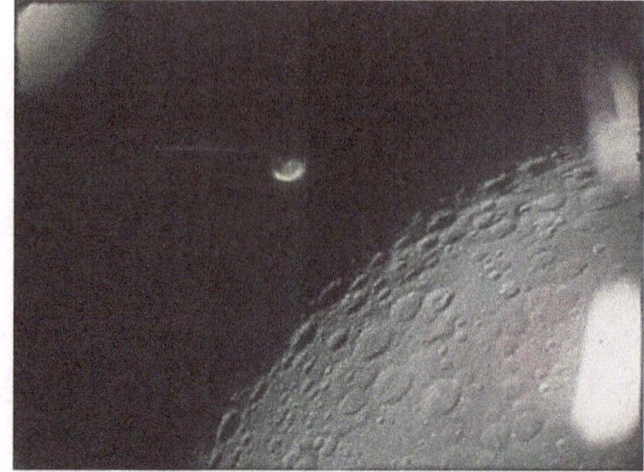

Fig.16.7: High-resolution, digital scan of a full frame from the original Apollo 16 film showing the object in question (top center) and its position relative to the moon. Reflections in the window are also visible (left and right). Credit: NASA

Beginning their return from the moon to an April 27, 1972, splashdown, Astronauts John Young, Thomas Mattingly and Charles Duke captured about four seconds of video footage of an object that seemed to look a lot like Hollywood's version of a spacecraft from another world.

The thing was described as "a saucer-shaped object with a dome on top." The images were captured with a 16mm motion picture camera shooting at 12 frames per second from a command/service module window. The object appears momentarily near the moon. As the camera pans, it moves out of the field of view. It reappears as the camera pans back. It appeared in about 50 frames, Figure 16.8.

Some very bright people recently worked hard to analyze that footage. Their conclusion was that the object wasn't at all what some observers thought it seemed to be. There is no indication the Apollo 16 crew ever thought the film showed anything special.

A group headed by Gregory Byrne of Johnson Space Center's Image Science and Analysis Group completed a report on its investigation earlier this year. They used a video copy of the film initially, then did a high-resolution digital scan of the original film for detailed analysis.

They stabilized images to correct for camera movement, and then aligned multiple frames in a sequence. One thing that showed them was that the object appeared to move slightly with respect to the moon, because of parallax brought about by slight camera motions and the nearness of the object to the camera, Figure 16.9.

Fig.16.8: Image Enhancement of the Object and Linear Feature - Curtsy: NASA

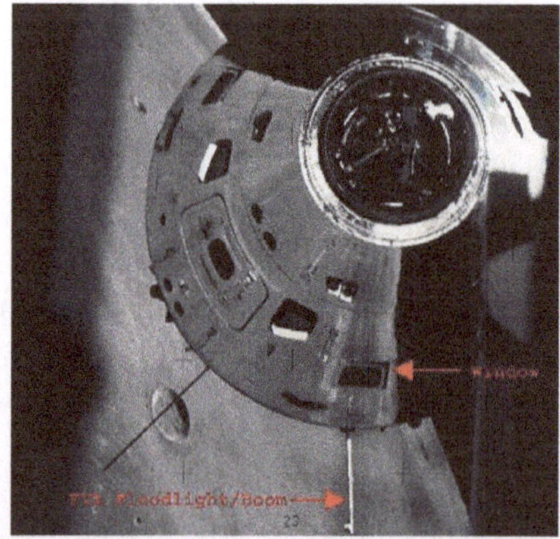

Fig.16.9: View of the Apollo Command/ Service Module from the Lunar Module during Apollo 17 showing the location of the EVA floodlight/boom. Credit: NASA

The investigators also combined several frames in a sequence, to give them higher resolution and greater contrast than individual frames. The combinations showed them more clearly a "linear feature" attached to one side of the object. They also looked at archived images from other Apollo missions. Bottom line: "All of the evidence in this analysis is consistent with the conclusion that the object in the Apollo 16 film was the EVA [spacewalk] floodlight/boom. There is no evidence in the photographic record to suggest otherwise."

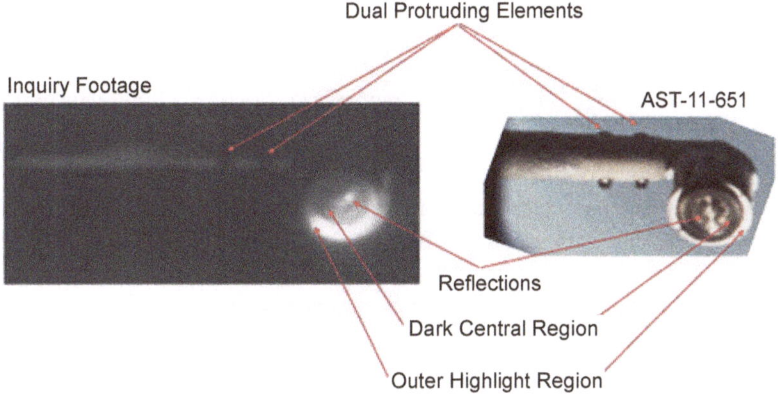

Fig.16.10: *Enhanced Apollo 16 image (left) compared with features of the EVA floodlight/boom from the perspective of a Command/Service Module window, Figure 16.10. Curtsy: NASA*

Fig.16.11: Unidentified Flying Object (UFO), *also Called* flying Saucer.

Figure 16.10 - Any aerial object or optical phenomenon not readily identifiable to the observer, UFOs became a major subject of interest following the development of rocketry after World War II and were thought by some researchers to be intelligent extraterrestrial life visiting Earth. Figure 16.11.

HISTORY - FLYING SAUCERS AND PROJECT BLUE BOOK

The first well-known UFO sighting occurred in 1947, when businessman Kenneth Arnold claimed to see a group of nine high-speed objects near Mount Rainier in Washington while flying his small plane. Arnold estimated the speed of the crescent-shaped objects as several thousand miles per hour and said they moved "like saucers skipping on water." In the newspaper report that followed, it was mistakenly stated that the objects were saucer-shaped, hence the term *flying saucer*.

Sightings of unidentified aerial phenomena increased, and in 1948 the U.S. Air Force began an investigation of these reports called Project Sign. The initial opinion of those involved with the project was that the UFOs were most likely sophisticated Soviet aircraft, although some researchers suggested that they might be spacecraft from other worlds, the so-called extraterrestrial hypothesis (ETH). Within a year, Project Sign was succeeded by Project Grudge, which in 1952 was itself replaced by the longest-lived of the official inquiries into UFOs, Project Blue Book, headquartered at Wright-Patterson Air Force Base in Dayton, Ohio.

From 1952 to 1969 Project Blue Book compiled reports of more than 12,000 sightings or events, each of which was ultimately classified as

1. "Identified" with a known astronomical, atmospheric, or artificial (human-caused) phenomenon or
2. "Unidentified." The latter category, approximately 6 percent of the total, included cases for which there was insufficient information to make an identification with a known phenomenon.

THE ROBERTSON PANEL AND THE CONDON REPORT

An American obsession with the UFO phenomenon was under way. In the hot summer of 1952, a provocative series of radar and visual sightings occurred near National Airport in Washington, D.C. Although these events were attributed to temperature inversions in the air over the city, not everyone was convinced by this explanation. Meanwhile, the number of UFO reports had climbed to a record high. This led the Central Intelligence Agency to prompt the U.S. government to establish an expert panel of scientists to investigate the phenomena. The panel was headed by H.P. Robertson, a physicist at the California Institute of Technology in Pasadena, California, and included other physicists, an astronomer, and a rocket engineer.

The Robertson Panel met for three days in 1953 and interviewed military officers and the head of Project Blue Book. They also reviewed films and photographs of UFOs. Their conclusions were that

1. 90 percent of the sightings could be easily attributed to astronomical and meteorologic phenomena (e.g., bright planets and stars, meteors, auroras, ion clouds) or to such earthly objects as aircraft, balloons, birds, and searchlights,
2. there was no obvious security threat, and
3. there was no evidence to support the ETH.

Parts of the panel's report were kept classified until 1979, and this long period of secrecy helped fuel suspicions of a government cover-up.

A second committee was set up in 1966 at the request of the Air Force to review the most interesting material gathered by Project Blue Book. Two years later this committee, which made a detailed study of 59 UFO sightings, released its results as *Scientific Study of Unidentified Flying Objects*—also known as the Condon Report, named for Edward U. Condon, the physicist who headed the investigation. The Condon Report was reviewed by a special committee of the National Academy of Sciences. A total of 37 scientists wrote chapters or parts of chapters for the report, which covered investigations of the 59 UFO sightings in detail. Like the Robertson Panel, the committee concluded that there was no evidence of anything other than commonplace phenomena in the reports and that UFOs did not warrant further investigation. This, together with a decline in sighting activity, led to the dismantling of Project Blue Book in 1969.

OTHER INVESTIGATIONS OF UFO'S

Despite the failure of the ETH to make headway with the expert committees, a few scientists and engineers, most notably J. Allen Hynek, an astronomer at Northwestern University in Evanston, Illinois, who had been involved with projects Sign, Grudge, and Blue Book, concluded that a small fraction of the most-reliable UFO reports gave definite indications for the presence of extraterrestrial visitors. Hynek founded the Center for UFO Studies (CUFOS), which continues to investigate the phenomenon. Another major U.S. study of UFO sightings was the Advanced Aviation Threat Identification Program (AATIP), a secret project that ran from 2007 to 2012. When the existence of the AATIP was made public in December 2017, the most newsworthy aspect of it was a report that the U.S. government possessed alloys and compounds purportedly attained from UFOs that were of unidentifiable nature, but many scientists remained skeptical about this claim.

Aside from the American efforts, the only other official and fairly complete records of UFO sightings were kept in Canada, where they were transferred in 1968 from the Canadian Department of National Defense to the Canadian National Research Council. The Canadian records comprised about 750 sightings. Less-complete records have been maintained in the United Kingdom, Sweden, Denmark, Australia, and Greece. In the United States, CUFOS and the Mutual UFO Network in Bellevue, Colorado, continue to log sightings reported by the public.

In the Soviet Union, sightings of UFOs were often prompted by tests of secret military rockets. In order to obscure the true nature of the tests, the government sometimes encouraged the public's belief that these rockets might be extraterrestrial craft but eventually decided that the descriptions themselves might give away too much information. UFO sightings in China have been similarly provoked by military activity that is unknown to the public.

POSSIBLE EXPLANATIONS FOR UFO SIGHTINGS AND ALIEN ABDUCTIONS

UFO reports have varied widely in reliability, as judged by the number of witnesses, whether the witnesses were independent of each other, the observing conditions (e.g., fog, haze, type of illumination), and the direction of sighting. Typically, witnesses who take the trouble to report a sighting consider the object to be of extraterrestrial origin or possibly a military craft but certainly under intelligent control. This inference is usually based on what is perceived as formation flying by sets of objects, unnatural—often sudden—motions, the lack of sound, changes in brightness or **color**, and strange shapes.

The unaided eye plays tricks is well known. A bright light, such as the planet Venus, often appears to move. Astronomical objects can also be disconcerting to drivers, as they seem to "follow" the car. Visual impressions of distance and speed of UFOs are also highly unreliable because they are based on an assumed size and are often made against a blank sky with no background object (clouds, mountains, etc.) to set a maximum distance. Reflections from windows and eyeglasses produce superimposed views, and complex optical systems, such as camera lenses, can turn point sources of light into apparently saucer-shaped phenomena. Such optical illusions and the psychological desire to interpret images are known to account for many visual UFO reports, and at least some sightings are known to be hoaxes. Radar sightings, while in certain respects more reliable, fail to discriminate between artificial objects and meteor trails, ionized gas, rain, or thermal discontinuities in the atmosphere.

UFO ABDUCTION

"Contact events," such as abductions, are often associated with UFOs because they are ascribed to extraterrestrial visitors. However, the credibility of the ETH as an explanation for abductions is disputed by most psychologists who have investigated this phenomenon. They suggest that a common experience known as "sleep paralysis" may be the culprit, as this causes sleepers to experience a temporary immobility and a belief that they are being watched.

THE 1970S AND 1990S: CIA - THE UFO ISSUE REFUSES TO DIE

Central Intelligence Agency CIA report, known as the Condon report did not satisfy many UFOlogists, who considered it a coverup for CIA activities in UFO research. Additional sightings in the early 1970s fueled beliefs that the CIA was somehow involved in a vast conspiracy.

On 7 June 1975, William Spaulding, head of a small UFO group, Ground Saucer Watch (GSW), wrote to CIA requesting a copy of the Robertson panel report and all records relating to UFOs. Spaulding was convinced that the Agency was withholding major files on UFOs. Agency officials provided Spaulding with a copy of the Robertson panel report and of the Durant report.

On 14 July 1975, Spaulding again wrote the Agency questioning the authenticity of the reports he had received and alleging a CIA coverup of its UFO activities. Gene Wilson, CIA's Information and Privacy Coordinator, replied in an attempt to satisfy Spaulding, "At no time prior to the formation of the Robertson Panel and subsequent to the issuance of the panel's report has CIA engaged in the study of the UFO phenomena. "The Robertson" panel report, according to Wilson, was "the summation of Agency interest and involvement in UFOs." Wilson also inferred that there were no additional documents in CIA's possession that related to UFOs. Wilson was ill informed.

In September 1977, Spaulding and GSW, unconvinced by Wilson's response, filed a Freedom of Information Act (FOIA) lawsuit against the Agency that specifically requested all UFO documents in CIA's possession. Deluged by similar FOIA requests for Agency information on UFOs, CIA officials agreed, after much legal maneuvering, to conduct a "reasonable search" of CIA files for UFO materials. Despite an Agency-wide unsympathetic attitude toward the suit, Agency officials, led by Launie Ziebell from the Office of General Counsel, conducted a thorough search for records pertaining to UFOs. Persistent, demanding, and even threatening at times, Ziebell and his group scoured the Agency. They even turned up an old UFO file under a secretary's desk. The search finally produced 355 documents totaling approximately 900 pages. On 14 December 1978, the Agency released all but 57 documents of about 100 pages to GSW. It withheld these 57 documents on national security grounds and to protect sources and methods.

Although the released documents produced no smoking gun and revealed only a low-level Agency interest in the UFO phenomena after the Robertson panel report of 1953, the press treated the release in a sensational manner. *The New York Times*, for example, claimed that the declassified documents confirmed intensive government concern over UFOs and that the Agency was secretly involved in the surveillance of UFOs. GSW then sued for the release of the withheld documents, claiming that the Agency was still holding out key information. It was much like the John F. Kennedy assassination issue. No matter how much material the Agency released and no matter how dull and prosaic the information, people continued to believe in a Agency coverup and conspiracy.

Stansfield Turner was so upset when he read *The New York Times* article that he asked his senior officers, "Are we in UFOs?" After reviewing the records, Don Wortman, Deputy Director for Administration, reported to Turner that there was "no organized Agency effort to do research in connection with UFO phenomena nor has there been an organized effort to collect intelligence on UFOs since the 1950s." Wortman assured Turner that the Agency records held only "sporadic instances of correspondence dealing with the subject," including various kinds of reports of UFO sightings. There was no Agency program to collect actively information on UFOs, and the material released to GSW had few deletions. Thus assured, Turner had the General Counsel press for a summary judgment against the new lawsuit by GSW. In May 1980, the courts dismissed the lawsuit, finding that the Agency had conducted a thorough and adequate search in good faith.

During the late 1970s and 1980s, the Agency continued its low-key interest in UFOs and UFO sightings. While most scientists now dismissed flying saucers reports as a quaint part of the 1950s and 1960s, some in the Agency and in the Intelligence, Community shifted their interest to studying parapsychology and psychic phenomena associated with UFO sightings. CIA officials also looked at the UFO problem to determine what UFO sightings might tell them about Soviet progress in rockets and missiles and reviewed its counterintelligence aspects. Agency analysts from the Life Science Division of OSI and OSWR officially devoted a small amount of their time to issues relating to UFOs. These included counterintelligence concerns that the Soviets and the KGB were using US citizens and UFO groups to obtain information on sensitive US weapons development programs (such as the Stealth aircraft), the vulnerability of the US air-defense network to penetration by foreign missiles mimicking UFOs, and evidence of Soviet advanced technology associated with UFO sightings.

CIA also maintained Intelligence Community coordination with other agencies regarding their work in parapsychology, psychic phenomena, and "remote viewing" experiments. In general, the Agency took a conservative scientific view of these unconventional scientific issues. There was no formal or official UFO project within the Agency in the 1980s, and Agency officials purposely kept files on UFOs to a minimum to avoid creating records that might mislead the public if released.

The 1980s also produced renewed charges that the Agency was still withholding documents relating to the 1947 Roswell incident, in which a flying saucer supposedly crashed in New Mexico, and the surfacing of documents which purportedly revealed the existence of a top secret US research and development intelligence operation responsible only to the President on UFOs in the late 1940s and early 1950s. UFOlogists had long argued that, following a flying saucer crash in New Mexico in 1947, the government not only recovered debris from the crashed saucer but also four or five alien bodies. According to some UFOlogists, the government clamped tight security around the project and has refused to divulge its investigation results and research ever since. In September 1994, the US Air Force released a new report on the Roswell incident that concluded that the debris found in New Mexico in 1947 probably came from a once top-secret balloon operation, Project MOGUL, designed to monitor the atmosphere for evidence of Soviet nuclear tests.

Fig.16.12: *UFO Sightings From around the World*

Circa 1984, a series of documents surfaced which some UFO logists said proved that President Truman created a top-secret committee in 1947, Majestic-12, to secure the recovery of UFO wreckage from Roswell and any other UFO crash sight for scientific study and to examine any alien bodies recovered from such sites. Most if not all of these documents have proved to be fabrications. Yet the controversy persists.

Like the JFK assassination conspiracy theories, the UFO issue probably will not go away soon, no matter what the Agency does or says. The belief that we are not alone in the universe is too emotionally appealing and the distrust of our government is too pervasive to make the issue amenable to traditional scientific studies of rational explanation and evidence.

RESEARCH CONDUCTED BASED ON WITNESSES

Research Category per Country as Guide to zones, Figure 16.12:

The lists below contain UFO reports from various countries areas.

Argentina	Australia
Belarus	Belgium
Brazil	Canada
Canary Islands	China
France	India
Indonesia	Iran
Italy	Mexico
New Zealand	Norway
Outer space	Russia
South Africa	Spain
Sweden	United Kingdom
United States	

LIST OF CREDIBLE REPORTED UFO SIGHTINGS

This is a list of notable reported sightings of unidentified flying objects (UFOs) arranged by date. It includes reports of close encounters and abductions. This list is not an endorsement of the extraterrestrial hypothesis, the *psychosocial hypothesis*, the interdimensional hypothesis, the field of ufology in general, or UFO religions.

The term UFO as stated before was coined by officer *Edward J. Ruppelt* who investigated aerial phenomena for the *United States Air Force* in his role as leader of *Project Blue Book* during the 1950s. Before Project Blue Book, UFOs were often referred to as "*flying saucers*" in reference to the widely publicized Kenneth Arnold UFO sighting. Ruppelt explained that "the term 'flying saucer' is misleading when applied to objects of every conceivable shape and performance."

What would later be called UFOs were reported before the 1947 flying saucer craze. Allied pilots during World War II described them as foo fighters. In Sweden, early 20th-century UFOs were called *Spökraketer* or ghost rockets. Mystery airships reported in the late 19th century, are seen as predecessors to UFO sightings. Unexplained aerial phenomena have been recorded stretching back into antiquity. Prior to the Scientific Revolution, they were often interpreted through the lenses of gods, ghosts, demons, and omens. Rice University professor of religion Jeffrey J. Kripal says of UFO encounters, "These are not especially rare events, nor are they restricted to any culture, race, religion, or time period." Incidents of UFO citation reports throughout Ancient and Recent History, Table 16,1

ANTIQUITY

Table 16.1: Incidents of UFO Citation Reports Throughout Ancient and Recent History – Curtsy Wikipedia

Date	Name	Location	Description
c. 1440 BC	Fiery disks	AF, Ancient Egypt; Lower Egypt	According to the disputed Tulli Papyrus, the scribes of the pharaoh Thutmose III reported that «fiery disks» were encountered floating over the skies. The Condon Committee disputed the legitimacy of the Tulli Papyrus stating, «Tulli was taken in and that the papyrus is a fake.»
218 BC	Ships in the sky	EU, Roman Republic; Rome, Italia	Livy records a number of portents in the winter of this year, including navium speciem de caelo adfulsisse ("phantom ships had been seen gleaming in the sky").
76 BC	Spark from a falling star	AS, Roman Republic; Asia	According to Pliny the Elder, a spark fell from a star and grew as it descended until it appeared to be the size of the Moon. It then ascended back up to the heavens and was transformed into a light.
7 BC	Flame-like wine-jars from the sky	AS, Roman Republic; Phrygia, Asia	According to Plutarch, a Roman army commanded by Lucullus was about to begin a battle with Mithridates VI of Pontus when «the sky burst asunder, and a huge, flame-like body was seen to fall between the two armies». Plutarch reports the shape of the object as like a wine-jar (pithos).[9][7]
AD 65	Sky army	AS, Roman Empire; Judaea	Romano-Jewish historian Flavius Josephus reported chariots «hurtling through the clouds» prior to the First Jewish–Roman War.[7]
AD 196	Angel hair	EU, Roman Empire; Rome, Italia	Historian Cassius Dio described «A fine rain resembling silver descended from a clear sky upon the Forum of Augustus." He used some of the material to plate some of his bronze coins, but by the fourth day afterwards, the silvery coating was gone.[7]

Date	Name	Location	Description
AD c. 740	Air ship of Clonmacnoise	EU, Ireland; Teltown in County Meath, and Clonmacnoise in County Offaly	Several sets of Irish annals, those of Ulster, Tigernach, Clonmacnoise, and the Four Masters, all have entries to the effect that "ships with their crews were seen in the air".[10]

16th–17th centuries

Date	Name	Location	Description
1561-4-14	1561 celestial phenomenon over Nuremberg	EU, Holy Roman Empire; Nuremberg, Bavaria	Residents of Nuremberg described an aerial battle, followed by the appearance of a large black triangular object, and then a crash outside of the city. A broadsheet recorded that witnesses observed hundreds of spheres, cylinders and other odd-shaped objects that moved erratically overhead.[11]
1566-8-7	1566 celestial phenomenon over Basel	EU, Switzerland; Basel, Basel	A broadsheet published in 1566 depicted numerous spherical objects appearing out of the sun.[11] The event was recorded and depicted by Samuel Coccius, "a student of the Holy Scripture and of the free arts, at Basel".[12]
1609-9-22	Gwanghaegun period UFO Turmoil	AS, Joseon (Korea); Gangwon Province	On September 22, 1609, multiple witnesses reported seeing UFOs in Goseong, Wonju Gangneung at 사시 (9am-11am), Chuncheon County at 오시 (11am-1pm) and Yangyang County at 미시 (1pm-3pm). They described a Halo or washbowl that was divided in two.[13]

19th century

Date	Name	Location	Description
1803-2-22 or 1803-3-24	Utsuro-bune at Haratono-hama	AS, Japan; Hitachi Province	In 1803, local fishermen reportedly found a closed vessel with small windows adrift. They said when they investigated it that "a beautiful young woman" with red and white hair and dressed in strange clothes emerged, holding a square box "that no one was allowed to touch" and that she spoke to them in a language they had never heard before.[14]

Date	Name	Location	Description
1883-8-12	Bonilla observation	NA, Mexico; Zacatecas Observatory, Zacatecas	On August 12, the astronomer José Bonilla counted over 400 dark, unidentified objects crossing the sun while observing sunspot activity at Zacatecas Observatory in Mexico. He was able to take photographs, exposing wet plates at 1/100 second. He published an account of the event three years later in L'Astronomie, a French astronomy journal.[15][16]
1896-11-17 to 1897-4-23	Mystery airships	NA, United States	Newspapers across California, and later other states, printed reports of strange airships. Common elements of the descriptions included flapping wings, a cigar-shaped body, and a metal hull.[17]
1897-4-17	Aurora, Texas, UFO incident	NA, United States; Aurora, Texas	Local correspondent S.E. Hayden reported the crash of an airship piloted by an alien. According to Hayden, the spaceman was buried in the local cemetery. Residents of Aurora embrace the story without taking it seriously.[18][19]

20th century

1909–1948[**edit**]

Date	Name	Location	Description
1909	*New Zealand airship sightings*	*OC, New Zealand; Otago*	In August 1909, moving and whirring lights were reported in the sky around Otago. In the following months, many sightings were reported across New Zealand with varying descriptions of the craft and crew.[20][21][22]
1917-8-13, 1917-9-13, 1917-10-13	*Miracle of the Sun*	*EU, Portugal; Fátima, Santarém District*	Thousands of people gathered in Fátima based on reported *Marian apparitions* and observed bizarre solar activity. Catholic bishop *José Alves Correia da Silva* declared the miracle «worthy of belief» on 13 October 1930, and the primarily Catholic witnesses viewed the event in religious terms. Later, *Jacques Vallée*, Joaquim Fernandes and Fina d'Armada interpreted it as a mass UFO sighting.[23][24][25]
c. 1940	*Foo fighters*	*IC*, Over *World War II* theaters	During World War II, *allied* fighter pilots above Europe reported colorful balls of light following their aircraft at high speeds.[26]
1942-2-24	*Battle of Los Angeles*	*NA, United States; Los Angeles, California*	Just months after the Japanese *Attack on Pearl Harbor*, U.S. radar stations picked up an unidentified aerial object in the early morning. For several hours, anti-aircraft artillery fired thousands of rounds into the searchlight-scoured sky. The *LA Times* reported that "the air over Los Angeles erupted like a volcano."[27][28]

Date	Name	Location	Description
1946	*The Ghost Rockets*	*EU*, *Scandinavia* and other parts of *Europe*	Thousands of UFO sightings were reported over Europe. Due in part to concerns that foreign governments were testing *recovered experimental German technology*, the Swedish and Greek governments investigated the reports separately.[29]
1946-5-18	*Ängelholm UFO memorial*	*EU*, *Sweden*; *Ängelholm*, *Kristianstads County*	Swedish entrepreneur, Gösta Carlsson, the founder and owner of Cernelle AB, attributes his success to a 1946 UFO encounter. Decades later, he erected a concrete monument in the clearing where he says the flying saucer landed.[30]
1947-6-21	*Maury Island incident*	*NA*, *United States*; *Puget Sound* near *Maury Island*, *Washington*	*Fred Crisman* mailed an account from employee Harold A. Dahl, along with a cigar box of metal wreckage, to *Raymond A. Palmer* who had previously published the *Shaver Mystery* stories. Dahl claimed that his dog was killed and his son was injured by debris in an encounter with six flying doughnut-shaped objects. He also reported that he was subsequently threatened by *Men in Black*. On July 31, 1947, Palmer arranged a meeting between Crisman, Dahl, Air Force investigators, and flying saucer witnesses *Kenneth Arnold* & *Emil Smith*. [31][32]
1947-6-24	*Kenneth Arnold UFO sighting*	*NA*, *United States*; North of *Mount Rainier*, *Washington*	Private pilot *Kenneth Arnold* was flying near *Mount Rainier* when he reported seeing a group of thin, reflective craft moving at high speeds and flashing in the sun like mirrors. Bill Bequette of the *East Oregonian*, who first interviewed Arnold, summarized the sighting as, "nine saucer-like aircraft flying in formation." This introduced the term flying saucers, and Arnold's sighting sparked an explosion of UFO reports around the country. [33][34][35][36][37]
c. 1947	*1947 flying disc craze*	*NA*, *United States*; *Washington* and other states	After the *Kenneth Arnold* sighting was reported in the news, over 800 similar sightings were reported throughout 1947.[38][39]
1947-7-4	*Flight 105 UFO sighting*	*NA*, *United States*; En route from *Boise*, *Idaho* to *Pendleton*, *Oregon*	A United Airlines crew including Captain Emil Smith, co-pilot Ralph Stephens, and stewardess Marty Morrow witnessed nine unidentified objects. Believing them to be aircraft, Smith flashed the plane's landing lights intending to alert the objects which he described as "smooth on the bottom and rough appearing on top".[40][41]

Date	Name	Location	Description
1947-7-8	*The Roswell Incident*	*NA*, *United States*; about 30 mi. north of *Roswell, New Mexico*	*Walter Haut*, a *United States Army Air Forces* spokesperson, issued a press release announcing the «capture» of a «*flying saucer*". Hours later, the Army announced that the find was a crashed weather balloon. In 1978, the case regained attention after *Jesse Marcel*, the Army Officer who recovered the wreckage, told UFO researchers that the weather balloon explanation was a cover story. In 1994, the Air Force attributed the incident to the previously classified *Project Mogul*.[42]
c. 1948	*The Green Fireballs*	*NA*, *United States*; *New Mexico* and other parts of the *Southwestern United States*	The US Air Force investigated reports of green flares streaking across the sky after an Air Force C-47 transport encountered a green ball of fire on 5 December 1948. The pilot, Captain Goede, described the object as larger than a meteor and not arching downward as a meteor would. The Air Force investigation was inconclusive.[43][44]
1948-1-7	*Mantell UFO incident*	*NA*, *United States*; *Kentucky*	The flight tower at *Godman Army Airfield* instructed Captain Thomas Mantell to pursue a UFO sighted over *Fort Knox, Kentucky*. His aircraft crashed while in pursuit and Mantell died in *Franklin, Kentucky*. According to the historical marker placed on *Interstate 65* near the site it «is still uncertain what Mantell was pursuing».[45]
1948-3-25	*Aztec, New Mexico UFO hoax*	*NA*, *United States*; *New Mexico*	Conmen Silas Newton and Leo Gebauer sold "magnetic oil-detecting machines" based on the story that they had replicated technology from a crashed spaceship. The pair were convicted of fraud in 1953. Elements of their story regarding a crashed ship with occupants were later entangled in the Roswell narrative.[46][47][48]
1948-7-24	*Chiles-Whitted UFO encounter*	*NA*, *United States*; *Montgomery, Alabama*	Clarence Chiles and John Whitted, American commercial pilots, reported that their airplane had nearly collided with a UFO near Montgomery. According to the pilots the object "looked like a wingless aircraft...it seemed to have two rows of windows through which glowed a very bright light, as brilliant as a magnesium flare."[49][50]

Date	Name	Location	Description
1948-10-1	*Gorman dogfight*	*NA, United States; North Dakota*	A US Air Force pilot sighted and pursued a UFO for 27 minutes over *Fargo, North Dakota*. According to US Air Force officer *Edward J. Ruppelt*, this was one of three cases, along with the Mantell incident and Chiles-Whitted encounter, that shifted the Air Force's attitude about UFO reports leading to the creation of *Project Blue Book.[51][52]*

1950–1974

Date	Name	Location	Description
1950-5-11	McMinnville UFO photographs	NA, United States; a farm near McMinnville, Oregon	A farmer took pictures of a purported "flying saucer". These were the first flying saucer photographs since the coining of the term.[53]
1950-8-15	Mariana UFO incident	NA, United States; Great Falls, Montana	The manager of Great Falls' pro baseball team took color film of two UFOs flying over Great Falls. The film was extensively analyzed by the US Air Force and several independent investigators.[54][55]
1951-8-25	Lubbock Lights	NA, United States; Lubbock, Texas	Several Lights in V-Shaped formations were repeatedly spotted flying over the city. Witnesses included W. I. Robinson, A. G. Oberg, and W. L. Ducker, professors of geology, chemical engineering, and petroleum engineering respectively. Teenage student Carl Hart Jr. photographed the lights[56]
1952-7-12 to 1952-7-29	1952 Washington, D.C. UFO incident	NA, United States; Washington, D.C.	A series of sightings in July 1952 accompanied radar contacts at three separate airports in the Washington area. The sightings made front-page headlines around the nation, and ultimately lead to the formation of the Robertson Panel by the CIA.[57][58]
1952-7-14	Nash-Fortenberry UFO sighting	NA, United States; Norfolk, Virginia	W. Nash and W. Fortenberry, pilots of a DC-4 airliner of Pan American Airways, sighted eight large, round, glowing red objects.[59]
1952-9-12	The Flatwoods Monster	NA, United States; Flatwoods, West Virginia	Three local boys followed a bright object into the forest to what they believed was a UFO landing. They went to the nearby home of Kathleen May who accompanied them back to the spot along with 2 other children and teenage National Guardsman Eugene Lemon. In the forest, they smelled a foul odor and saw what May described as a tall figure with claws and "a head that resembled the ace of spades".[60]

Date	Name	Location	Description
1954-10-27	Fiorentina Stadium Mass Sighting	EU, Italy; Stadio Artemio Franchi in Florence	A football game between Fiorentina and Pistoiese was under way at the Stadio Artemio Franchi when a group of UFOs traveling at high speed abruptly stopped over the stadium. The stadium became silent as the crowd of around 10,000 spectators witnessed the event and described the UFOs as cigar shaped.[61][62]
1955-8-21 to 1955-8-22	Kelly–Hopkinsville encounter	NA, United States; A farmhouse near Hopkinsville, Kentucky	After the sighting of a disc-shaped aircraft a group of strange, goblin-like creatures are reported to have repeatedly approached a farm house and looked inside through the windows. Members of the two families present shot at them several times with little or no effect. The encounter lasted from evening to dawn.[63]
1956-7-24	Lakenheath-Bentwaters incident	EU, United Kingdom; Suffolk, England	United States Air Force (USAF) and Royal Air Force (RAF) radar operators (from Lakenheath RAF Station, Bentwaters RAF Station, and Sculthorpe RAF Station) detected up to 15 objects over Suffolk. An RAF pilot was sent out from Waterbeach RAF Station in a de Havilland Venom, a jet aircraft with Airborne Interception radar. The pilot reported spotting the object on radar and visually observing a luminous white object that moved behind his craft when he attempted to intercept.[64][65]
1956-8-13	Elizabeth Klarer	AF, South Africa; Drakensberg	A series of photos depicting a supposed UFO, were taken on 24 July near Rosetta in the Drakensberg region. The photographer, meteorologist Elizabeth Klarer, claimed detailed adventures with an alien race including having had an alien lover, Akon, who would have fathered her son Ayling.[66]
1957-5-3	Gordon Cooper UFO Sightings	NA, United States; Edwards Air Force Base, California	Gordon Cooper, one of the original Project Mercury astronauts, witnessed a type of metallic craft without wings flying over Germany in the 1950s. At the time, Cooper believed these to be Soviet aircraft. His attitude later changed after an incident at Edwards Air Force Base. Cooper sent a crew of James Bittick and Jack Gettys out to a dry lake bed to set up data-recording photography equipment. Cooper said the two men, both familiar with experimental aircraft, came back shaken and talking about witnessing a wingless aircraft with retractable legs silently land and take off near them. Cooper reported the incident to The Pentagon which asked for all photographs of the craft. Cooper looked at the photos before sending them off and felt that the government covered up a UFO encounter.[67][68]

Date	Name	Location	Description
1957-5-20	Milton Torres 1957 UFO Encounter	EU, United Kingdom; East Anglia	U.S. Air Force fighter pilot Milton Torres reports that he was ordered to intercept and fire on a UFO displaying "very unusual flight patterns" over East Anglia. Ground radar operators tracked what was believed to be an unidentified aircraft for some time before Torres' plane was scrambled to intercept.[69][70][71][72]
1957-10-15	Antônio Vilas-Boas Abduction	SA, Brazil; Near São Francisco de Sales, Minas Gerais	Law student, Antônio Vilas-Boas, described being abducted by humanoid aliens and taken aboard their egg-shaped craft. He also said that he was confined within a small round room where he was compelled to have sex with a four foot tall alien woman.[73]
1957-11-2	Levelland UFO case	NA, United States; Levelland, Texas	Numerous drivers observed glowing objects hovering over the highway. They described them as approximately 200 feet long and shaped like eggs or cigars. The appearance of these lights was immediately followed by electrical failures including ignition failures in their cars.[74][75]
1957-11-4	Kirtland AFB UFO sighting	NA, United States; Albuquerque, New Mexico	Two Civil Aeronautics Administration tower operators observed an egg-shaped object emitting a white light beneath it come down to the Kirtland AFB runway as if landing. At seven times magnification through binoculars, they could see neither wings, tail, nor fuselage. They watched the object come to a stop several feet above the ground, hover, and then rapidly ascend. The operators then called CAA Radar Approach Control who tracked the object on radar.[64][65]
1961-9-19	Betty and Barney Hill abduction	NA, United States; South of Lancaster on Route 3, New Hampshire	The Hills reported the first alien abduction experience to be widely spread in English-language publications. While driving home, they observed a light move through the sky and land ahead of them. Barney Hill said that, against his will, he turned the car down a side road towards the light, where he found six small humanoid beings waiting for them. Betty Hill reported that they inserted a needle through her navel among other vaguely medical tests.[76]
1964-4-24	Lonnie Zamora incident	NA, United States; Socorro, New Mexico	Police officer Lonnie Zamora investigated a roaring sound. Zamora and a nearby tourist found a craft that took off shortly after their arrival. The craft left impressions in the ground that did not aid in identification.[77]

Date	Name	Location	Description
1965-6-4	Project Gemini UFO	OS, Low Earth orbit; Above Hawaii	During Gemini 4, astronaut James McDivitt spotted a white cylinder with a protruding arm traveling in his orbit. McDivitt has said that it was impossible for him to assign scale to the object against the black background of space, saying that it could have been small enough to hold in his hands or "the size of the Empire State Building."[78]
1965-9-3	The Incident at Exeter	NA, United States; Exeter, New Hampshire	Numerous people reported pulsating UFOs in Exeter, New Hampshire. The events were the subject of *The Incident at Exeter* by John G. Fuller.[79]
1965-7-1	The Valensole UFO incident	EU, France; Valensole, Provence-Alpes-Côte d'Azur	French farmer, Maurice Massé, witnessed a spherical vehicle in his lavender field. He noticed and approached two individuals that he observed near the vehicle, but after one pointed a tube in his direction, he stood still feeling paralyzed. He described the beings as child-sized, pale, large-headed, with only holes where a mouth should be. Massé said that after they left in their vehicle, he was never again able to grow a healthy plant in the area where the craft had landed. He did not personally comment on the effect this had on him, saying, "One always says too much." His wife reported that the man was plagued with exhaustion for months, that he had confessed some type of communication to her, and that it was a "spiritual experience" for her husband.[80][81]
1965-12-9	Kecksburg UFO incident	NA, United States; Kecksburg, Pennsylvania	Local newspapers and newscasts reported a fireball observed in the skies over Kecksburg. According to some residents, they found an acorn-shaped object in the woods. The U.S. military closed off the area to investigate and reported no evidence of a crash. A model built for *Unsolved Mysteries* is kept on display by the fire department, which leads an annual UFO celebration.[82][83][84]
1966-4-6	Westall UFO	OC, Australia; Clayton South, Victoria	Several hundred students and school faculty watched an object land at the Grange Reserve for horses, lift off, and vanish. Witnesses of "The Clayton Incident" still gather for reunions.[85]
1967-8-29	Close encounter of Cussac	EU, France; near Cussac, Auvergne	A 13-year-old boy and his younger sister reported an incident to their father, local police, and investigators. According to police reports, they witnessed a brilliant sphere and four small black occupants while herding cattle outside their village.[86]

Date	Name	Location	Description
1967-10-4	Shag Harbour UFO incident	NA, Canada; Gulf of Maine near Shag Harbour, Nova Scotia	A large illuminated object was reported to have crashed into waters near Shag Harbour. A Canadian naval search followed, and officially referred to the object as a UFO. There were other UFO sightings in the area at the time.[87][88]
1969-1-6	Jimmy Carter UFO incident	NA, United States; Leary, Georgia	Jimmy Carter (US President 1981–1977) reported seeing an unidentified flying object while at Leary, Georgia in 1969.[89]
1969-4-12	Finnish Air Force sighting	EU, Finland; Pori	Seven disc-shaped objects hovering in formation around 1500–3000 meters were spotted by the air traffic controller and two pilots. When fighter pilot Tarmo Tukeva investigated, they accelerated away.[90]
1969-9-1	Berkshire UFO sightings	NA, United States; Berkshire County, Massachusetts	Four unrelated families alleged being abducted by a UFO and moved by a ray of light.[91]
1972-6-27	1972 UFO sightings in the eastern Cape	AF, South Africa; Fort Beaufort	A farmer near Fort Beaufort in the eastern Cape fired shots at what he said was an unknown light in the sky. The incident received coverage by international press, and led to businesses capitalizing on the incident, with a tavern calling itself the «UFO Bar» and painting flying saucers on the walls.[92][93]
1973-10-11	Pascagoula Abduction	NA, United States; Pascagoula, Mississippi	Charles Hickson and Calvin Parker were fishing from a pier on the Pascagoula River when they say that they heard whirring sounds and witnessed a craft over 30 feet long with flashing lights. Both men say they were paralyzed and then taken by humanoids with «robotic slit-mouths» and «crab-like pincers».[94][95]
1974-1-23	Berwyn Mountain UFO incident	EU, United Kingdom; Llandrillo, Merionethshire, North Wales	An alleged UFO crash involving lights in the sky moments before a large impact shock. The cause of the incident was however soon revealed as a 3.5 magnitude earthquake.[96][97]
1974-8-23	John Lennon UFO incident	NA, United States; New York City, New York	Musician John Lennon and then-assistant May Pang report seeing a craft emitting lights that changed color in the night sky above their Manhattan penthouse. Lennon would later reference this experience in his song «Nobody Told Me".[98][99]

1975–2000

Date	Name	Location	Description
1975-11-05	Travis Walton	NA, United States; Near Turkey Springs in Apache-Sitgreaves National Forest, Arizona	Logger Travis Walton disappeared for over 5 days resulting in a police investigation of his coworkers. When questioned on where he had been, Walton said that he had been taken aboard a spacecraft by nearly human creatures.[100] Walton›s alien abduction account is the basis for the book *The Walton Experience* (1978), the film *Fire in the Sky* (1993), and the documentary «Alien Abduction: Travis Walton» (2022).[101]
1976-9-17	*1976 Tehran UFO incident*	*AS*, *Iran*; *Tehran*, *Tehran province*	The 1976 Tehran UFO Incident was a radar and visual sighting of a UFO over the capital of Iran, during early morning hours. Two Imperial Iranian Air Force F-4 Phantom II jet interceptors reported losing instrumentation and communications as they approached the object.*[102]*
1977-9-20	*Petrozavodsk phenomenon*	*EU*, *Soviet Union*, *Finland*, *Lithuania*, and *Denmark*	Residents of *Petrozavodsk* reported a giant glowing «*jellyfish*" of light (visible for over ten minutes) looming in the early morning sky. The light was seen and photographed in several *Baltic Sea countries*. In response to the phenomenon, the Soviet Union created a government program to study anomalous atmospheric phenomena. This program would later attribute the Petrozavodsk sightings to the secret night launch of the *Kosmos 955* spy satellite. According to Soviet astrophysicist, Yuli Platov, sunlight can cause the giant plumes of gas and dust produced by rockets to glow, especially "in *twilight hours*, when the rocket streaks through sunlit regions and the observer is on the nighttime side of the Earth."*[103]*
1978-5-10	*Emilcin Abduction*	*EU*, *Poland*; *Emilcin*, *Lublin Voivodeship*	Polish farmer Jan Wolski reported that while returning home, two "short, green-faced humanoid entities" wearing black overalls jumped onto his horse-drawn cart and started speaking an incomprehensible language. After about 1000 ft (300 m), he reported seeing a white flying object, from which an alien creature came out and invited Wolski inside. The farmer said that they examined him once inside.*[104]*

Date	Name	Location	Description
1978-10-21	*Valentich disappearance*	*OC*, Australia; *Victoria*	Frederick Valentich left *Moorabbin Airport* in a *Cessna 182 Skylane*, a single-engined light aircraft. At 7:06 PM, he began reporting a strange craft to *Melbourne* air traffic control. Valentich›s last words were, «That strange aircraft is hovering on top of me again ... it is hovering and it›s not an aircraft.»*[105]* Neither the pilot nor the plane were ever found.[106]
1978-12-6	*Zanfretta UFO Incident*	*EU*, *Italy*; *Torriglia*, *Genoa*	Italian nightwatchman Pier Fortunato Zanfretta perceived a red, oval object and phoned his supervisor. During the call, he described non-human creatures that he said were attacking him. He was later found in a state of shock and his experience was adapted into a stage play.[107]
1979-11-11	*Manises UFO incident*	*EU*, *Spain*; *Valencia*, *Valencian Community*	En route to *Las Palmas*, commercial pilot Francisco Javier Lerdo de Tejada radioed air traffic control regarding a pair of red lights approaching his *TAE Supercaravelle*. Neither air traffic control in Barcelona nor the military identified the object. Tejada made an emergency landing at the nearby airport in Manises.[108]
1980-4-11	Arequipa UFO incident	*SA*, *Peru*; *Arequipa Region*	Early in the morning of April 11, La Joya *Air Force* Base ordered fighter pilot Oscar Santa María Huertas to intercept an object in restricted air space. Huertas pursued the object in a *Sukhoi Su-22* and fired a barrage of *30mm shells* into it. According to Huertas the object did not seem damaged and rose to 19,200 meters. He described it as similar in shape to an incandescent lightbulb with a much wider circular silver base and said that it «lacked all the typical components of aircraft. It had no wings, propulsion jets, exhausts, windows, antennae, and so forth». [109][110][111][112]
1980-12-24 to 1980 -12-24	*Rendlesham Forest incident*	*EU*, *United Kingdom*; *Rendlesham Forest*, *Suffolk*, *England*	*United States Air Force* personnel reported various unusual observations at RAF Woodbridge and RAF Bentwaters, two American air bases located in England. Their reports included lights in the sky, a metallic triangular object in the forest, multi-colored lights moving through the forest, and higher levels of radiation.[113]

Date	Name	Location	Description
1981-1-8	*Trans-en-Provence case*	*EU, France; Trans-en-Provence, Provence-Alpes-Côte d'Azur*	Retired contractor, Renato Nicolai reported the landing of a flying object near his home to local police. Renato believed it to be a military craft. Groupe d'Étude des Phénomènes Aérospatiaux Non-identifiés (*GEPAN*), a branch of French Space Agency created to investigate UFOs, conducted an investigation, photographed circular impressions on the ground, and took samples of the area.[114] Skeptics have been critical of the GEPAN investigation which took place 40 days after the initial sighting.[115][116]
1986-5-19	*"Night of the UFOs"* in Brazil	*SA, Brazil; São Paulo, Rio de Janeiro, Minas Gerais* and *Goiás*	Radar and visual contacts were obtained with multiple 'bright colorful objects' in the sky across several states. *Mirage 2000* and *F-5* fighters were scrambled but failed to intercept, with pilots describing the objects as capable of impossible maneuvers and rapidly accelerating to as much as Mach 15 once approached.[117]
1986-11-17	*Japan Air Lines Cargo Flight 1628 incident*	*NA, United States; Alaska*	While piloting a *Japanese Boeing 747-200F cargo aircraft* on a *polar route* from France to *Narita International Airport* in Japan, the flight crew witnessed several unidentified objects over eastern Alaska. Captain Kenju Terauchi (Terauchi Kenju), *co-pilot* Takanori Tamefuji (Tamefuji Takanori), and *flight engineer* Yoshio Tsukuda (Tsukuda Yoshio) reported rectangular arrays of what Captain Terauchi described as glowing *nozzles* or thrusters.[118][119]
1987-12-1	*Ilkley Moor UFO incident*	*EU*, United Kingdom; *Ilkley Moor*	Retired police officer, Philip Spencer, took a photograph of what he said was a strange being on the moor. According to Spencer, the being fled after being photographed and left in a domed craft.[120][121]
1994-9-16	*Ariel School UFO incident*	*AF, Zimbabwe; Ruwa*	Over sixty students reported seeing a silver craft land in a field near the school. They described occupants dressed in all black that exited the silver object.[122]
1996-01-20	*Varginha UFO incident*	*SA, Brazil; Varginha, Minas Gerais*	In 1996, various individuals reported possibly unrelated incidents with what they described as UFOs, creatures, and the Brazilian military.[123] The events are the inspiration for saucer-shaped Varginha water tower, the *Nave Espacial de Varginha*; the 1998 Brazilian video game *Incidente em Varginha*; and 1996, a film by director Rodrigo Brandão.[124][125]

Date	Name	Location	Description
1997-3-13	*Phoenix Lights*	*NA, United States; Phoenix, Arizona*	Hundreds of people witnessed a series of lights in a "V" pattern moving through Arizona. The incident is unusual for the large number of photographs.[126]

21st century

Date	Name	Location	Description
2000-10-5	Bonsall UFO sighting	EU, United Kingdom; Bonsall, Derbyshire	Resident, Sharon Rowlands, filmed a luminous object in the night sky starting around 9:15 PM and continuing for several minutes. Various other residents reported strange lights including a man who described a "pink glow, vertically shaped like a shoe box".[127][128][129]
c. 2004	Tinley Park Lights	NA, United States; Illinois	At 8 PM on Halloween night (2004), residents were outside trick-or-treating on the streets of Tinley Park and other Chicago suburbs, when thousands of people watched, photographed, and filmed a formation of red lights hanging in the October sky above them. Similar formations of lights drifted over the area the following autumn.[130][131][132]
2004-3-5	2004 Mexican UFO incident	NA, Mexico; Campeche	Mexican Air Force pilots filmed (initially unidentified) lights in the sky using infrared cameras while searching for drug-smuggling planes. Multiple subsequent investigations identified these as massive burn-off flares from a cluster of off-shore oil platforms in the Bay of Campeche.[133][134][135]
2004-11-14	USS Nimitz UFO incident	NA, United States; Off the coast of San Diego, California	Several pilots from VFA-41 squadron flying Super Hornets from the USS Nimitz, were directed by the USS Princeton to intercept one of several unidentified flying objects detected by radar. The pilots reported a visual encounter and recorded an infrared video. The Navy has verified that the video was taken by Navy personnel and has stated that it has not yet identified the nature of the sightings which they classify as unexplained aerial phenomena.[136][137][138]
2006-11-7	2006 O'Hare International Airport UFO sighting	NA, United States; Chicago, Illinois	United Airlines employees and pilots reported sightings of a saucer-shaped, unlit craft hovering over a Chicago O'Hare Airport terminal, before appearing to leave with a rapid vertical rise.[139]

Date	Name	Location	Description
2007-4-23	2007 Alderney UFO sighting	EU, Bailiwick of Guernsey; Alderney	Two airline pilots on separate flights reported a massive object at an altitude of around 3,500 feet off the coast of Alderney.[140]
2007-11-28 to 2011-12-13	Dudley Dorito	EU, United Kingdom; West Midlands conurbation	The Dudley Dorito were multiple sightings of a black triangle over the West Midlands conurbation of the United Kingdom which began in November 2007. The phrase was coined by the local press after hearing witness descriptions of the object.[141][142]
2008-6-20	Wales UFO sightings	EU, United Kingdom; Various cities, Wales	According to media reports, a police helicopter was almost hit by a UFO, before it tried to pursue it. Hundreds of people reported to have witnessed a UFO on the same or preceding days, from different areas of Wales.[143]
2009-1-5	Morristown UFO hoax	NA, United States; Morristown, New Jersey	In the evening, citizens in Morristown and other town in Morris County, New Jersey saw five red lights in the sky. After three months, two men from the Morristown area announced that they had organized a UFO hoax, meant as a «social experiment».[144]
2010-1-25	Harbour Mille incident	NA, Canada; Harbour Mille, Newfoundland and Labrador	At least three UFOs that looked like missiles but emitted no noise were spotted over Harbour Mille.[145]
2014-6-2 to 2015-3-10	USS Theodore Roosevelt UFO incidents	NA, United States; East Coast of the United States	Multiple UFO radar-visual encounters by United States Navy pilots over a period of nine months. Videos of two of the encounters were released. The Navy has verified that the videos were taken by Navy personnel and has stated that it has not yet identified the nature of the sightings which they classify as unexplained aerial phenomena.[146][147]
2021-2-21	Pilot's UFO sighting over New Mexico	NA, United States; Clayton, New Mexico	The pilot, at an altitude of 37,000 feet (11,278 m) reported seeing a long cylindrical object that almost looked like a "cruise missile type of thing" moving really fast right over the top of them according to published audio. American Airlines confirmed that the radio transmission came from flight 2292. FAA a few days later stated: «A pilot reported seeing an object over New Mexico shortly after noon local time on Sunday, Feb. 21, 2021. FAA air traffic controllers did not see any object in the area on their radarscopes.»[148]

Date	Name	Location	Description
2023-1-28 to 2023-2-13	List of high-altitude object events in 2023	IC, United States, Canada, Colombia, Costa Rica, and Venezuela	Multiple airborne objects, sometimes reported in mainstream media as UFOs, were observed and sometimes shot down by military aircraft. Many of the objects were reported as meteorology or espionage balloons.

MATHEMATICAL ANALYSIS APPLIED TO USA – (SIMPLE)

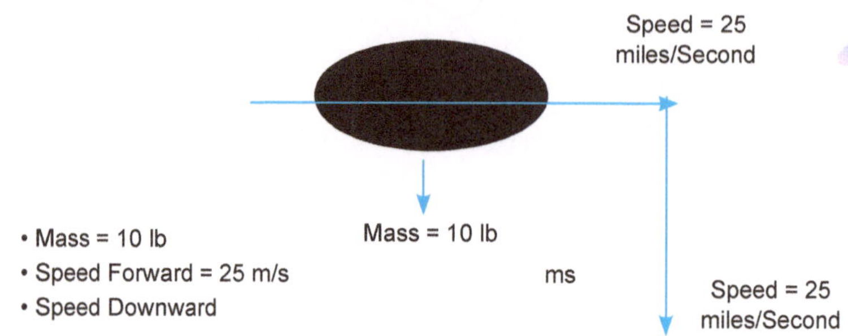

Fig.16.13: *Mathematical Analysis of the UFO Object*

Observation

- The mass of the UFO Disk travels forward at a speed of 25m/s. Figure 16.13
- The same mass instantaneously travels downward at the same speed of 25 m/s
- This is exactly what was observed in all the UFO recorded when they start motion from zero speed reaching any speed, they desire up to 186,282 m/s.

Analysis

According to Sir Isaac Newton's Laws of dynamic motions, the UFO's appearance, disappearance, and traveling speeds, contradicts these laws of dynamic motion.

The reason of the violation and the impossibilities:

a. Any object with any mass cannot reach the desired speed without gradual acceleration
b. The sudden stop of any object with any mass must have a deceleration until it reached zero speed
c. Any normal object with a mass will acquire inertia during forward acceleration
d. Thus, the UFO object violate the laws of physics of the universe

Results

1. The UFO is visible in all sighting and observations
2. The recorded data of speed and dynamic motions are accurate and verifiable
3. The environment has no effect on the data or the UFO object during citation

Conclusion

1. The UFO object is massless (Zero Mass)

2. The UFO object has the perpetual power to appear and disappear at will – (Quantum Mechanics), according to Albert Einstein Theory of Relativity
3. The object can make any tun, in any direction, including a U-turn successfully without losing Inertia
4. The UFO object does not have the same Earthly physical properties such as carbon, oxygen, and hydrogen, as these materials still have mass to be considered
5. The object can only be made Ethereal of LIGHT

Since the object is made of light or similar entity of physics, such as holographs, it opens a new dimension of our investigation and analysis. It is alarming to tap into the Ethereal dimensions.

It is conceivable to experience an entity attempting to masquerade itself as an entity of light or alike. Upon searching the books, we remembered that the New Testament has referred to other being transformed into a form of light or a messenger of light. 2 Corinthians 11:14, states: '***And it is no wonder, for Satan himself masquerades as an angel of light.***' Our research could not stop here!

THE CAUSE OF THE MASQUERADE

The eminent question is why the masquerade in a form of light?

The whole Earth and all its inhabitants are posing the same question, what is the message the UFOs are attempting to convey, Figure 16.14. The schism that separates us from them has been crossed unless those UFOs have never been isolated from Earth dwelling. They cannot be the spirits of the dead. The dead no matter what kind of people they were, they depart the Earth the instant their souls depart their bodies. Therefore, who are those entities that masquerade in light form, and why?

Fig.16.14: UFO Masquerade in for of Light

Upon reading further to gain the knowledge of life on Earth as well as life beyond, I realized we are not dealing with elusion. We are dealing with entities, which may take whatever ethereal it desired to serve their cause.

THE INTENT OF UFOS

There were several incidents in the past, where unknown entities appeared and disappeared serving very specific reasons. Then they depart after explicit message they leave behind. Let us state several of those incidents where messengers ascended from Heaven to Earth:

1. **Genesis 18:2-15**

   When he [Abraham] lifted up his eyes and looked, behold, three men [Angels] were standing opposite him; and when he [Abraham] saw them, he [Abraham] ran from the tent door to meet them and bowed himself to the earth, and said, "My Lord [Jesus], if now I have found favor in Your sight, please do not pass Your servant by. Please let a little water be brought and wash your feet, and rest yourselves under the tree

2. **Genesis 16:7-14**

   Now the angel of the Lord [Jesus] found her by a spring of water in the wilderness, by the spring on the way to Shur. He [Jesus] said, "Hagar, Sarai's maid, where have you come from and where are you going?" And she said, "I am fleeing from the presence of my mistress Sarai." Then the angel of the Lord said to her, "Return to your mistress, and submit yourself to her authority."

3. **Genesis 19:1-22**

   Now the two angels [2 Angels] came to Sodom in the evening as Lot was sitting in the gate of Sodom. When Lot saw them, he rose to meet them and bowed down with his face to the ground. And he said, "Now behold, my lords, please turn aside into your servant's house, and spend the night, and wash your feet; then you may rise early and go on your way." They said however, "No, but we shall spend the night in the square." Yet he urged them strongly, so they turned aside to him and entered his house; and he prepared a feast for them, and baked unleavened bread, and they ate.

4. **Genesis 28:10-12**

   Then Jacob departed from Beersheba and went toward Haran. He came to a certain place and spent the night there, because the sun had set; and he took one of the stones of the place and put it under his head, and lay down in that place. He had a dream, and behold, a ladder was set on the earth with its top reaching to heaven; and behold, the angels of God were ascending and descending on it.

5. **Genesis 16:7-14**

   Now the Angel of the Lord [Jesus] found her by a spring of water in the wilderness, by the spring on the way to Shur. He said, "Hagar, Sarai's maid, where have you come from and where are you going?" And she said, "I am fleeing from the presence of my mistress Sarai." Then the angel of the Lord said to her, "Return to your mistress, and submit yourself to her authority."

6. **Exodus 3:1**

   Now Moses was pasturing the flock of Jethro his father-in-law, the priest of Midian; and he led the flock to the west side of the wilderness and came to Horeb, the mountain of God.

Numbers 22:31-35

Then the Lord opened the eyes of Balaam, and he saw the Angel of the Lord [Jesus] standing in the way with his drawn sword in his hand; and he bowed all the way to the ground. The angel of the Lord [Jesus] said to him, "Why have you struck your donkey these three times? Behold, I have come out as an adversary, because your way was contrary to me. But the donkey saw me and turned aside from me these three times. If she had not turned aside from me, I would surely have killed you just now, and let her live."

Joshua 5:13-15

Now it came about when Joshua was by Jericho, that he lifted up his eyes and looked, and behold, a man was standing opposite him [angel]with his sword drawn in his hand, and Joshua went to him and said to him, "Are you for us or for our adversaries?" He said, "No; rather I indeed come now as captain of the host of the Lord [angel]." And Joshua fell on his face to the earth, and bowed down, and said to him, "What has my lord to say to his servant?" The captain of the Lord's host said to Joshua, "Remove your sandals from your feet, for the place where you are standing is holy." And Joshua did so.

Judges 2:1-4

Now the angel of the Lord [Jesus] came up from Gilgal to Bochim. And he said, "I brought you up out of Egypt and led you into the land which I have sworn to your fathers; and I said, 'I will never break My covenant with you, and as for you, you shall make no covenant with the inhabitants of this land; you shall tear down their altars.' But you have not obeyed Me; what is this you have done? Therefore, I also said, 'I will not drive them out before you; but they will become as thorns in your sides and their gods will be a snare to you

Judges 6:11-24

Then the Angel of the Lord came and sat under the oak that was in Ophrah, which belonged to Joash the Abiezrite as his son Gideon was beating out wheat in the wine press in order to save it from the Midianites. The angel of the Lord appeared to him and said to him, "The Lord is with you, O valiant warrior." Then Gideon said to him, "O my lord, if the Lord is with us, why then has all this happened to us? And where are all His miracles which our fathers told us about, saying, 'Did not the Lord bring us up from Egypt?' But now the Lord has abandoned us and given us into the hand of Midian."

Judges 13:6-21

Then the woman came and told her husband, saying, "A man of God [Jesus] came to me and his appearance was like the appearance of the angel of God, very awesome. And I did not ask him where he came from, nor did he tell me his name. But he said to me, 'Behold, you shall conceive and give birth to a son, and now you shall not drink wine or strong drink nor eat any unclean thing, for the boy shall be a Nazirite to God from the womb to the day of his death.'" Then Manoah entreated the Lord and said, "O Lord, please let the man of God whom You have sent come to us again that he may teach us what to do for the boy who is to be born."

2 Samuel 24:16

When the angel [Angel] stretched out his hand toward Jerusalem to destroy it, the Lord relented from the calamity and said to the angel who destroyed the people, "It is enough! Now relax your hand!" And the angel of the Lord was by the threshing floor of Araunah the Jebusite.

Daniel 6:21

Then Daniel spoke to the king, "O king, live forever!

Zechariah 2:3

And behold, the angel who was speaking with me was going out, and another angel was coming out to meet him,

Psalm 34:7

The **angel** of the LORD encamps around those who fear him, and delivers them.

Judges 13:2-22

There was a certain man of Zorah, of the tribe of the Danites, whose name was Manoah. And his wife was barren and had no children. And the **Angel** of the LORD appeared to the woman and said to her, "Behold, you are barren and have not borne children, but you shall conceive and bear a son. Therefore, be careful and drink no wine or strong drink, and eat nothing unclean, for behold, you shall conceive and bear a son. No razor shall come upon his head, for the child shall be a Nazirite to God from the womb, and he shall begin to save Israel from the hand of the Philistines." Then the woman came and told her husband, "A man of God came to me, and his appearance was like the appearance of the angel of God, very awesome. I did not ask him where he was from, and he did not tell me his name, ...

2 Kings 6:15-17

When the servant of the man of God rose early in the morning and went out, behold, an army with horses and chariots was all around the city. And the servant said, "Alas, my master! What shall we do?" He said, "Do not be afraid, for those who are with us are more than those who are with them." Then Elisha prayed and said, "O LORD, please open his eyes that he may see." So, the LORD opened the eyes of the young man, and he saw, and behold, the mountain was full of horses and **chariots of fire** all around Elisha.

Fig.16.15: *The Seraphim Angel Achieves God's Will*

Isaiah 6:1-3

In the year that King Uzziah died I saw the Lord sitting upon a throne, high and lifted up; and the train of his robe filled the temple. Above him stood the seraphim [**angel**], Figure 16.15. Each had six wings: with two he covered his face, and with two he covered his feet, and with two he flew. And one called to another and said: "Holy, holy, holy is the LORD of hosts; the whole earth is full of his glory!"

Daniel 10:13

The prince of the kingdom of Persia withstood me twenty-one days, but Michael [**angel**], one of the chief princes, came to help me, for I was left there with the kings of Persia,

Daniel 8:16

And I heard a man's voice between the banks of the Ulai, and it called, "Gabriel [**angel**], make this man understand the vision."

Fig.16.16: *Daniels Facing Heavens ignoring the roaring Lions Behind*

2 Kings 6:17

Then Elisha prayed and said, "O LORD, please open his eyes that he may see." So the LORD opened the eyes of the young man, and he saw, and behold, the mountain was full of horses and chariots of fire all around Elisha.

Daniel 9:21

While I was speaking in prayer, Figure 16.16, the man Gabriel [**angel**], whom I had seen in the vision at the first, came to me in swift flight at the time of the evening sacrifice.

Isaiah 7:14

Therefore, the Lord himself will give you a sign. Behold, the virgin shall conceive and bear a son, and shall call his name Immanuel, Figure 16.17.

Fig.16.17: *Isaiah 7:14 - Therefore, the Lord himself will give you a sign. Behold, the virgin shall conceive and bear a son, and shall call his name Immanuel.*

Numbers 22:22

But God's anger was kindled because he went, and the **Angel** of the LORD took his stand in the way as his adversary. Now he was riding on the donkey, and his two servants were with him.

Genesis 18:1-33

And the LORD appeared to him by the oaks of Mamre, as he sat at the door of his tent in the heat of the day. He lifted up his eyes and looked, and behold, three men [**3 angels**] were standing in front of him. When he saw them, he ran from the tent door to meet them and bowed himself to the earth and said, "O Lord, if I have found favor in your sight, do not pass by your servant. Let a little water be brought, and wash your feet, and rest yourselves under the tree, while I bring a morsel of bread, that you may refresh yourselves, and after that you may pass on—since you have come to your servant." So, they said, "Do as you have said." ...

TOP OF FORM

Luke 1:19

And the **angel** answered him, "I am Gabriel. I stand in the presence of God, and I was sent to speak to you and to bring you this good news.

Hebrews 13:2

Do not neglect to show hospitality to strangers, for thereby some have entertained **angels** unawares.

Colossians 2:18

Let no one disqualify you, insisting on asceticism and worship of **angels**, going on in detail about visions, puffed up without reason by his sensuous mind,

Revelation 4:1-11

After this I looked, and behold, a door standing open in heaven! And the first voice, which I had heard speaking to me like a trumpet, said, "Come up here, and I will show you what must take place after this." At once I was in the Spirit, and behold, a throne stood in heaven, with one seated on the throne. And he who sat there had the appearance of jasper and carnelian, and around the throne was a rainbow that had the appearance of an emerald. Around the throne were twenty-four thrones, and seated on the thrones were twenty-four elders, clothed in white garments, with golden crowns on their heads. From the throne came flashes of lightning, and rumblings and peals of thunder, and before the throne were burning seven torches of fire, which are the seven spirits of God, ...

Hebrews 1:14

Are they not all ministering **spirits** [angels] sent out to serve for the sake of those who are to inherit salvation?

Acts 5:19

But during the night an **angel** of the Lord opened the prison doors and brought them out, and said,

Acts 5:16

The people also gathered from the towns around Jerusalem, bringing the sick and those afflicted with unclean spirits, and they were all healed.

Luke 1:35

And the **angel** answered her, "The Holy Spirit will come upon you, and the power of the Most High will overshadow you; therefore, the child to be born will be called holy—the Son of God.

Jude 1:9

But when the **archangel** Michael, contending with the devil, was disputing about the body of Moses, he did not presume to pronounce a blasphemous judgment, but said, "The Lord rebuke you, Figure 16.18."

Matthew 26:53

Do you think that I cannot appeal to my Father, and he will at once send me more than twelve legions of **angels**? Figure 16.19

Fig.16.18: The ***Archangel*** *Michael, Contending with the Devil*

Fig.16.19: Twelve Legions of ***Angels***

Hebrews 13:1-2

Let brotherly love continue. Do not neglect to show hospitality to strangers, for thereby some have entertained angels unawares.

Colossians 1:16

For by him all things were created, in heaven and on earth, visible and invisible, whether thrones or dominions or rulers or authorities—all things were created through him and for him. Figure 16.20

Fig.16.20: *Colossians 1:16 - For by him all things were created, in heaven and on earth,* ***visible and invisible****, whether thrones or dominions or rulers or authorities—all things were created through him and for him.*

John 20:12

And she saw two **angels** in white, sitting where the body of Jesus had lain, one at the head and one at the feet.

1 John 4:1

Beloved, do not believe every **spirit**, but test the **spirits** to see whether they are from God, for many false prophets have gone out into the world.

Galatians 1:8

But even if we or an **angel** from heaven should preach to you a gospel contrary to the one, we preached to you, let him be accursed.

Matthew 10:1

And he called to him his twelve disciples and gave them authority over unclean **spirits**, to cast them out, and to heal every disease and every affliction.

Matthew 4:11

Then the devil left him, and behold, **angels** came and were ministering to him.

John 1:1

In the beginning was the Word, and the Word was with God, and the Word was God.

Luke 1:26

In the sixth month the **angel** Gabriel was sent from God to a city of Galilee named Nazareth,

Hebrews 12:22

But you have come to Mount Zion and to the city of the living God, the heavenly Jerusalem, and to innumerable **angels** in festal gathering,

Acts 12:5-17

So, Peter was kept in prison, but earnest prayer for him was made to God by the church. Now when Herod was about to bring him out, on that very night, Peter was sleeping between two soldiers, bound with two chains, and sentries before the door were guarding the prison. And behold, an **angel** of the Lord stood next to him, and a light shone in the cell. He struck Peter on the side and woke him, saying, "Get up quickly." And the chains fell off his hands. And the angel said to him, "Dress yourself and put on your sandals." And he did so. And he said to him, "Wrap your cloak around you and follow me." And he went out and followed him. He did not know that what was being done by the angel was real, but thought he was seeing a vision. ...

Acts 1:11

And said, "Men of Galilee, why do you stand looking into heaven? This Jesus, who was taken up from you into heaven, will come in the same way as you saw him go into heaven."

Luke 2:13

And suddenly there was with the **angel** a multitude of the heavenly host praising God and saying,

Matthew 28:5

But the **angel** said to the women, "Do not be afraid, for I know that you seek Jesus who was crucified.

Matthew 2:13

Now when they had departed, behold, an **angel** of the Lord appeared to Joseph in a dream and said, "Rise, take the child and his mother, and flee to Egypt, and remain there until I tell you, for Herod is about to search for the child, to destroy him."

Luke 1:26-38

In the sixth month the **angel** Gabriel was sent from God to a city of Galilee named Nazareth, to a virgin betrothed to a man whose name was Joseph, of the house of David. And the virgin's name was Mary. And he came to her and said, "Greetings, O favored one, the Lord is with you!" But she was greatly troubled at the saying, and tried to discern what sort of greeting this might be. And the angel said to her, "Do not be afraid, Mary, for you have found favor with God. ...

Matthew 18:10

"See that you do not despise one of these little ones. For I tell you that in heaven their **angels** always see the face of my Father who is in heaven.

Matthew 2:13-15

Now when they had departed, behold, an **angel** of the Lord appeared to Joseph in a dream and said, "Rise, take the child and his mother, and flee to Egypt, and remain there until I tell you, for Herod is about to search for the child, to destroy him." And he rose and took the child and his mother by night and departed to Egypt and remained there until the death of Herod. This was to fulfill what the Lord had spoken by the prophet, "Out of Egypt I called my son."

Matthew 25:41

"Then he will say to those on his left, 'Depart from me, you cursed, into the eternal fire prepared for the devil and his angels.

THE ETHEREAL ENTITY OF A SPIRIT (UFO)

The ethereal entity is a spirit being that does not have mass and is able to move even faster than speed of light. The speed of thought is faster than the speed of light. It can transport any spirit being to whatever location desired.

The reading of scripture above highlights the spirit being as angels. Also, there are two types of angels. One type is pure and wholesome intended to serve God's will through serving mankind. The other type of angels is the rebellious against the Lord God, while they have devoted their loyalty and pledged their allegiance to Satan the prince of darkness, evil and mischief.

The cause of the UFOs is not clearly manifested. Nevertheless, none of the UFO has demonstrated compassion and wholesome intent to mankind. They demonstrated instead, confusion, strife, disagreements, utter anxieties.

We may speculate a logical intent deduced from their increased appearances, in particularly in recent years. Their past history was dim and chaotic up to Noah's era.

Prior to Noah, the rebellious spirits knew in a general term, that the Messiah will come, as promised to redeem mankind from the curse of death caused by Adam's rebellious in violating Gods order. However, the exact timing was hidden from them. As a result, they schemed a plan to destroy God's pool of human genes. The rebellious spirits, similar to UFOs, took upon themselves the task to possess human bodies and produce a superior human form in strength and beauty. They have seduced women and were able to impregnate them. Consequently, the human gene pool was defiled by combining the falling mankind with the rebellious spirits beings called the Nephilim.

Jesus Christ, the Son of God, God kept for Himself only 8 people unpolluted by the scheme of Satan, the prince of the UFO, the falling angels. As the Universal Flood swept out the polluted beings whom they defiled God's plan of salvation. They perished in the flood along with all the corrupted race of mankind.

Similarly, nowadays the rebellious falling angels with their leader the prince of darkness are aware of the great moment, where Christ will appear in the sky to rapture His Pride, the collective company of universal believers, the Universal Church. At the same time the prince of darkness, the Beast together with his two prominent devoted agents, one is called the Anti-Christ and the other is called the False Prophet. It is likely, once the church is raptured to Heaven, the Anti-Christ and the False Prophet will claim that the UFOs have abducted the believers, to maintain their grip on the rest of mankind who were left behind.

LIFE ON OTHER PLANETS?

Why is the Universe so Big?

Many people, Christian or otherwise, struggle with the notion that the earth is the only inhabited planet in this enormous universe. In short, is there life on other planets? Figure 16.21.

Fig.16.21: *Why the Universe is So Big? Curtsy - NASA*

Those who believe life evolved on the earth usually see it as virtual 'fact' that life has evolved on countless other planets. Discovering life on other planets would in turn be seen as confirming their evolutionary belief.

But even many Christians think, 'God must have created life elsewhere, otherwise this enormous universe would be an awful waste of space.' In my experience, this seems to be the major underlying reason why people think that there must be other life 'out there,' However, our thinking should be based on what God said He did (the Bible), and not what we think He would, should or might have done.

Firstly, since God is the one who made the universe, it can scarcely be 'big' to Him. Humans struggle with its vastness because our comprehension is limited to the created time/space dimensions within which we exist, and it is mind-bending to try and comprehend anything beyond our dimensional existence. Size is only relative to us as inhabitants of this universe. And size and time are related somewhat. Because the universe is big to us we consider how long it would take us to travel across it, for example.

However, time itself began with the creation of the physical universe, so how can we comprehend what eternity is, or might be?

What was 'Before' the Universe?

Similarly, how do we imagine how 'big' God is?

We cannot use a tape measure that is made of the very atoms He made to measure Him. One example of this might be if you were asked to build a small house and you did. Now you are asked to build a large house. In our dimensions, for you to build the larger house it would require more effort and take more time. So, is it harder, or does it take longer for God to build a big universe compared to a smaller one (according to our perspective on what constitutes large or small of course)? Of course not, because He isn't bound by time and space (which He created). *Isaiah 40:28* says; '*… the everlasting God, the LORD, the Creator of the ends of the earth, does not grow weak nor weary …*'.

We are impressed that God made trillions of galaxies with billions of stars in them and suitably so, because that is one of the reasons for making them. But as mentioned, size is not an issue for God. Stars are relatively simple structures as they are just great big balls of gas. It would take more 'creative input', in that sense, for Jesus' miracle of feeding the five thousand than for the creation of countless quasars (there is immense genetic complexity in the structure of even a dead fish).

THE BIBLE AND ET'S

It is often asked, 'Just because the Bible teaches about God creating intelligent life only on Earth, why *couldn't* He have done so elsewhere?' After all, Scripture does not discuss everything, e.g., motorcars. However, the biblical objection to ET is not merely an argument from silence. Motor cars, for example, are not a salvation issue, but we believe that sentient, intelligent, moral-decision-capable beings is, because it would undermine the authority of Scripture. In short, understanding the big picture of the Bible/gospel message allows us to conclude clearly that the reason the Bible doesn't mention extraterrestrials (ETs) is that there aren't any.

Surely, if the earth were to be favored with a visitation by real extraterrestrials from a galaxy far, far away, then one would reasonably expect that the Bible, and God in His sovereignty and foreknowledge, to mention such a momentous occasion, because it would clearly redefine man's place in the universe.

1. The Bible indicates that the *whole creation* groans and travails under the weight of sin (*Romans 8:18–22*). The effect of the Curse following Adam's Fall was *universal.* Otherwise, what would be the point of God destroying this *whole creation* to make way for a new heavens and Earth—*2 Peter 3:13, Revelation 21:1* ff? Therefore, any ETs living elsewhere would have been (unjustly) affected by the Adamic Curse through no fault of their own—they would not have inherited Adam's sin nature.
2. When Christ (God) appeared in the flesh, He came to Earth not only to redeem mankind but eventually the *whole creation* back to Himself (*Romans 8:21, Colossians 1:20*). However, Christ's atoning death at Calvary cannot save these hypothetical ETs, because one needs to be a physical descendant of Adam for Christ to be our 'kinsman-redeemer' (*Isaiah 59:20*). Jesus was called 'the last Adam' because there was a real first man, Adam (*1_Corinthians 15:22,45*)—not a first Vulcan, Klingon etc. This is so a sinless human Substitute takes on the punishment all humans deserve for sin (*Isaiah 53:6,10; Matthew 20:28; 1 John 2:2, 4:10*), with no need to atone for any (non-existent) sin of his own (*Hebrews 7:27*).
3. Since this would mean that any ETs would be lost for eternity when this present creation is destroyed in a fervent heat (2 Peter 3:10, 12), some have wondered whether Christ's sacrifice might be repeated elsewhere for other beings. However, Christ died *once for all* (*Romans 6:10, 1 Peter 3:18*) on the *earth.* He is not going to be crucified and resurrected again on other planets (*Hebrews 9:26*). This is confirmed by the fact that the redeemed (earthly) church is known as Christ's bride (*Ephesians 5:22–33; Revelation 19:7–9*) in a marriage that will last for eternity. Christ is *not* going to be a polygamist with many other brides from other planets.
4. The Bible makes no provision for God to redeem any other species, any more than to redeem fallen angels (*Hebrews 2:16*).

ET'S - NO BIBLICAL REFERENCE!

One attempt to fit ETs in the Bible is on the basis of a word in *Hebrews 11:3*: 'Through faith we understand that the ***worlds*** were framed by the word of God, so that things which are seen were not made of things which do appear.'

The word 'worlds' appears in the KJV translation and some others, and some claim that it refers to other inhabitable planets. However, the word is αἰῶν (*aiōn*), from which we derive the word 'eons.' Thus, modern translations render the word as 'universe' (entire space-time continuum) because it correctly describes 'everything that exists in time and space, visible and invisible, present and eternal.' Even if it was referring to other planets, it is an unwarranted extrapolation to presume intelligent life on them.

It should also be remembered that expressions like "the heavens and earth" (*Genesis 1:1*) are a figure of speech known as a *merism.* This occurs when two opposites or extremes are combined to represent the whole or the sum of its parts. For example, if I said "I painted the whole building from top to bottom." One would understand this to mean everything in the whole building. Similarly, biblical Hebrew has no word for 'the universe' and can at best say 'the all', so instead it used the merism "the heavens and the earth." It is clear that New Testament passages like the aforementioned *Romans 8:18–22 and Hebrews 11:3* are pointing back to the Genesis ("heavens and earth") creation, and thus, everything that God made and when time as we know it began.

Another is the passage in *John 10:16* in which Jesus says, 'I have other sheep, which are not of this fold; I must bring them also, and they will hear My voice; and they will become one flock with one shepherd.' However, even an ET-believing astronomer at the Vatican (thus a 'hostile witness' to the 'no ETs cause'), a Jesuit priest by the name of Guy Consalmagno, concedes, 'In context, these "other sheep" are presumably a reference to the *Gentiles, not extraterrestrials.*' Jesus' teaching was causing division among the Jews (vs. 19), because they always believed that salvation from God was for them alone. Jesus was reaffirming that He would be the Savior of *all* mankind.

NOVEL METHOD

A more recent idea to allow for ETs arose out of a perceived need to protect Christianity in the event of a real alien visitation to Earth. Michael S. Heiser is an influential Christian UFOlogist/speaker with a Ph.D. in Hebrew Bible and Ancient Semitic Languages. He claims that the arguments put forward earlier might not apply to God-created aliens. Because they are not descendants of Adam, they have not inherited his sin nature, and thus, are not morally guilty before God. Just like 'bunny rabbits' on the earth, they do not need salvation—even though they will die, they are going to neither heaven nor hell.

On the surface this seems a compelling argument; after all, fallen angels are intelligent but are beyond salvation ("For surely it is not angels he helps, but Abraham's descendants." *Hebrews 2:16*). Angels are immortal and not of our corporeal dimension. And Heiser's ETs in spaceships require a level of intelligence not found in rabbits. This acutely highlights the injustice of their suffering the effects of the Curse, including death and ultimately extinction when the heavens are 'rolled up like a scroll' (*Revelation 6:14*). It also seems bizarre to assign no moral responsibility for the actions of highly intelligent beings.

Heiser also claims that vastly intelligent ETs would not displace mankind's position as being made in the image of God because 'image' just means humans have been placed as God's representatives on the earth.

However, the Bible says we are made in God's image *and* likeness (*Genesis 1:26*). Man was immediately created a fully intelligent being about 6,000 years ago and was involved in craftsmanship shortly thereafter (*Genesis 4:22*). Since that time, even we have not been able to develop technologies advanced enough to travel to other star systems. If aliens were capable of developing incredible faster-than-light spaceships needed to get here, one would presume they must have been created with vastly superior intellect to ours—which would make them even more in God's likeness in that sense than we are. Or, their creation is much older than the 6,000 years of the biblical six-day timeframe; the aliens were created before man and had sufficient time to develop their technologies. However, God created Earth on Day 1 and later the heavenly bodies on Day 4.

INFLUENCE BY UNBIBLICAL VIEWS

Although Heiser does not promote theistic evolution, he is sympathetic to a universe billions of years old, as proposed by the progressive creationist Dr Hugh Ross. In theory, this could allow the time necessary for any unseen ETs to develop the almost science-fiction-like technologies required to get here. However, this is circular reasoning.

UNBIBLICAL LONG AGE

There is a huge problem for the Gospel in these long ages. First, it's important to understand that modern scientific idea of long ages (i.e., millions and billions of years) derived from the belief that sedimentary rock layers on Earth represent eons of time. This in turn derived from the dogmatic assumption that there were no special acts of creation or a global Flood, so that Earth's features must be explained by processes seen to be happening now. This philosophy of uniformitarianism seems to amply fulfil the Apostle Peter's prophecy recorded in *2 Peter 3:3–7.*

The conflict with the Gospel is that these very same rock layers contain fossils—a record of dead things showing evidence of violence, disease and suffering. Thus, taking a millions-of-years view, even without evolution, places death and suffering long before the Fall of Adam. This undermines the Gospel and the very reasons that Christ came to the earth—such as reversing the effects of the Curse. *Romans 5:12* clearly states that sin and death entered into the creation as a result of Adam's actions. There was no death before the Fall.

RANKING THE CREATED ORDER

Psalm 8:5 says that man was made a little lower than the God [Angels in other translations] and crowned with glory and honor. Heiser has said that salvation is based upon ranking, not intelligence. If so, where in the Bible (which omits to mention them) would ET sit in this pecking order? Would they be higher than man, and lower than angels, for example? If these advanced ETs were capable of visiting the earth, mankind would now be subject to *their* dominion. (Even if the ETs were friendly, potentially they would be much more powerful due to their intelligence and technology.) This would be in direct contravention to God's ordained authority structure when he ordered mankind to 'subdue' the earth—also known as the dominion mandate (*Genesis 1:28*).

Be Inspired

Psalm 19:1 tells us a major reason that the universe is so vast: 'The heavens declare the glory of God; and the firmament shows His handiwork.' There are many similar passages in Scripture. They help us understand who God is and how powerful He is.

It reminds us that the more we discover about this incredible universe, the more we should be in awe of the One who made it all. In short rather than looking up and wondering 'I wonder what else is out there?' and imaginary aliens we've never seen. We should instead be considering the very One that made it all.

Fig.16.22: *Two identical Mars Rovers traverse the surface searching for evidence of water. Evolutionary researchers are eagerly looking for past or present signs of (even) microscopic life. Curtsy - NASA*

Could There be 'Simple Life' Else-where in Space?

The Bible's 'big picture' seems to preclude *intelligent* life elsewhere in God's universe. But what if bacteria were found on other planets, for example? This is exceedingly unlikely, but 'God-made' bacteria would not violate the Gospel. And in any case, any 'microbes on Mars' were likely as a result of human contamination. What would be their purpose? The entire focus of creation is mankind on this

Earth; the living forms on Earth's beautifully balanced biosphere are part of our created life support system, Figure 16.22.

If bacteria are found elsewhere in the solar system, it will be hailed as proof that life can 'just evolve.' However, we have previously predicted in print that in such an unlikely event, the organisms will have earth-type DNA, etc., consistent with having originated from here as contaminants—either carried by recent man-made probes, or riding fragments of rock blasted from Earth by meteorite impacts.

ALIENS - POPULAR FANTASY

Fig.16.23: Most Aliens - Popular Fantasies

News-media reported a scene on this glowing halo in the skies over Moscow Russia in October, 2009, Figure 16.23. It was a weather anomaly caused by several weather fronts passing through Moscow as the sun shined from the west.

SUBSTITUTED CHRISTIANITY

Belief in extraterrestrial life has effectively become a new religion for many, replacing traditional belief. For example, the sociologist Gerald Eberlein stated:

Mundane objects like these lenticular clouds above Skaftafell glacier in Iceland are mistaken as flying saucers due to our cultural conditioning to believe in such things.

" ... research has shown that people who are not affiliated with any church, but who claim that they are religious, are particularly susceptible to the possible existence of extraterrestrials. For them, UFOlogy is a substitute religion."

A close examination of the phenomenon shows that it also provides a perfect substitute 'replacement theology' for the big picture Creation, Fall, Redemption, and Restoration aspects of Christianity. Much of the UFO literature seeks to:

a. Replace the Creator God with aliens, even claiming the Bible is actually referring to these ETs and not God.

b. Substitute the Fall for an environmentalist catastrophe or message, i.e., the aliens will not allow us to destroy ourselves and 'mother earth' as we are allegedly doing.

c. The ETs will be the saviors of mankind and will rescue the earth in the process. They even have rapture-type events predicted for the future.

d. There will be a restoration of the Earth or the 'true (UFO) believers' will be transported to a new Edenic or Nirvana type planet that has not yet been corrupted.

Please observe that many of the UFO stories have been received and channeled by contacts who claim to be receiving special revelation from our 'space brothers'. This is an occult activity that masquerades as an enlightened practice.

Any Life Out There?

Not that long ago the idea that older (on the evolutionary scale), more technologically advanced (than humans), wiser and benevolent extraterrestrials have visited or are currently visiting the Earth would have seen you carted off to the local psych ward. But not anymore. Popular polls have even suggested that up *to 20 million Americans have seen a UFO and that 4* million claims to have been abducted by *aliens*. Such ideas have become very mainstream and children in particular are very susceptible and are openly targeted. Many seemingly credible scientists now believe that life on Earth originated from outer space via either directed or undirected *panspermia* ('seeds from space'—from the Greek words pas/pan [all] and sperma [seed]). Some very famous scientists have even suggested that our DNA may contain revelation from our alien creators once we are smart enough (like them) to decipher it. Ironically, this makes many of them 'creationists' (sort of). It appears that they just don't want the God of the Bible to be responsible.

The movie *The Fourth Kind* claimed to be a representation of 'real' alien abduction episodes in Nome, Alaska.

The strength of such ideas also presents a challenge to the traditional Christian worldview, because they are in stark contrast to the idea that mankind is the central focus of God's creation. A straightforward understanding of Scripture leaves no place for similarly sentient and intelligent morally aware beings elsewhere in the universe. The primary role of the creation was to bring forward a bride for Christ.

Such a seemingly 'dogmatic' stance has led to shouts of 'arrogance' aimed at Christians. However, these are strange accusations, coming from those who believe that the universe is widely inhabited by ETs, because to date we do not have a shred of indisputable evidence that ET has visited us or is living out there somewhere.

We have often been challenged by Christians who have no problem with the concept of alien life. In our experience, this 'Christian' view is driven by an anthropomorphic (or human-like) view of the abilities and purposes of God. In short, many conclude that God would not have made the universe so big just for mankind, otherwise it would seem an awful waste of space.

First, God is God and there in none like Him. A quick search of the Scriptures reinforces this at least a dozen times using terminology like:

There is **none like** you among the gods, O Lord, nor are there any works like yours. *Psalm 86:8*, (emphasis ours).

In *Isaiah 55:9*, note that He specifically uses the universe as an example of His awesome power, God tells us not to attribute fallible human reasoning to Him:

"For as the heavens are higher than the earth, so are my ways higher than your ways and my thoughts than your thoughts."

In Evolution and the science of fiction many writers highlighted that the most popular entertainment genre today is science fiction and how popular cultural ideas are shaping our worldviews. It involves the majority by far of the highest grossing movies of all time. See our Avatar movie review and Avatar Aspirations Sci fi helps create an alternate belief system.

The Bible is notably silent about any extraterrestrials, yet if an ET visit in their technologically advanced hyperdrive spaceships was imminent, it would be reasonable to presume that our sovereign and prescient

Creator would let us know about such things. This is not an 'argument from silence,' but rather from 'conspicuous absence,' After all, a commissioned report by NASA as far back as 1960 concluded:

"While the discovery of intelligent life in other parts of the universe is not likely in the immediate future, it could nevertheless, happen at any time. Discovery of intelligent beings on other planets could ... could send sweeping changes or even the downfall of civilization, ... societies sure of their place have disintegrated when confronted by a superior society."

There is no need to invoke a size argument to explain why God made the universe so big—we need to remember that it is only big to *us* as mortal finite beings. The Bible indicates that He made all things in heaven and on Earth in six days—for a Creator capable of this in the first place, it is not hard for Him to make it whatever size He chooses.

Again, in Evolution and the science of fiction we read:

"Apologist John Whitcomb writes: 'It must be recognized ... that it required no more exertion of energy for God to create a trillion galaxies than to create one planet'. *Isaiah 40:28* says; ' ... the everlasting God, the LORD, the Creator of the ends of the earth, does not grow weak nor weary ... .' Actually, stars are rather simple structures—they have been described as 'glowing balls of gas'. There is far more complexity in the genetic code of the simplest organism than in a thousand galaxies. It would thus take more 'creative input,' in that sense, for Jesus' miracle of feeding the five thousand than for the creation of countless quasars (there is immense complexity in the structure of even a dead fish). And presumably, if God had created only our solar system, sceptics would ask why, if He were so great, did He not create something on a larger scale."

Isaiah 45:18 tells us specifically that the Earth was made for a special purpose:

"For this is what the LORD says—he who created the heavens, he is God; he who fashioned and made the earth, he founded it; **he did not create it to be empty, but formed it to be inhabited**—he says: 'I am the LORD, and there is no other' (emphasis ours)."

And lastly, *Psalm 19:1* reminds us that:

"The heavens declare the glory of God; and the expanse proclaims His handiwork".

In short, instead of looking at the awesomeness of the universe and thinking "I wonder what else is out there", we should be reminded of the awesomeness of the One who made it all (*Romans 1:20*) for surely the universe provides us with some indication of how incredible the Creator really is.

It should also be noted that many people do see seemingly inexplicable things in the sky and some even have experiences with alleged alien beings. We do not and have never denied that such experiences can occur. However, we do dispute the source (cause) of many experiences, particularly when people think that just because they have seen a light in the sky, it must automatically be extraterrestrial in origin. Christians, too, have fallen into this trap because of the integration of popular culture into their worldview. We recommend reading. God warns us many times in Scripture to have nothing to do with mediums, diviners, those who tell the future or the spirits of the dead.

Please observe that because the Bible, in these cases, mentions such things, it does not mean that they are true. The warnings are due to the source behind such manifestations. Secular UFO researcher John Keel noted:

"The UFOs do not seem to exist as tangible, manufactured objects. They do not conform to the natural laws of our environment. They seem to be nothing more than transmogrifications tailoring themselves to our abilities to understand. The thousands of contacts with the entities indicate that they are liars and put-on artists. ***The UFO manifestations seem to be, by and large, merely minor variations of the age-old demonological phenomenon***" (Emphasis ours).

EVOLUTIONISM AT ITS CORE

The main advocates of extraterrestrial life, however, are evolutionists. In The UFO phenomenon—growing and not going away! Is it the next great challenge for the church? :

"The interest in the search for extraterrestrial life is huge—mainly fueled by the enormous popularity of science fiction and its depiction of advanced alien life on other planets. Many evolutionists are acutely aware of this. In fact, in the minds of many young people, the idea of alien life 'proves' the theory of evolution itself. It has almost become a circular self-serving hypothesis!

1. Evolutionists believe:
   a. Life evolved on the earth
   b. The universe is an enormous place
   c. Life must have evolved elsewhere
2. Because of evolutionary teaching, many people now believe:
   a. Since the universe is an enormous place, we cannot be the only ones)
   b. Aliens must exist [if they evolved here, why not elsewhere, as the conditions must be right on one of the many billions of Earth-like planets they presume to exist.] (Science fiction can't be *all* wrong!)
   c. If aliens exist, they must have evolved
   d. It 'proves' that life must have evolved on the earth also."

UFO commentator and author Ronald Story also notes:

" ... science fiction has become our myth, and science has become our religion. Due mainly to media influences and a hideously complicated world, most people are finding it increasingly difficult to distinguish fantasy from reality."

In a big picture sense, the whole realm of UFO beliefs is truly a religious idea built upon yet another religious idea (evolution):

"The common denominator of both camps is the belief that evolution has occurred for countless eons on the earth and all over the universe. This point cannot be emphasized strongly enough—it is the basis for virtually all belief in alien life, whatever form one thinks that life may take."

THE LATEST CLAIMS

Many 'true believers' in UFOs become very frustrated and impatient at the lack of information emanating from official sources about UFOs. Because there are so many sightings, and indeed, because of their own experiences, they feel that the government must be complicit in hiding the truth. This was particularly the case for *The Roswell Incident*. The truth behind the speculations that something otherworldly crash-landed on a ranch in Roswell, New Mexico, was not aided by the UFO true believers. An overlay of deliberate misinformation helped this event become a defining moment in UFO lore.

LATEST CLAIMS AND SPECULATIONS

Vatican Astronomer Wants to Baptize ET

In late 2009 the **Vatican** held a conference to discuss how Catholic theology is affected by the existence of extraterrestrials. It should be worthy of note that no one ever really questioned whether ET exists or not. That 'fact' seemed to be a foregone conclusion. Now the Pope's astronomer Dr Guy Consolmagno, said he would even be happy to ***baptize an alien***, but "Only if they asked". It was reported that "A self-confessed science fiction fan, he said he was 'comfortable' with the idea of alien life."

Robert Jamieson, former Combat YouTube Missile Targeting Team Commander for the US Air Force, speaking at the National Press Club Conference on 27 September, 2010

He is also an avowed fan of evolutionism and is quick to dismiss the creationist viewpoint and intelligent design. He seems to be ignorant of both, and it would appear than he has simply never heard a good creation scientist presentation (like so many who openly ridicule). His comment that "Intelligent aliens may be living among the stars and are likely to have souls" also displays a warped view of Scripture, unfortunately. Although professing to be Christian, his views, like those of many theistic evolutionists, can hardly be held up as a bastion of biblical truth.

Former Air Force Officers Made Sworn Testimonies

In September, 2010 ex-US Air Force officers and one enlisted man gave a lengthy presentation at the US National Press Club. As such, it was widely reported on in the media.

Some of these men are certainly specialists in their respective fields, and one would think that they are not all prone to flights of fancy or delusional behavior, nor do they have some collective agenda to fabricate stories. Some of the reports are very powerful, such as strange lights hovering over nuclear missile silos which are seen descending into the silos. On some such occasions the actual missiles were deactivated.

The behaviors of the strange objects certainly would give the impression that there was some intelligent force guiding them. They also testified that on many occasions CIA agents came out and took over the investigations and hushed up the events without further explanation being offered to the officers.

These events are nothing new and have been known amongst UFO researchers for many years. Robert Muller, a high-ranking nuclear physicist who served as one of the US government's top advisors on nuclear secrets, was asked if there was a cover-up about alien visitations. He said (laughingly) and denying any conspiracy:

"If there is, I'm part of it. There's no cover up. If there is, I would certainly know about it."

He said that the officers have simply seen things that they don't understand. And that:

" ... the paradigm of the era was that things you don't understand are flying saucers, [then] that's how you reported them."

He offered no explanation though as to *what* they were, but we concur with his assessment that it is the cultural overlay that drives the interpretation of the evidence (just like evolution theory). This was reinforced when some of the officers attributed motive to the actions of the UFOs. How can they know such things when a UFO by definition is merely an 'unidentified flying object,' and in these cases they still remain unidentified?

The Air Force officers did say that on some occasions they were told that because these were matters of national security (after all it did involve the nation's nuclear arsenal), the public would panic and that's why it was being hushed up.

LIGHT APPEARANCE

Falling Angels – Deceiving Mankind

The UFO phenomenon is the age-old spiritual deception in a modern and very believable guise. It is being performed by ***fallen angels*** with the intent of deceiving mankind as to our place in the universe as per the

Bible. The Bible records the activities of angels. They appear in a myriad of forms, sometimes appear physically and can most certainly affect our world, even leading people to killing on occasions.

However, governments along with their scientific advisors and researchers do not officially recognize the supernatural realm, and much less that such phenomena could be malevolent angels. Their aim is to explain everything naturalistically:

"So why does the government have nothing to say in this matter?

This remains a mystery, and perhaps it is the religious connotations that prohibit it from 'official' classification. Because the occultic and spiritual nature of such claims is not scientifically verifiable, perhaps governments do not want to admit that they cannot explain what is really happening.

One could imagine the public disquiet if its leaders actually admitted that ***'something is happening, we don't know what, and we are powerless to do anything about it.'*** [Besides widespread panic] It would be open season for every bizarre claim of the 'UFOnuts' and hoaxers professing to have the answer, and could lead to a serious, although unwitting, endorsement of those self-professed UFO messiahs who claim to be in contact with the ETs. It would be even more difficult to weed out the apparently genuine and serious claims from those of the frauds and fame-seekers. Hence, it seems, the blanket statement, ***"UFOs pose no threat to national security."***

Thus, seeming silence on UFO events only adds to the conspiracy idea, that the government must know something and is covering it up—particularly when former NASA Apollo astronauts like Edgar Mitchell have made similar claims that the aliens are with us.

'United Nations Appoints an ET Ambassador'—Hoax Report

The internet was 'on fire' as news that the United Nations was set to appoint an ambassador to greet the aliens when they 'eventually do land' spread virally across the globe. However, this was a hoax. The reports were so specific that they even named Malaysian astrophysicist Dr Mazlan Othman as the appointee to allegedly head the Office for Outer Space Affairs (UNOOSA), a department that really exists. Many of the major news services were taken in by the claim without properly checking the facts. This aptly demonstrates how credible such a move would be regarded as, due to the public perception that aliens are a 'fact.' Such calls for an appointee are nothing new. The fear was that 'primitive earthlings' are likely to shoot first and ask questions later, which could result in a galactic confrontation when they are 'only here to help us.' The belief in the reality of alien visitations to the earth is so strong that it is:

" ... is demanding governments worldwide disclose and use secret alien technologies obtained in alleged UFO crashes to stem climate change."

It is said:

" ... would like to see what (alien) technology there might be that could eliminate the burning of fossil fuels within a generation ... that could be a way to save our planet.'

The belief that aliens possess the technology to save mankind from the effects of alleged global warming is nothing new. The religion of UFOlogy has many looking to the stars for our 'saviors from outer space.' A writer stated that in July 1978, UFO enthusiast and then prime minister of Grenada, Sir Eric Gairy, was petitioning the U.N. to create a special group for the purpose of investigating UFOs. He believed that America was keeping quiet despite knowing the 'truth' about them. Famous NASA Mercury/Gemini Astronaut Gordon Cooper supported his idea, and during a lengthy speech, he was quoted as saying:

"I believe that these extraterrestrial vehicles and their crews are visiting the planet from other planets that are a little more technically advanced than we are on earth. I feel that we need to have a top-level coordinated program to scientifically collect and analyze data from all over the earth concerning any type of encounter and to determine how best to interfere with these visitors in a friendly fashion … ."

At the time of writing an internet search still reveals hundreds of web pages still claiming that the UN appointment is accurate. However, Ms. Othman has apparently personally denied the claim although she apparently recently stated in a speech to fellow scientists about what to do when we make contact with the ETs:

"When we do, we should have in place a coordinated response that takes into account all the sensitivities related to the subject. The U.N. is a ready-made mechanism for such coordination."

Such elaborate hoaxes are common in UFOlogy. So strong is the UFO believers' preexisting belief that alien visitations have occurred on the earth, that for them the end justifies the means in order to reveal the truth widely.

As one can realize, UFO beliefs are strongly held in all quarters of the populace. We would not be surprised if in the future such an appointment could be made by the UN or various governments, which would only further embellish the idea of ET being with us.

THE DISCLOSURE PROJECT

This is something we have been asked about on several occasions. A worldwide movement known as *The Disclosure Project* has been operating since 1993 to help 'reveal the truth' about alien contacts. It was commenced by UFOlogist Steven Greer, who claims governments have been covering up ET's calling cards. His movement tapped into the public consciousness when he also stated that governments have been concealing advanced energy technologies that could ultimately benefit mankind. Such ideas are usually called 'Top Secret Black Projects' that are not revealed because they could upset the socioeconomic or politically stability of the world.

Greer is a former '5th Kind' contactee (mutual contact with aliens), which would go a long way to explaining his passion about the subject. After all he believes he knows and has experienced the 'truth' by virtue of his ET conferences. He dropped out of college to study as a teacher at the Maharishi International University. This is a 'college' started by Maharishi Mahesh Yogi—the famous guru for the pop group the Beatles back in the 1970s. Yogi is also known as a pioneer of new age transcendental meditation. It is no surprise how people's exposure to UFOs leads them into the occult as was shown time and again, and has also been demonstrated by psychologists and other professionals.

Greer has since qualified to become a medical doctor, but also uses his talents to help people make contact with ETs using meditation, remote viewing and channeling. He is one of the most well-known UFOlogists in the world and has appeared regularly on the media to express his views. The media loves nothing more than a good conspiracy after all. In recent months there has been much speculation that he has even met with government officials to discuss Project Disclosure. Greer's star is rising and he is being regarded as somewhat of an expert in this area.

However, it is quite easy for most Christians to see the occult factor in his thinking. This is quite worrying as one could almost expect official sources to state that there has been alien contact in an effort to explain simply something they don't understand and misinterpret—the same way as the former US Air Force officers mentioned earlier.

NUCLEAR PHYSICIST - UFO COVERUPS

Another report reached us that a 'credible' nuclear physicist who has worked for major American companies claimed to have proof of cover-ups like the aforementioned Greer. However, the scientist in question is well-known to us. He is Stanton Friedman, a 'true believer' who has been investigating UFOs for over 50 years.

Back in 1978, Friedman was introduced to Captain Jesse Marcel, the Air Force officer who claimed to have found a wrecked alien craft on a ranch in Roswell back in 1947. It was Friedman's initial enquiries that led authors Berlitz and Moore to pen the massively popular book *The Roswell Incident*.

Friedman is no innocent bystander. He is one of the most outspoken proponents of the extraterrestrial hypothesis and has even accused the SETI project of being complicit in the cover-ups. As mentioned, everybody loves a good conspiracy theory. A 2002 Roper poll, commissioned by the Sci-Fi Channel, noted that 72 percent of Americans believed the U.S. government isn't telling all it knows about UFOs, and 68 percent thought the government knows more about extraterrestrial life than it cares to disclose. Friedman stated:

"I don't know of any government on this planet that wants its citizens to owe their primary allegiance to the planet. Nationalism is the only game in town."

Again, his religious zeal for aliens to exist was demonstrated when he added:

"I'm still optimistic that, within my lifespan—and I'm 75—we'll get at least a part of the story, that we're not alone in the universe."

However, as we have often suggested, besides the desire for such things to be real, there are often more mundane agendas at work. UFOs are big business (just ask Hollywood), and this latest claim by Friedman coincided with the release of his latest book *Science was Wrong*. Claims of government cover-ups are usually a sure-fire way of getting media attention.

BRITISH SCHOOL CHILDREN TRAINED DURING UFO CRASH DRILLS

Perhaps one of the more comical claims, yet sadly also very disturbing, was that 8–10-year-old schoolchildren in the UK have been conducting UFO crash drills. On further investigation these appear to be isolated incidents in just a few schools and not education department curriculum (not yet anyway). The scenario is that children are told that a UFO (common vernacular for alien spaceship/flying saucer etc.) has crashed in the school grounds. They are then told to secure the scene and collect wreckage until the authorities arrive. The local police are also complicit in the stunt as they eventually turn up and take over the 'investigation'. The children are then asked to write about their experiences. A teacher at the Sandford Primary School stated:

"The children didn't know what was going on … As they approached the crash site, we could see how amazed and perplexed they were. It was a fantastic first reaction.

"Police constable Gary Densham, who took part in one of the staged UFO crashes last year at the Lanchester Endowed Parochial Primary, said, 'The older pupils were asking questions about the crash site, like whether it was safe, but the younger children were convinced they'd seen the crash happen. Their imaginations were brilliant.'"

Disassociation between Reality and Fantasy

One wonders how most ordinary people can accept the brainwashing of their children in the school stunt just mentioned. It creates a disassociation between reality and fantasy for them so that it will make it increasingly difficult for them to discern the truth in the future. Particularly when the authority figures in their lives deceive them so openly.

The many media claims dealt with above are just a few from the constant stream of such items that come to our attention on a regular basis. If there is a conspiracy or a plot afoot, then it is emanating from spiritual forces that are seeking to undermine God's rightful place as Creator, and the veracity of His Word—undermining the Gospel in the process.

Every day, people see strange things in the sky that they can't explain. It does not automatically follow that such sightings are alien spaceships. We have even had many Christians approach or contact us to claim that what they have seen *must* have been extraterrestrial in nature. This is due to a kind of cultural conditioning that people are exposed to via their education and popular media. The Apostle Paul said:

"The Spirit clearly says that in later times some will abandon the faith and follow deceiving spirits and things taught by demons" (*I Timothy 4:1*).

We realize that taking a stand against the idea of ET life will polarize Christians in this modern culture. But do not be deceived by powerful manifestations emanating from the spiritual realm. The Lord Jesus warned:

"For false messiahs and false prophets will appear and perform ***great signs and wonders to deceive***, if possible, even the elect. See, I have told you ahead of time" (*Matthew 24:24–25*) (emphasis added).

As we have often stated, we should make a stand on the truth and not be concerned that the Bible might somehow be falsified (e.g., by a 'real ET') when it comes to big picture issues like God being the Creator. Our strong view is that there is simply no room in His Word to squeeze in a benevolent alien or two. As sightings and alleged contacts increase, and governments take official positions to seemingly endorse the notion, it is worth remembering that mankind's inclination is towards rebellion and rejection of God, seemingly looking to anything else rather than accept Him as Creator. *Jeremiah 17:9* reminds us:

"The heart is deceitful above all things, and desperately wicked: who can know it?"

Many Christians have already been led astray by accepting another manmade philosophy (evolution) that had its origin with the lies of the evil one when he asked the very first humans:

"Did God really say?"

And told them:

"You shall be as Gods" (*Genesis 3*).

The entirety of the UFO phenomenon has its genesis in the origins issue, via the promotion of evolution on other planets. A belief in biblical creation, and using the Bible as our worldview filter, helps us to critically think about all of reality, including the supernatural.

UFOlogy—Religion or Science

Fig.16.24: Searching for Extra Terristial Life - courtesy Wikipedia.org, released under the GFDL

The Allen array will increase SETI's capacity to search for extraterrestrial life. It is named after Paul Allen, the co-founder of Microsoft. He is just one of many prominent industry and media leaders who have contributed millions of dollars to SETI's programs, Figure 16.24.

Supporters of 'SETI—the Search for Extraterrestrial Intelligence'—have reacted strongly against George Basalla, professor emeritus of history at the University of Delaware, who says in his book *Civilized Life in the Universe* (Oxford University Press, 2006) that SETI is more of a faith than a science.

Basalla highlights SETI's failure to make 'contact' despite over 40 years of trying and says its continuing efforts in the absence of any evidence indicate that SETI relies more on 'religious zeal' than anything else.

In response, SETI Institute's David Darling argues that the quest for ET is 'true to the methodology and spirit of science.' In an article at *Space.com* he writes: 'It isn't an unreasonable hypothesis that if intelligence has come about on one planet [Earth] that it may also have arisen elsewhere, especially given the vast number of stars in this and other galaxies.'

What he's talking about, of course, is evolution—i.e., if life evolved here, why not elsewhere? Darling refers to the debate going on at SETI about 'how often primitive life, such as bacteria, serves as the precursor of complex, multicellular life, and, ultimately advanced intelligence.' He adds: 'Forecasting how intelligence will evolve is a hazardous business.'

EVOLUTION AND THE SCIENCE FICTION

Denial of Awesome Creator

Science fiction and ETs are hugely popular genre now accounts for 12 of the 15 highest-grossing movies of all time. 'Sci-fi' has replaced the war movie and the western as the modern version of 'good guys vs bad guys' and 'shoot 'em-up' escapism.

Science Fact or Fantasy?

Although enjoyable, we need to be mindful that the majority do not view sci-fi through 'Biblical glasses'—that is, they don't have a worldview based on the Bible and its foundational account of Creation.

How many have stopped to consider that the popularization, success, and even worship of science fiction has been underpinned by an enormous cultural change that has gripped our planet—the overwhelming acceptance of the theory of evolution?

The whole concept of searching uncharted space for alien life-forms is clearly based on the premise that, if evolution occurred on the Earth, then it must have occurred elsewhere in the universe.

Yet, the whole hypothesis of evolution is itself based on unobservable events happening in an unobservable past, and much of it (especially 'chemical evolution') denies experimental reality. One could argue, then, that evolution is 'the science of fiction'.

Science fiction beliefs are so 'mainstream' that governments today have even funded schemes like the SETI project (**S**earch for **E**xtra **T**errestrial **I**ntelligence). The reality, though, is that there has never been one single documented observation of an alien encounter.

No Scriptural Basis

Many wonder (even some Christians)—inspired by the wonder of special effects— 'Could there be life on other planets?' But a straightforward reading of Genesis gives us no indication that God created intelligent, alien life-forms elsewhere in the universe. Would such questions even be asked if science fiction were not so popular?

Romans 8:22 also tells us that the *whole* creation has been groaning (because of sin and the subsequent Curse) right up to the present time. Therefore, it would not make sense that intelligent beings on another planet (part of this creation) had, because of the sin of Adam on the Earth, been subjected to the Curse.

Also, God the Creator of the universe (in particular, the Second Person of the Trinity) took on human nature (Jesus) as the 'last Adam' (*John 1:1–18*).

He came to *this* Earth to fulfil His plan for the redemption of the human race (offspring of the 'first Adam') which would redeem the entire universe as well. Redeemed humanity will be Christ's bride throughout eternity—and Christ will only have one bride, effectively eliminating the notion of 'other races' in the universe (*Eph. 5:22–33, Rev. 19:7–9*).

Science fiction has helped people grasp the enormous size of the universe. But many go on to ask, 'Why would God go to all the trouble of creating billions of galaxies and stars?' Apologist John Whitcomb writes: 'It must be recognized … that it required no more exertion of energy for God to create a trillion galaxies than to create one planet'. *Isaiah 40:28* says; '… the everlasting God, the LORD, the Creator of the ends of the earth, does not grow weak nor weary …'.

Actually, stars are rather simple structures—they have been described as 'glowing balls of gas.' There is far more complexity in the genetic code of the simplest organism than in a thousand galaxies.

Scripture explains that the purpose of stars—created on Day 4 of Creation Week (*Genesis 1:14*)—was to divide day and night, as signs and seasons, and for days and years. In other words, the focus of the entire creation, even those stars that are mega-distances away, is for humankind, on this Earth.

Incidentally, scientific evidence on redshifts now suggests strongly that our galaxy, the Milky Way, is at or near the physical center of the universe.

Faith Based on 'Unrealism'

Due to evolutionary beliefs, many individuals and churches have rejected the Creation account in Genesis as the true history of the universe. This rejection, plus a fat 'diet' of science fiction, has 'immunized' many against the Gospel.

How many science fiction stories do you know of that glorify God? Many subtly portray aliens as spiritual beings with a belief system, or with an all-powerful 'force' controlling their destiny. Others depict 'highly evolved' civilizations in a positive light for having rejected the supernatural beliefs of their ancestors.

The Genesis account of God as the Creator is completely incompatible with evolution-based science fiction beliefs, many of which are just plain 'occultic' in their portrayal of spirituality. Noticeably, there is even a proliferation of cults that believe a benevolent alien race will return and 'beam' them away to a better place. This apparent desire to escape reality shifts the focus of salvation from Christ the Creator to hope in a fantasy.

THE UNIVERSE MAGNIFIES ITS CREATOR

Although through science fiction we can marvel at (and enjoy) the imaginativeness of mankind, and its ability to 'transport' us to strange new worlds, we need to be mindful that science fiction, with its constant evolutionary overtones, undermines the witness of Creation (remember to wear your Biblical glasses). *Psalm 19:1* emphatically state: 'The heavens declare the glory of God; and the expanse proclaims His handiwork.' When we look at the awesome wonder of the stars set in space, including those we are only just now finding out about via powerful telescopes, we should be reminded of the Awesome One who created them (*Romans 1:20*).

APPENDIX

a. Furthermore, Jesus dying for alien beings makes no sense, since Jesus took on human nature, and remains the God-man forever as our Savior. If He were to atone for Vulcans, say, He would need to become a Vulcan. The whole purpose of creation is focused on the race on Earth, of which some will be Christ's 'bride' throughout eternity. Christ will not have multiple 'brides.'

b. This means there is a small chance of hitting one in each linear kilometer travelled, but over such vast distances, a hit is almost certain. The Appendix gives calculations of the damaging effects of dust at such high speeds.

c. The devil and his evil angels are fallen created beings. Satan's kingdom will exist only as long as God permits.

d. William Alnor, cult expert and award-winning journalist, studied the UFO phenomenon for many years. His book, UFOs in the New Age, Baker Book House, US, 1992, documents his investigations that led to the conclusion that some UFO phenomena have an occult source. Gary Bates came to a similar conclusion in his book, Alien Intrusion.

e. Of course, there are angelic beings. These were made early in Creation Week—referred to as 'sons of God' and 'morning stars' in the poetry of the book of Job, they rejoiced and sang at the formation of the earth's 'foundations' (Job 38:7).

f. The church was bought with the blood of its Savior from the wound in His side, a clear analogy to the first woman being born from a 'wound' in Adam's side.

g. Ross believes in soulless man-like creatures before Adam, similar in spiritual status to Heiser's hypothetical ETs. For a complete refutation of Ross's ideas see *Refuting Compromise* by Jonathan Sarfati Master Books, Arkansas, USA, 2004.

h. Consolgmagno, G., Humans are not God's only intelligent works, 3 January 2006. He actually took the affirmative side in a debate with CMI's Dr Jonathan Sarfati (they didn't see each other's arguments before publication in the liberal *Science and Theology News*).

REFERENCES

1. Parts of this chapter are based on Gitt, W., God and the extraterrestrials, Creation 19(4):46–48, 1997; creation.com/god&et. See also Grigg, R., Did life come from outer space? Creation 22(4):40–43, 2000; creation.com/lifefromspace. For detailed treatment of this topic, see Bates, G., Alien Intrusion: UFOs and the Evolution Connection, Creation Book Publishers, US, 2010; creation.com/ai.
2. Sarfati, J., Life on Mars? Creation 19(1):18–20, 1996; creation.com/marslife, Sarfati, J., Life from Mars? Journal of Creation 10(3):293–296, 1996; creation.com/lifefrommars.
3. Anon., Another blow to Mars 'life' claim, Creation 20(2):8, 1998; creation.com/marsblow.
4. Holmes, B., Death knell for Martian life, New Scientist 152(2061/2):4, 1996.
5. Out there—Readers Digest exclusive poll, July 2005.
6. Morris, H.M., The Genesis Record, Baker Book House, US, p. 169, 1976.
7. Leupold, H.C., Exposition of Genesis, Volume 1, Baker Book House, US, p. 250, 1942.

8. Bates, G., Designed by aliens? Creation 25(4):54–55, 2003; creation.com/alien-made.
9. For articles on the origin of life, see: Batten, D., Origin of life: An explanation of what is needed for abiogenesis; creation.com/ool, 26 Nov. 2013; Sarfati, J., Self-replicating enzymes? Journal of Creation 11(1):4–6, 1997; creation.com/replicating, Thaxton, C.B., Bradley, W.L. and Olsen, R.L., The Mystery of Life's Origin, Philosophical Library Inc., US, 1984. See more at creation.com/origin.
10. Erdling, H., UFOlogie als Ersatzreligion, Focus Magazin 45:254, 1995; focus.de/kultur/ leben/modernes-leben-ufologie-als-ersatzreligion_aid_154571.html.
11. Compare Grigg, R., *Did life come from outer space? Creation* 22(4):40–43, 2000; creation.com/life-from-space, Bates, G., *Alien Intrusion: UFOs and the evolution connection*, Master Books, Arkansas, USA, 2004.
12. Sarfati, J., *Conclusive evidence for life from Mars? Remember last time!* creation.com/mars, 15 May 2002.
13. Matthews, M., *Space life? Answering unearthly allegations, Creation* 25(3):54–55, 2003; creation.com/ space-life.
14. Sarfati, J., The Fall: a cosmic catastrophe: Hugh Ross's blunders on plant death in the Bible, *Journal of Creation* **19**(3):60–64, 2005; creation.com/plant-death. Return to text.
15. Henry, J.F., An old age for the earth is the heart of evolution, *Creation Research Society Quarterly* **40**(3):164–172, December 2003; creationresearch.org/crsq/articles/40/40_3/Henry.htm. Return to text.
16. Mortenson, T., *The Great Turning Point*, Master Books, Arkansas, USA, 2004.
17. Hallo Erdling, Ufologie, *Focus* 45:254, 6 November 1995.
18. Cited in Chuck Missler and Mark Eastman, Alien Encounters (Indianapolis, IN: Koinonia House), p. 193, 2003.
19. Cited in Gary Bates, *Alien Intrusion: UFOs and the Evolution Connection*, Creation Book Publishers, Powder Springs, GA, 2009, p. 158-159.
20. Story, R.D., editor, The Mammoth Encyclopedia of Extraterrestrial Encounters (London: Constable & Robinson), p. 679, 2002.
21. Gary Bates, *Alien Intrusion: UFOs and the Evolution Connection*, Creation Book Publishers, Powder Springs, GA, p. 36, 2009.
22. I'd love to baptise ET, says Vatican's stargazer, dailymail.co.uk, 25 January, 2011.
23. Disclosure Conference, youtube.com, 25 January, 2011.
24. Air Force Officers' Sworn Testimony About UFO Encounters, video.foxnews.com, 25 January, 2011.
25. Gary Bates *Alien Intrusion: UFOs and the Evolution Connection*, Creation Book Publishers, Powder Springs, GA, 2009, p. 160.
26. United Nations to appoint space ambassador to act as first contact for aliens visiting Earth, dailymail.co.uk, 2 February, 2011.
27. U.N. Denies Appointing 'First Contact' for Visiting Space Aliens, foxnews.com, 2 February, 2011.
28. UFO science key to halting climate change: former Canadian defense minister, physorg.com, 5 March 2007.
29. Cited in Gary Bates, *Alien Intrusion: UFOs and the Evolution Connection*, Creation Book Publishers, Powder Springs, GA, p. 23, 2009.

30. Nuclear Physicist Describes Vast UFO Cover-Up, aolnews.com, 2 February, 2011.
31. UK Schools Hold UFO Crash Drills, aolnews.com, 2 February, 2011.
32. TNG Trivia, ugcs.caltech.edu, 12 July 2002.
33. Whitcomb, J., *The Bible and Astronomy*, BMH Books, Indiana, p. 28, 1984.
34. Humphreys, R., Our galaxy is the center of the universe, 'quantized' red shifts show, *J. Creation* 16(2):95–104, 200.

THE STRING THEORY AND THE ORIGIN OF THE UNIVERSE

WHAT IS THE STUFF IN THE UNIVERSE MADE OF?

Our high-school physics tells us atoms, with 90 natural elements building the world around us. But what are these atoms made of?

Over the last century, we've ripped atoms apart, eventually finding electrons and quarks, allegedly, the fundamental building blocks of stuff. But are they really fundamental? Have we truly reached the bottom?

Not according to the hypothetical string theory.

String theory is allegedly, the idea that everything in the universe, every particle of light and matter, is comprised of miniscule vibrating strings, Figure 17.1.

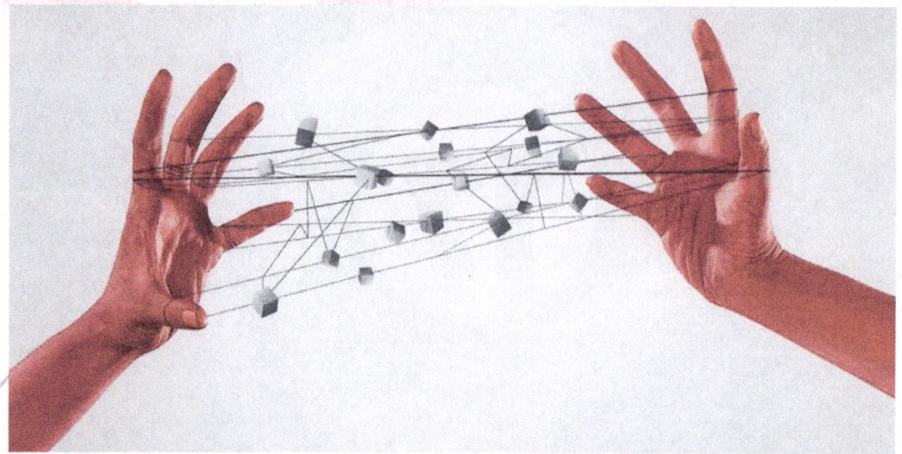

Fig.17.1: String Theorists Say the Universe is Built of Infinitesimal Vibrating Strings — but How do You Prove that? - Curtsy - Getty Images

These strings are truly tiny, many billions of times smaller than an individual proton within an atomic nucleus. They vibrate at countless billions of times per second in ***ten dimensions*** of space, or maybe ***eleven***. Or it might be ***twenty-six*** dimensions.

Why would physicists, a seemingly sensible bunch, think this is the way the universe operates?

To understand why, we need to step back more than 100 years.

TWO INCOMPATIBLE THEORIES

At the end of the 1800s, physics was riding high. Scientists thought they understood gravity, electricity, magnetism, heat and gases.

In 1900, Lord ***Kelvin*** apparently retorted that *"there is nothing new to be discovered in physics."*

However, soon after the dawn of the 20th century, this cozy situation began to fall apart.

Einstein rewrote the very notions of space and time with his special and general theories of relativity, whilst Planck, Bohr and Heisenberg revealed that the world of the very small obeyed the apparently nonsensical rules of ***quantum mechanics***.

By the mid-1900s, our view of the universe had been utterly revolutionized. And these new strange physical theories were yielding incredibly accurate predictions for experimental tests.

However, despite this success, physicists were unhappy. The problem was with forces. It was realized that there were four fundamental forces that underpin the universe: gravity, electromagnetism, strong nuclear force and weak nuclear force.

Of these, the strong and weak nuclear forces only operate at subatomic scales. At a larger scale, the rest of the universe is a never-ending battle between gravity and electromagnetism.

Gravity is described by Einstein's mathematics of curved space and time. But the other three forces are written in the language of quantum mechanics. And these two methods are completely incompatible.

This situation frustrates physicists as they have to remember two independent sets of mathematics to describe the physics of the universe.

It also worries them, as they are unable to describe physical processes in situations when the fundamental forces are battling for dominance, such as at the birth of the universe or in the center of black holes.

So, they have searched for one set of mathematics to describe all of the forces, a ***Grand Unified Theory*** that will mean that there is less to remember, Figure 17.2.

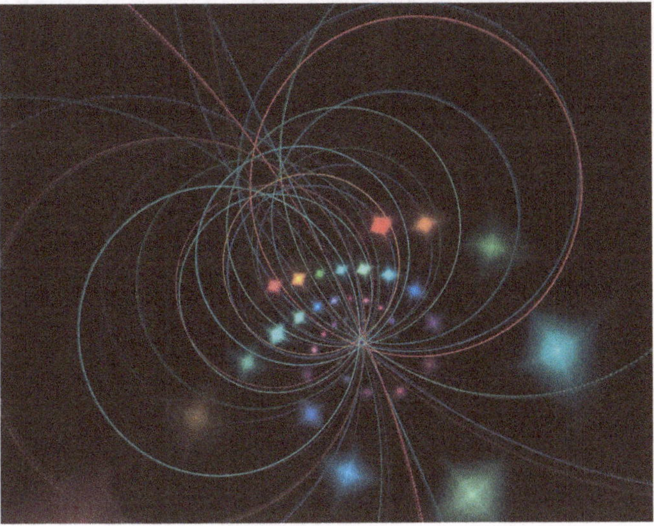

Fig.17.2: *String theory turns the page on the standard description of the universe by replacing all matter and force particles with just one element: tiny vibrating strings that twist and turn in complicated ways that, from our perspective, look like particles. Curtsy - Shutterstock*

The search for this Grand Unified Theory is not new, and many have tried and failed. Einstein was searching for a way to unify gravity and electromagnetism until his dying days.

This search for grand unification brings us to string theory.

VIBRATING STRINGS PLAYING A COSMIC NOTE

Like most scientific discoveries, the birth of string theory was very messy. It was born in the post-war explosion in particle physics which led to the discovery that the universe appeared to be built of a small family of fundamental particles — ***quarks***, ***leptons*** and ***force-carrying bosons***.

This helped make sense of the ever-growing zoo of particles flung out of high-powered accelerators, but physicists asked whether the apparently fundamental quarks, leptons and bosons were themselves made of similar stuff?

Scrabbling within the mathematics, physicists started to find similarities in the particles, representing them as one-dimensional loops of stringy-stuff, Figure 17.3.

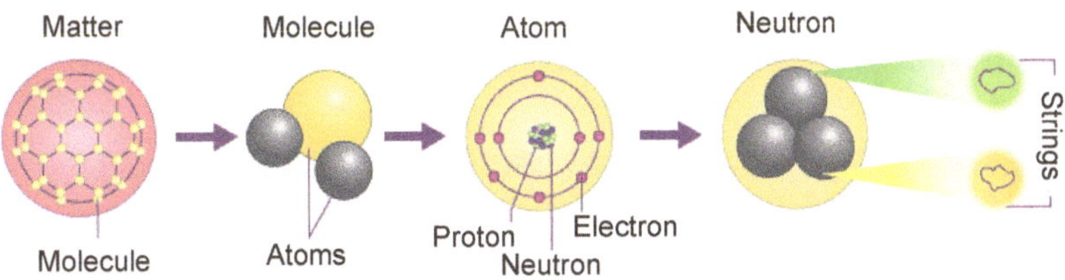

Fig.17.3: *String Theory is a Theoretical Framework that Tries to Solve the Problems of How and Why the Universe is Like*

Different vibrations of this stringy stuff correspond to each of the fundamental particles; one note played on a fundamental string is an electron, another note is a quark, another is a photon, the particle of light.

The strings themselves are not made of anything smaller —they are the true fundamental pieces of the universe.

But the mathematics of string theory is a little strange, and in putting the pieces together, physicists needed to add more and more dimensions of space to make their theories work, many more than the three we experience in our everyday lives.

If string theory is correct, more convoluted mathematical trickery is required to hide these extra dimensions from us.

String theorists are built of stern stuff and working with complex vibrations in multiple dimensions didn't daunt them.

With its simplicity as the underlying idea that can explain everything in the universe, string theory has proven very seductive.

Since its crystallization in the 1980s, it has continued to grow and evolve into a group of ideas — known as "M-Theory" — although no-one seems to know what M stands for.

M-theorists are confident that they are on the correct road towards grand unification, and soon will be able to pull together all of the individual threads and declare victory over the fundamental forces, Figure 17.4.

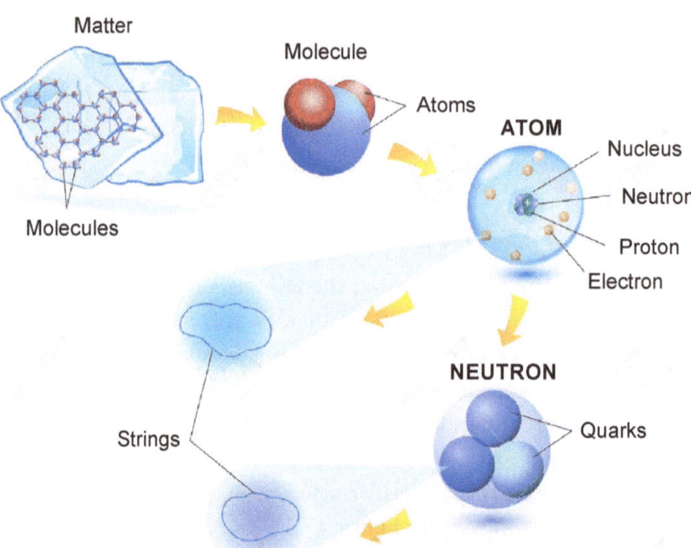

Fig.17.4: String theory, Superstrings M-Theory On an example of a matter, molecules, atoms, electrons, protons, neutrons and quarks, Microcosm and Macrocosm

MATHEMATICALLY ELEGANT - IMPOSSIBLE TO PROVE

Not Everyone is Convinced

Whilst scientific journals contain many thousands and thousands of pages of mathematical investigations of string theory, they all lack one important thing — **predictions** that can be tested through experiments.

To probe the tiny scale of strings we would need to build a huge version of the Large ***Hadron Collider*** — at least as large as our *Milky Way galaxy*, if not the observable universe.

As you can imagine, unless something radically changes in the mathematics of string theory, experimental verification will remain forever out of reach, Figure 17.5.

Fig.17.5: *Large Hadron Collider Smashes Beam Intensity Record, Inches Closer to Discovering God Particle*

Again, string theorists remain undaunted. Even if we cannot test the theory, they say, we should continue our efforts on ***M-theory*** as the idea is so beautiful, it just cannot be wrong.

With a bit more effort, they say, we will hold in our hands the theory to describe everything.

But to others, this is going too far. Science without predictions is not science, it's simply mathematics.

And if string theory is never tested against nature, then it will never be science.

String theory has made other physicists grumpy, with its fame overshadowing their own search for a ***Grand Unified Theory***

But their approaches can seem as strange and weird as string theory, with ideas such as "Loop Quantum Gravity" which suggests even space and time are built from fundamental bits.

These physicists feel that the theorists on the string theory bandwagon are heading down a dead-end path in the search for ultimate physics, Figure 17.6.

Fig.17.6: *Stream Grand Unified Theory*

Is string theory, or its many descendants in "M-theory", on the right track towards grand unification?

In truth, we simply don't know. We don't know if any of our ideas are truly inching a way towards the ultimate theory, or if a completely different approach is required.

In fact, we simply don't know if a grand unified theory even exists.

But this will not stop physicists from continuing their search.

What is String Theory?

What if our universe is made of vibrating strings? What if these invisible strings vibrate to create all sorts of matter and energy in this universe?

String theory is a concept and or a pipe dream in physics that states the universe is constructed by tiny vibrating strings, smaller than the smallest subatomic particles.

It states the universe is constructed by tiny vibrating strings, smaller than the smallest subatomic particles.

String theory is perhaps the most high-profile candidate for what physicists call a theory of everything – a single mathematical framework capable of describing the entirety of the known universe.

Hypothetically, as these fundamental strings twist, fold and vibrate, scientists imagine the vibrating strings create matter, energy and all sorts of phenomena like electromagnetism, gravity, etc.

String theory is one of the most famous ideas in modern physics, but it is also one of the most ever confusing concepts in science.

At its heart is the idea that the fundamental particles we observe are not point-like dots, but rather tiny strings that are so small that our best instruments cannot tell that they are not points.

It also predicts that there are extra dimensions to space beyond the obvious length, breadth and depth, but we do not experience them because they are bunched up in tiny spaces, Figure 17.3.

At present, physicists have to rely on two such frameworks. Quantum theory, which accurately describes the physics of the very small, and general relativity, developed by Albert Einstein, which describes the physics of the enormously large. The trouble is, the two theories contradict each other and don't get along.

The trouble is centered around '***gravity***.' It's the only one of the four fundamental forces of nature described by general relativity (Gravity, Electromagnetism, Strong Nuclear Forces, and Weak Nuclear Forces), and the only one that quantum theory cannot address. Coming up with a model that ties up all four forces in one neat package is a long-standing dream for theoretical physicists, named 'Theory of Everything.' Figure 17.7.

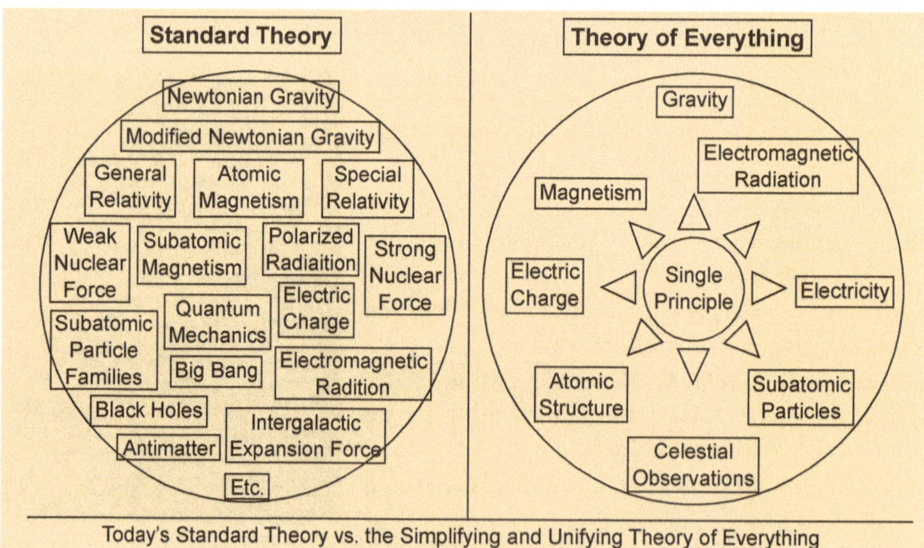

Fig.17.7: *Theories in Physics: The Theory of Everything*

String theory claims to make that dream a reality. In simple terms, it does this by reimagining what reality is made of. Instead of treating subatomic particles as the fundamental building blacks of matter, string theory says that everything is made of unbelievably tiny strings, whose vibrations produce effects that we interpret as atoms, electrons and quarks.

String theory is a theoretical framework that tries to tackle the problems of how and why the universe is like what it is now, Figure 17.3. In string theory, point-like matter particles are replaced by one-dimensional entities called strings.

String theory explains how these infinitesimal strings travel and interact with each other. In normal classical distance scales (larger than the string scale), a string is very similar to a normal particle with its *charge*, mass and other characteristics. The characteristics of strings are directly controlled by their vibrational states.

One of the vibrational conditions of the strings matches the states of the graviton. Graviton is a quantum particle that regulates gravitational force. In this regard, string theory is also called the *theory of quantum gravity*.

STRING THEORY AS THEORY OF EVERYTHING

The unification of ***quantum physics*** and ***relativity*** has been the biggest hurdle since the dawn of modern physics. Many famous physicists regarded the String theory as the Theory of Everything. It is considered the ultimate framework that could merge general relativity and quantum physics, Figure 17.8. We cannot go beyond without a unified theory, because both mechanics control almost everything modern physics deals with.

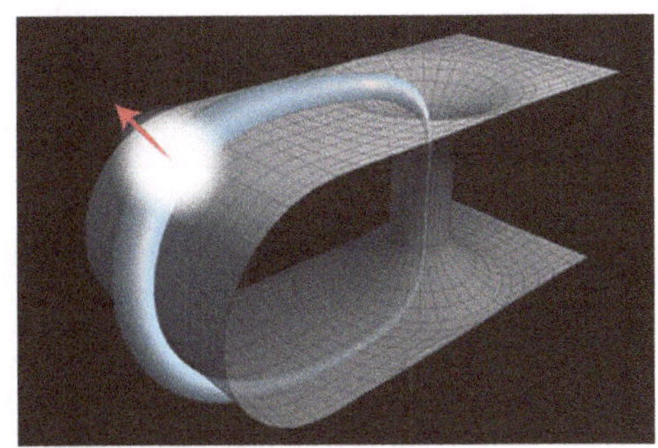

Fig.17.8: "Time-traveling photons Connect General Relativity to Quantum Mechanics" Curtsy - sciencesprings

Quantum mechanics regulates all phenomena at the subatomic level, and general relativity controls gigantic interstellar activities in the universe. The weird part is both of them do not merge well together. Flag bearers of the string theory propose that string theory could solve the disparities between the two physical realms.

STRING THEORY ORIGIN

In general relativity, gravity is a force that bends and warps space-time around supermassive bodies. Even though gravity is one of the four fundamental forces in nature, it is very weak compared to the other three forces (electromagnetism, weak force and strong force). So, it can't be observed or identified on the scale of subatomic particles. However, *gravity* is very dominant in long-distance scenarios. It controls the structure of the macro universe (galaxies, planets, stars, moons), Figure 17.9.

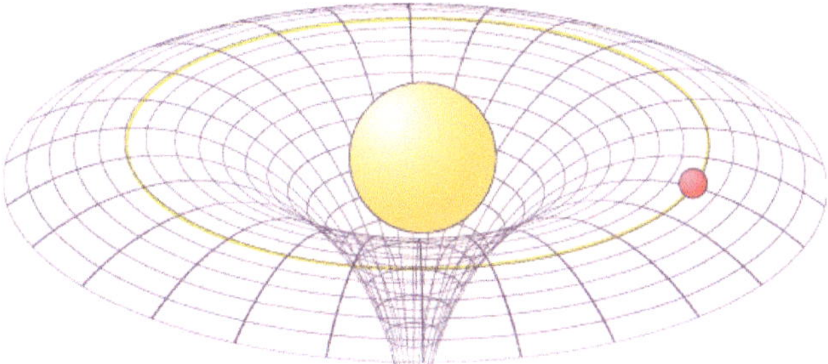

Fig.17.9: Space/Time Wrapping by Gravitational Force

As far as quantum mechanics is concerned, gravity doesn't have much effect. The probable nature of the quantum realm also poses a significant challenge for the induction of gravity in the quantum realm.

Generally, gravity does not act as a particle as its own. Even if a hypothetical model is introduced to explain the particle nature of a gravity particle, it violates fundamental energy laws.

In the 1970s, theorists tried to discard the self-destructive idea of point-like gravity particles. Instead of point particles, ***strings*** were introduced. Even if strings collide, there will be no infinite energy problem. Strings can smoothly smash and rebound without implying any physically nonsense infinities.

Bosonic string theory was one of the earliest versions of string theory. It only included fundamental particles called bosons. It was then evolved into superstring theory. It predominantly introduced a relation called supersymmetry between bosons and fermions. Later, physicists developed five concrete versions.

STRINGS IN STRING THEORY

String theory was developed from the idea of remodeling point particles to one-dimensional entities called strings. The interplay of strings can simply be stated by generalizing the theory of perturbation from quantum field theory.

Since string theory has no full non-perturbative explanation, most hypothetical and theoretical questions are still out of their domain.

Fig.17.10: *Open Type String Vs Closed Type String*

A string can be open or closed, Figure 17.10. An open string has two open ends, and a closed string forms a closed loop. Strings have a tiny dimension, approximately the order of Planck length, which is 10^{-32} meters. This is the scale at which ***quantum gravity*** is believed to become significant.

In classical or general relativity scales, strings are not very different from zero-dimensional point particles. The vibrational nature of strings determines the characteristics of the particle.

Even though bosonic string theory was the first concrete version of string theory, it was only concerned about bosons (a group of particles that transfer forces between fermions or matter). In order to negotiate such limit, bosonic string theory was replaced by a class of theories called superstring theories, Figure 17.11. These theories explain both fermions and bosons. So, the whole standard model came under the string theory hypothesis. The most striking development was the introduction of supersymmetry. It allows each boson to have a counterpart fermion.

There are many versions of superstring theory.

In order for that to work, string theory has to make one more radical assumption. That instead of living in a universe with three dimensions of space and one of time, we live in one with either 9, 10 or 25 dimensions of space, plus time. These extra dimensions are then curled up so tightly that we don't notice them – much like a silken thread appears one-dimensional until you get close enough to notice its width.

String theory, often called the "**Theory of Everything**," is a relatively young science that includes such unusual concepts as superstrings, *branes*, and extra dimensions. Scientists are hopeful that string theory will unlock one of the biggest mysteries of the universe, namely how gravity and quantum physics fit together.

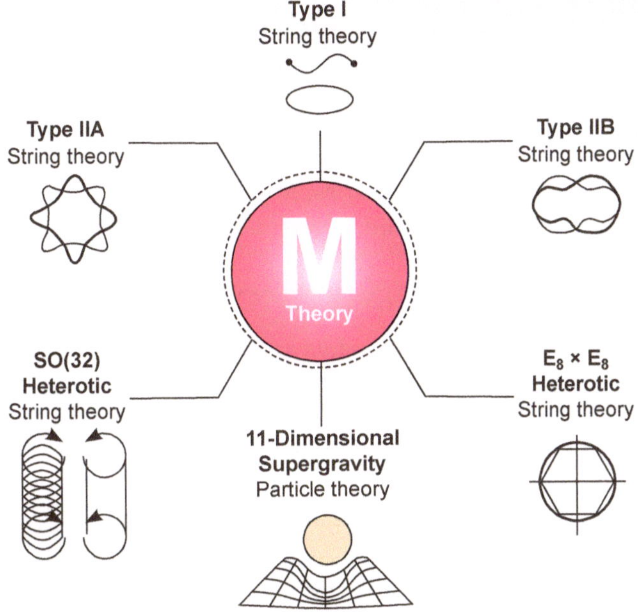

Fig.17.11: Different Versions of String Theory

DIFFERENT VERSIONS OF STRING THEORY

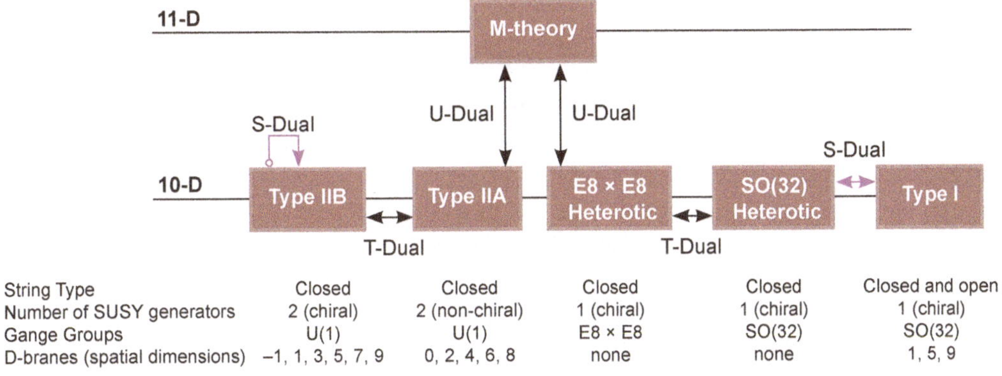

	Type IIB	Type IIA	E8 × E8 Heterotic	SO(32) Heterotic	Type I
String Type	Closed	Closed	Closed	Closed	Closed and open
Number of SUSY generators	2 (chiral)	2 (non-chiral)	1 (chiral)	1 (chiral)	1 (chiral)
Gange Groups	U(1)	U(1)	E8 × E8	SO(32)	SO(32)
D-branes (spatial dimensions)	–1, 1, 3, 5, 7, 9	0, 2, 4, 6, 8	none	none	1, 5, 9

Fig.17.12: Superstring Theory

1. Type I
2. Type IIA
3. Type IIB
4. Heterotic string theory (SO(32))
5. Heterotic string theory (E8×E8)

The various versions have different sets of strings. The particles that form at low energy levels show varied symmetries, Figure 17.12.

Heterotic, TypeIIB and TypeIIA theory consists of only closed strings. However, Type I consists of both closed and open strings.

M-THEORY

In 1995 Edward Witten stated that all the string theory versions are just different limiting cases of one super theory.

It is called the M theory, Figure 17.12. It is based on eleven dimensions. Edward formulated this conclusion, based on the findings of Michael Duff, Chris Hull, Ashok Sen and Paul Townsend.

MULTIPLE DIMENSIONS IN STRING THEORY

We, as humans, can only perceive and control three dimensions (length, height and width). With the inception of general relativity, time is considered the fourth dimension. The only difference is that we can control the first three spatial dimensions, but we cannot handle the time dimension, Figure 17.13.

In general relativity, space and time are not separate; space-time is one entity.

Gravity is considered as the byproduct of space-time geometry.

String theory must incorporate far more space-time dimensions than standard 3d or 4d space-time for a solid mathematical base.

There are twenty-six and ten dimensions in bosonic string and superstring theory, respectively. However, M-theory points to space-time with 11-dimensions.

Fig.17.13: *Multiple Dimensions in String Theory*

Branes (Membranes) in String Theory

- Branes are objects which can travel through space-time fabric, as per the constraints of quantum mechanics, Figure 17.14.
- Branes possess mass and can have properties like charge.
- Brane is an entity that is extended in multiple dimensions. The brane concept arises in string and universal theories like general relativity and *quantum mechanics*.

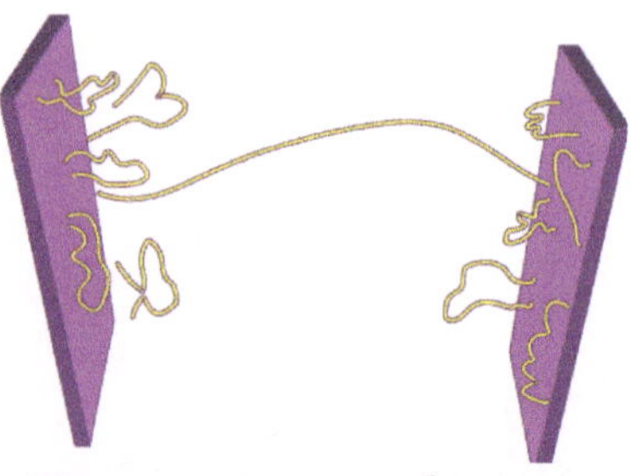

Fig.17.14: *Membranes (Branes) String Theory*

Here are some of the confusing technical sides of different branes.

- A 0-brane is a zero-dimensional entity.
- Point: 1-brane is a one-dimensional entity.
- String: 2-brane is a two-dimensional entity.
- Membrane: p-brane is a p-dimensional entity. As few types of string theory have nine spatial dimensions, p-branes may have dimension values of p up to 9. Time is an additional dimension.

Limitations of Hypothetical String Theory

- The major challenge of string theory is that the entire theory does not have a consistent definition in all scenarios. Contributed to its confusing concepts and definitions.
- Even though string theory is often hailed as the theory of everything, it still struggles to combine quantum physics and general relativity into a single universal theory.
- Another hurdle string theory puts forth is the prediction of multiple universes. The landscape of numerous universes has tangled the efforts to create particle physics theories under string theory. This is truly a pipe dream and would not ever become a valid theory.
- In quantum mechanics, the principle of observer-dependence is a fundamental characteristic. Obviously, string theory is still not able to integrate and resolve this quantum nature.
- String theory has not been able to explain dark energy and dark matter. They constitute almost 95% of the known universe.
- String theory cannot state whether the forces and particles forces in quantum physics can be merged or unified.
- String theory does not have a stable explanation of how the constants are created by nature in the standard model (particle physics).

UNDERSTANDING THE STRING THEORY HYPOTHESIS

Attempt to Define the Visible Universe

The Theory of Everything

Fig.17.15: String theory is the idea that the fundamental particles we observe are not point-like dots, but rather tiny strings - Curtsy clix, stock.xchng

String Theory Work in Progress to Define the Universe.

String theory is a work in progress, so trying to pin down exactly what the science is, or what its fundamental elements are, can be kind of complicated, Figure 17.15. The key string theory features include:

1. All objects in our universe are composed of vibrating filaments (strings) and membranes (branes) of energy.
2. String theory attempts to reconcile general relativity (gravity) with quantum physics.
3. A new connection (called *supersymmetry*) exists between two fundamentally different types of particles, *bosons* and *fermions.*
4. Several extra dimensions to the universe must exist.

SUPER-PARTNERS IN STRING THEORY

String theory's concept of *supersymmetry* is a fancy way of saying that each particle has a related particle called a *super-partner.* Keeping track of the confusing names of these super-partners can be tricky, so here are the rules in a nutshell. Even the rules are a laughing matter, Figure 17.16.

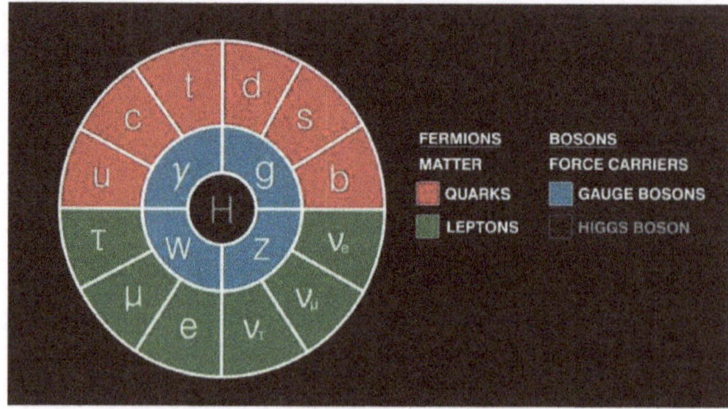

Fig.17.16: Superpartner of String Theory

a. The super-partner of a fermion begins with an *s*, so the super-partner of an electron is the selectron and the super-partner of the quark is the squark.
b. The super-partner of a boson ends in *ino*, so the super-partner of a photon is the photino and of the graviton is the gravitino.

Use the following table to see some examples of the super-partner names, Table 17.1.

Table 17.1: Some Superpartner Names

Standard Particle	Superpartner
Higgs boson	Higgsino
Neutrino	Sneutrino
Lepton	Slepton
Z boson	Zino
W boson	Wino
Gluon	Gluino
Muon	Smuon
Top quark	Stop squark

KEEPING TRACK OF STRING THEORY'S MANY NAMES

The Insanely and confusing 'String Theory' has gone through many name changes over the years, contributed to highly eccentric scientists and academic theorists. The list below provides an at-a-glance look at some of the major names for different types of string theory, Figure 17.17.

Some versions have more specific variations, which are shown as subentries. (These different variants are related in complex ways and sometimes overlap, so this breakdown into subentries is based on the order in which the theories developed.) Now, if you hear these names, you'll know they're talking about the most confusing theory on Earth today, the 'String Theory!'

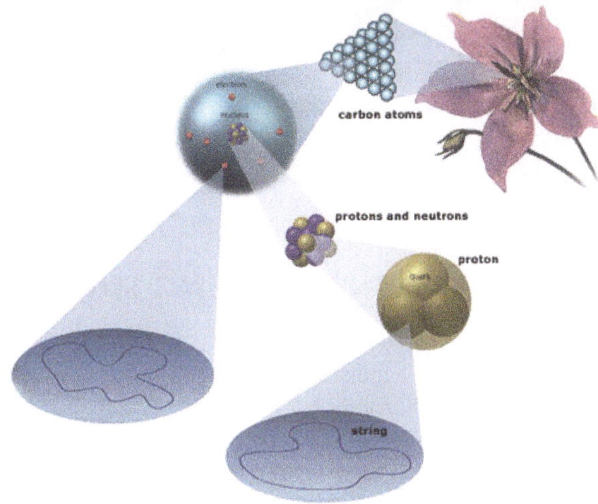

Fig.17.17: Strings to Carbon Atoms

KEY EVENTS IN STRING THEORY

The History of String Theory

Although string theory is a young science, it has had many notable achievements. What follows are some landmark events in the history of string theory.

1968: Gabriele Veneziano originally proposes the dual resonance model.

1970: String theory is created when physicists interpret Veneziano's model as describing a universe of vibrating strings.

1971: Supersymmetry is incorporated, creating superstring theory.

1974: String theories are shown to require extra dimensions. An object similar to the graviton is found in superstring theories.

1984: The first superstring revolution begins when it's shown that anomalies are absent in superstring theory.

1985: Heterotic string theory is developed. Calabi-Yau manifolds are shown to compactify the extra dimensions.

1995: Edward Witten proposes M-theory as unification of superstring theories, starting the second superstring revolution. Joe Polchinski shows branes are necessarily included in string theory.

1996: String theory is used to analyze black hole thermodynamics, matching earlier predictions from other methods.

1997: Juan Maldacena shows how it is possible to realize the holographic principle in string theory, which relates theories with gravity and theories without gravity living in one dimension-less.

STRING THEORY AND THE ORIGIN OF THE UNIVERSE

Perhaps the most serious problem with the **big bang** theory is the ***singularity problem***, which involves the original cause of the universe and the origin of matter and energy.

Big bang theorists can attempt to describe the early universe, but thus far, they have not explained why there is a universe to describe. However, there have been numerous attempts at such an explanation, and recently Paul J. Steinhardt has suggested yet another idea, this one based on string theory.

String theory is a new physical theory that seeks to solve the conflict between general relativity and quantum mechanics, thus providing a unified, overall description of physics as it is now known. But despite the advanced physics involved, Steinhardt's model does not overcome, or even address, the old problems that have plagued all previous naturalistic origins theories.

UNCONFUSING INTRODUCTION TO STRING THEORY

Quantum mechanics and general relativity are both enormously successful scientific theories. Quantum mechanics describes the bizarre workings of matter and energy on the smallest imaginable scales; general relativity describes the force of gravity, and becomes useful when gravitational fields are so extreme that they begin to depart from Isaac Newton's much simpler description.

The problem is that these two theories seem to be mutually exclusive. Both appear to be accurate descriptions of the world, but they cannot both be true. General relativity assumes that space-time is a smooth continuum which is curved by gravity. Quantum mechanics, by contrast, involves random, violent distortions of space-time in what has been called 'quantum foam' at the smallest scales. This is usually not a serious drawback, since strong gravitational fields, the domain of general relativity, usually occur on large scales, where the quantum effects are not noticed. But in some cases, such as the ***singularity*** inside a black hole, there are strong gravitational fields (requiring general relativity) coupled with very small scales (requiring quantum mechanics), Figure 17.18. If one simply joins the equations of general relativity to those of quantum mechanics, they give totally nonsensical answers to calculations, such as infinity.

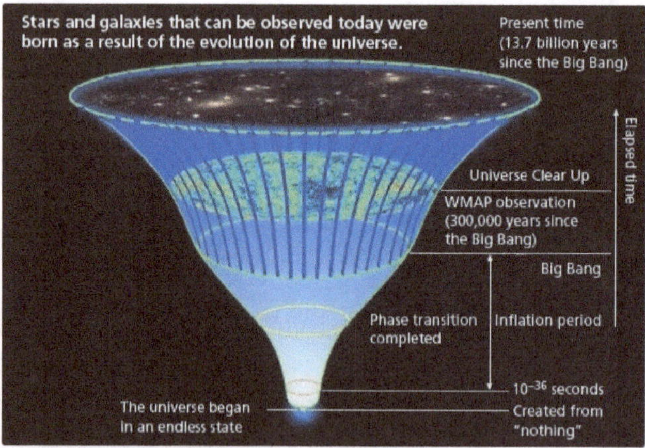

Fig.17.18: Black Hole Singularity - Curtsy Quora

String theory is the most successful, and the most popular, attempt to unify these apparently conflicting theories in a coherent description of reality. A critical part of this new theory is the idea that elementary particles such as the electron are not points, as is generally believed, but oscillating loops of 'string'. This apparently unrelated suggestion actually has profound implications for the conflict between general relativity and quantum mechanics.

When you want to probe something at very small scales (and assuming for the moment that you cannot see it directly), you can shoot very small projectiles at it and see how they are deflected, thereby reconstructing its image. Large projectiles will reveal only the largest-scale structure of the object in question, giving only a crude picture. Very small projectiles, however, can be deflected by the smallest irregularities in the target, revealing the finer details. In common experience we do this with photons, particles of light, without even thinking about it. The pattern of light and shadow, revealing an object's shape, is a result of the way light is deflected by the surface.

It takes small objects to probe small scales, and no object can be used to probe a scale smaller than its own size. Therefore, if all elementary particles are actually loops of string, then no scale smaller than these strings can possibly be probed. The problem of the ultra-small quantum foam that gives relativists a headache arises on precisely that scale. String theory therefore eliminates the problem and allows a unified description of nature.

String theory involves a number of very strange ideas. For example, the universe consists of nine or ten spatial dimensions, along with the one dimension of time. Furthermore, all forces arise from the same underlying force (just as electricity, magnetism and the weak force are already known to be interrelated), Figure 17.19. The properties of elementary particles are believed to arise from the patterns of vibrations in strings. String theory has not yet been completely developed or understood at the mathematical level, but so far it seems to have great potential.

Fig.17.19: *Multi Dimensions Universe – Curtsy - MAGGIE CHIANG/QUANTA MAGAZINE*

Despite its apparent theoretical promise, string theory does have the problem that there is not yet any experimental evidence for or against it. That will have to wait until either the theorists or the experimenters (or both) develop their techniques and ideas further. In the meantime, however, string theory seems to be the most promising solution, even if speculative.

STRING THEORY AND THE ORIGIN OF THE UNIVERSE

As might be expected of such a new physical theory, string theory has important implications in the field of cosmology, and it has been applied to the singularity problem.

Paul J. Steinhardt and Neil G. Turok have claimed that string theory allows for the possibility of an eternal universe. In their model, called the ***ekpyrotic*** model, our universe is only one of two.

These two universes, or 'branes,' lie side-by-side in a higher dimension (an analogy would be two flat sheets lying parallel in the third dimension) and are attracted by a 'new kind of force arising from quantum interactions' between the two branes. This attraction causes the branes to collide, with the energy of the collision producing a big bang, Figure 17.20. The branes then separate, space expands in each brane, stars and galaxies form, life evolves and asks where it came from, etc.

Eventually, this attractive force pulls the two branes back together in another big bang, and so it goes, in an eternal cycle with no beginning, no end, and ***no need for a cause or a creator***.

Fig.17.20: ***ekpyrotic*** *model Universe – Curtsy Tony Cradock Science*

Unfortunately, despite all the advanced (if unproven) physics involved in this theory, these two scientists have made a very elementary mistake that could have been perceived long ago. The ekpyrotic model, like any other theory that claims the universe is infinitely old, ignores the simple fact that the Second Law of Thermodynamics disallows any such possibility.

According to the Second Law, any isolated system must tend toward randomization and disorder or, more precisely, it must increase in entropy. Obviously, nothing that is wearing out can last forever, so this well-demonstrated law of physics requires that the universe cannot be infinitely old.

This argument has been made many times before, and big bang theorists will sometimes admit to it when it is convenient. But when an otherwise apparently promising origins theory contradicts the Second Law, that law is ignored, as in this case. (Sometimes it is claimed that this argument is not correct, since the Second Law applies only to isolated systems, whereas in the ekpyrotic model, the universe we live in is interacting with another, and thus is not isolated. However, the two universes can still be counted *together* as an isolated system. This *overall system* must still increase in entropy.)

Naturalistic theories of the origin of the universe generally fall into two categories:

1. those that claim the universe appeared from nothing with no cause, thus denying causality and the conservation laws; and
2. those that claim the universe is infinitely old, thus denying the Second Law of Thermodynamics.

Something Fundamentally Wrong with the Naturalistic Assumptions

The recent idea based on string theory makes no more progress than the previous attempts, indicating that there is something fundamentally wrong with the naturalistic assumptions behind such ideas. If we simply take the laws of physics at face value, however, it becomes obvious that the universe must have had a beginning, and that it must have been caused by some outside agent. Thus, there is no scientifically valid way to avoid the conclusion that '**In the beginning God created the heavens and the earth**.' Figure 17.21.

Fig.17.21: In The Beginning God Created – Genesis 1:1

COSMIC EVOLUTION HANGING BY A THREAD

Most people have heard of the expression '***the big bang.***' Its usage is so prevalent among mainstream scientists and the media that it has become the accepted 'fact' for how the universe began. However, there are an increasing number of secular scientists who are skeptical of this theory of cosmic evolution, and much of their skepticism has been caused by increasing discoveries that fly in the face of big bang theory. In May 2004 'An Open Letter to the Scientific Community' signed by dozens of secular scientists was advertised in the renowned *New Scientist.* At the time of writing this article, the total number of scientists signing the letter who are skeptical of the big bang has increased to over 400.

One of the great problems for those who believe that the universe came into existence by itself is that the universe *does not* present itself in such a manner. For example, the complexity of the observable universe fits better with the idea that it has been specially created *ex nihilo.* The 'first cause' problem is one of the great stumbling blocks for evolutionary cosmology. For example, consider the following logically valid argument.

1. Everything which has a beginning has a cause.

2. The universe has a beginning.
3. Therefore, the universe has a cause.

In short, if the universe had a beginning, then it must have had a cause.

DESIGN IS THE ANSWER

The so-called 'Anthropic Principle' derives from the observation that many aspects of the universe give at least the *appearance* of having been designed specifically for human life. There are an abundance of design features that sit uncomfortably for those who believe that we exist as a result of some giant cosmic lottery.

Accordingly, it is often claimed that there must be a myriad of other universes, with different features, even different laws of physics. And in our own universe the cosmic dice just happened to fall in such a way that the laws were those that would allow humans to evolve. Which is why we are here in the first place to ask such 'anthropic' questions. In other words, we are not here by design, it just looks that way, perhaps due to 'evolution by natural selection' among various (unobserved) universes.

Even though the big bang is passionately held to by some advocates of design, it also is used to attempt to avoid a Designer. The theory claims that all the matter, space and energy of the entire universe pre-existed in a particle no bigger than the head of a pin. Then, for no reason, it suddenly expanded, and the energy became matter that formed galaxies, stars and ultimately people.

Attempts to prop up this theory in the face of increasing problems have led to many weird hypotheses invoking mysterious but factual unseen forces known as 'dark matter' and 'dark energy' to basically hold the universe together or to push it apart. All of these speculations attempt to avoid the obvious spiritual implications of design (*Colossians 1:16–17*); For by him all things were created, in heaven and on earth, visible and invisible, whether thrones or dominions or rulers or authorities—all things were created through him and for him. And he is before all things, and in him all things hold together.

Theory of Everything?

One should consider that many of these 'cosmic convolutions' lack experimental support and are speculative ideas dressed in convoluted mathematics. One such idea is ***string theory***.

Although highly controversial, it has been gaining in popularity, and research is well-funded, particularly by those trying to prop up big bang ideology, Figure 17.22.

Fig.17.22: *Theory of Everything does not Make Sense – Curtsy : agsandrew / Adobe Stock*

Although String Theory not seem like it, it is a '*string theory for dummies*' type of explanation. ***But don't worry if you can't understand it. No one really does, including the author, as we will see.*** **And there is not a single shred of experimental evidence to support the claims.** String theory is presently completely unobservable and untestable. However, its advocates would also claim that it is not falsifiable, and therefore, it might be correct. This is the nature of such 'elegant' theories. One not infrequently hears similar claims about evolution in general.

The Founder Flounders!

A commentator in a popular science journal noted:

'The suspicion is that they're just floundering around doing ever more esoteric mathematics, but not really making any progress at all. It's a suspicion confirmed … by David Gross, (as we stated earlier) the Nobel Prize-winning American Theorist and one of the founders of string theory. At a prestigious conference of the best and brightest in physics he admitted: "***We don't know what we are talking about.***"'

One of the reasons that many are excited about the potential of string theory is that they hope it might evolve to become the elusive '*theory of everything*'—the holy grail that has eluded physicists for decades. The fundamental forces that govern our universe are gravity, electromagnetism, and the strong and weak nuclear forces (that hold atoms together). String theory is an attempt to unify these forces into one mathematical theory, an attempt that many physicists now believe to be impossible anyway.

DANGER APPEAL TO CHRISTIANS

Like the expression 'the big bang,' string theory is becoming part of the common language of cosmology, and sadly, many professing evangelical Christians, like the old-earth creationist Dr Hugh Ross, have jumped onto the string theory bandwagon. Some, always eager to find a way of not appearing too 'out of kilter' with the secular scientific community, propose that the other dimensions invoked by string theory could be the 'other' spiritual dimensions mentioned in Scripture.

Of course, it is reasonable to assume that there is another realm or dimension in which ***angels and/or God may exist***, and Paul spoke about a 'third heaven' (2 Corinthians 12:2), "*I know a man in Christ who fourteen years ago was caught up to the third heaven—whether in the body or out of the body I do not know, God knows.*" These may be compartments or areas of a spiritual realm or dimension, though not necessarily extra dimensions in themselves. The Bible does not speak about other dimensions specifically, and speculating on such matters in too much detail can lead to positions bordering on the heretical. This is one of the problems of taking the currently popular views of secular evolutionary scientists and trying to adapt them to Scripture.

INTERMARRY SECULAR GUESSWORK

Evolutionary ideas like string theory start from a worldview framework that there is no God. Therefore, it is dangerous practice to intermarry secular guesswork with the Bible. For example, like the big bang, there are many versions of string theory. Therefore, which one should we hang our theology on? If the secular theory changes tomorrow, do we then revise our theology? Finally, when the Bible cannot be twisted sufficiently to incorporate man-made ideologies that we assume to be correct, what happens to one's faith?

It's a much more reliable practice to start with Scripture when trying to understand our world and universe. The reality is --there is still so much we don't understand. But we can trust God. After all, He is the One who created it all to begin with. He encourages us to 'Trust in the Lord with all your heart, and lean not on your own understanding; In all your ways acknowledge Him, And He shall direct your paths' (Proverbs 3:5–6).

Many today propose that other universes may exist in parallel with our own. Some believe that aliens may exist in these other universe(s) and have harnessed the technology to flit between their own dimension and ours, and that this could account for the myriad of UFO sightings and alien encounters over the years, Figure 17.23.

Fig.17.23: *People Imagine - Aliens may Exist in Parallel Univer with Ours - Curtsy - iStockphoto*

Diagram - String Theory

Classical physics treats particles basically as zero-dimensional point-like objects, and quantum mechanics recognizes that they can also behave in filament wave-like fashion. But completely distinct from this physics, string theory proposes that they may instead exist as one-dimensional loops or 'strings,' Figure 17.24, that cut across or exist in nine spatial (and one time) dimensions compared to the three spatial (and one time) dimensions we understand and use in relativity theory.

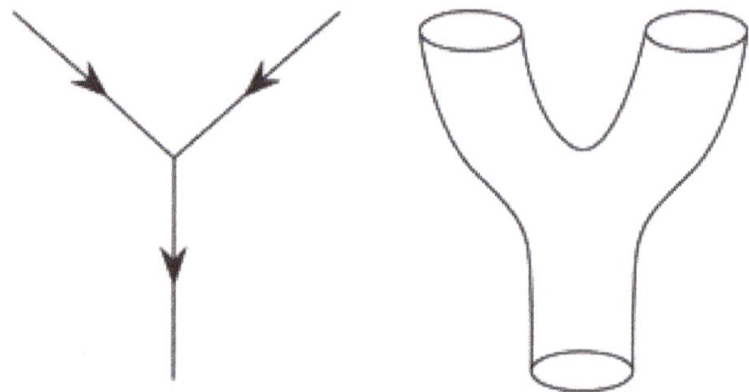

Fig.17.24: *Conventional point-like particles as compared to the theory that subatomic particles exist as closed strings. Left picture shows the merging of two point like particles, where the right picture shows two loops merging to form a single loop particle.*

Some suggest that these strings exist as higher dimensional objects called branes (membrane-like structures) that act like vibrating flat sheets or tubes etc. These ideas can even be extended to support the idea of multi-universes coexisting with our own, and that our own universe (the big bang) was the result of the collision of two of these branes in some higher-dimensional hyperspace. ***The goal is to get the universe to create itself***—that way, ***no first cause would be needed, hence no God.***

THEORY OF EVERYTHING

The Notable Questions in Physics

Theoretical physicists like to ask big questions. How did the universe begin?

What are its fundamental constituents? and

What are the laws of nature that govern those constituents?

If we look back over the 20th century, we can identify two pillars on which our current theories rest.

The first is **quantum mechanics**, which applies to the very small: atoms, subatomic particles and the forces between them. The second is **Einstein's general theory of relativity**, which applies to the very large: stars, galaxies and gravity, the driving force of the cosmos.

The problem we face is that the two are mutually incompatible. On the subatomic scale, Einstein's theory fails to comply with the quantum rules that govern the elementary particles. And on the cosmic scale, black holes are threatening the very foundations of quantum mechanics, Figure 17.25. Something has to give.

An all-embracing theory of physics that unifies quantum mechanics and general relativity would solve this problem, describing everything in the universe from the big bang to subatomic particles. We now have a leading candidate. Is it the much anticipated "theory of everything"? However, both Einstein and Hawking had failed to accomplish such task.

Fig.17.25: Distorted Space-time Bends Light from Distant Galaxies – Curtsy - Andrew Fruchter and the ERO Team NASA)

STRING THEORY UNSTRUNG

A Nobel laureate recently said that physics is in 'a period of utter confusion' and '***we don't know what we're talking about.***' Dr David Gross made these revealing admissions at a physicists' conference on the quantum structure of space and time, about the quest for a 'theory of everything.' That is, one that can describe every force and particle in nature.

In particular, string theory is billed as 'the great hope,' and has generated lots of mathematical equations. 'But these equations **tell us nothing about where space and time came from and describe nothing we would recognize**,' laments an editorial in *New Scientist*. (Emphasis added.)

The editorial's subtitle was 'The hunt for a theory of everything is going nowhere fast. It goes on to explain that when astronomers discovered the accelerating expansion of the universe (which string theory fails to account for), string theorists excused their theory by saying their equations describe all possible universes. But *New Scientist* rightly challenges that notion, querying that ***if string theory doesn't fit the known universe, 'is it science***?'

And *New Scientist* relays this joke circulating on physics blogs: '[W]e can after all, call our universe unique. Why? Because it is the only one that string theory cannot describe.'

This should be a lesson to would-be apologists, such as 'progressive creationist' Hugh Ross, who try to use string theory to prove the Bible (or at least his billions-of-years distortion of it).

Also, even Stephen Hawking has finally realized that a 'theory of everything' is a fantasy that founders on *Gödel's incompleteness proof*: that in any theoretical system as complex as arithmetic or above, there would always be true statements that cannot be proven within the system.

STEPHEN HAWKING: *KEY TO THE COSMOS*

Much of the material presented by Stephen Hawking has appeared in the 2010 book, *The Grand Design* by Hawking and Leonard Mlodinow.

Unfortunately, Hawking's scientific views are contaminated by his determination to exclude God from any rational involvement in the universe. A thorough review of Hawking books, his atheopathy goes beyond the evidence.

In "*Key to the Cosmos*" *book,* Hawking begins by tracing the development of scientific laws by Newton, Maxwell, Einstein, and Kaluza, the concept of quantum mechanics, the idea that electrons obey the laws of chance enunciated by Feynman, and Michaelson and Morley's experiment which showed that the hypothetical luminescent aether does not exist. Dr. Russell Humphreys Totally disagreed.

STRING THEORY

Hawking tells us what he believes about quarks as a parameter in the String Theory. These are proposed invisible particles inside neutrons and protons. According to him, quarks exist only insofar as they remain the best explanation of protons. Many creation physicists agree—for example, "***Has the 'God particle' been found*** ?"

Fig.17.26: Musician Energize the Strings - Curtsy iStockphoto

"Quarks are themselves made from something we physicists call strings, which are ever-more intricate distortions of space and time. ... Just as a violin string can vibrate to produce different musical notes, each subatomic string also vibrates, producing a different kind of fundamental particle. These tiniest particles give shape to the universe around us. ... String theory suggests that the vibrations of the strings produce tiny distortions in space/time at a microscopic scale. And they do so in a mind-boggling ten dimensions." Hawking explains this further in *The Grand Design* book as follows:

"According to string theory, particles are ... patterns of vibration that have length but no height or width. ... String theories ... are consistent only if space-time has ten dimensions, instead of the usual four"

"If the string vibrates in one way, it produces a certain kind of fundamental particle, say a quark. And if it vibrates in another way, it creates a neutrino, which is another kind of particle. String theory has the potential to explain why these particles interact with each other in the precise way they do—just like the harmony in a piece of music, Figure 17.26.

"And this is where the laws of physics come from—the laws that control everything in the universe, from the behavior of black holes to the life and death of stars, and something as simple as a roll of paper falling to the floor, or the flickering of a magnetic compass needle—the simplest and most fundamental of actions—all governed by the rules of string theory."

However, string theory has many problems as stated in an editorial in *New Scientist* lamenting: "But these equations **tell us nothing about where space and time came from** and describe nothing we would recognize."

SOURCE OF THE LAWS OF NATURE

For someone who goes out of his way to deny the existence of God and hence God as the creator of the laws of nature/physics, Hawking is quite vague as to his alternative source of these laws. His alleged simile between string theory and music is hardly proof of anything, and in fact he offers only "the potential" of string theory. Then, in his book *The Grand Design* he says (without elucidation): "the laws of nature in our universe arose from the big bang." Hawking quoted from his book saying: "The results described the book indicate that that our universe itself is also one of many, and that its apparent laws are not uniquely determined."

This inability to provide a cogent replacement for God as the source of scientific law is hardly surprising. Once you dismiss the concept of a Creator God who is not only a living supernatural being, but one who is also omnipotent, omniscient, and omnipresent, it certainly is difficult to contrive an adequate substitute. His 'multiverse' idea is *ad hoc*.

MIRACLES

Hawking is not nearly so vague about his denial of miracles. Three times in his book he refers to this:

1. "A scientific law is not a scientific law if it holds only when some supernatural being decides not to intervene."
2. "This book is rooted in the concept of scientific determinism which implies that … there are no miracles, or exceptions to the laws of nature."
3. "These laws should hold everywhere and at all times: otherwise, they wouldn't be laws.

There could be no exceptions or miracles. Gods or demons couldn't interfere in the running of the world."

Readers are obviously meant to make the mental jump: No miracles means no God, even if we've lost the origin of the laws of nature in the process and have to resort to string theory to get them back. Hawking, of course, is entitled to his axiom that God does not exist; at the same time we are entitled to our axiom that God *does* exist. So which axiom better fits the observable data? Let's examine this.

Consider an airplane. There are two forces operating on an airplane that keep it from flying. These are:

1. gravity (which tends to keep it on the ground), and
2. drag (which tends to stop it going forward).

So how does an airplane fly?

It flies when lift (generated by the angle of attack of the wings) overcomes gravity, and when thrust (generated by the propeller(s) or jet engine) overcomes drag, Figure 17.27. Therefore, when an airplane flies, there are (in this simple illustration) four forces acting on it, not just two. All four forces are valid. What happens to the airplane is determined by which forces are stronger at any time. Thus, science is not violated because there are more than two forces in play.

Fig.17.27: An airplane flies when lift overcomes gravity, and thrust overcomes drag. Curtsy iStockphoto

Applying this Principle to Hawking's logic:

1. It would obviously be nonsense for anyone to claim that the law of gravity is not a scientific law because there are times when lift can intervene and counteract it.
2. Lift is not an exception to the laws of nature. There is a force of gravity and there is a force of lift. Both forces operate together on an airplane in flight: otherwise, the airplane would shoot up to the edge of space.
3. Forces which balance other forces do not interfere in the running of the world. It is not a matter of there being no other laws than the ones Hawking defines; it is the one he denies that falsifies his premise.

The Christian philosopher Norman Geisler stated:

"Natural law is a description of the way God acts regularly in and through creation (Psalm 104:10–14), whereas a miracle is the way God acts on special occasions. So, both miracles and natural law involve the activity of God. The difference is that natural law is the regular, repeatable way God acts, whereas a miracle is not.

"Natural law describes the gradual activity of God in the world, whereas miracles manifest his immediate actions."

There is thus no basis for disallowing miracles, unless you could prove that God doesn't exist, but you can't prove a universal negative. Contrary to Hawking's wishful thinking, in any situation God can certainly bring other forces into play without violating science.

Indeed, seeing that God is the author of scientific law, a better conclusion is that God *does* intervene in the world when He wants to for His purposes, and in fact we should even *expect* Him to do so.

Hawking's error is that he confuses law with agency. He fails to understand that science can only *describe* things which are observable and repeatable; it cannot *prescribe* what cannot happen. Treating scientific laws as the cause of observed regularities is like claiming that the outline of a map is the cause of the shape of a coastline. It is therefore *unscientific* for Hawking to claim that God does not exist and that He cannot perform miracles.

M-THEORY AND MULTI-UNIVERSES

Hawking introduces us to the concept of multi-or parallel universes. He informed us:

"Currently there are several different versions of this string theory, which all together I call '***M-theory***'. Nobody seems to know what the 'M' stands for; it could be 'master', 'miracle' [Oops! how did that term get in there? —*Ed.*] or 'mystery', or perhaps all three. ... M-theory is making one remarkable prediction—that ours is not the only universe ... there should be hundreds of billions of billions of other universes, perhaps more universes than there are stars in the known cosmos."

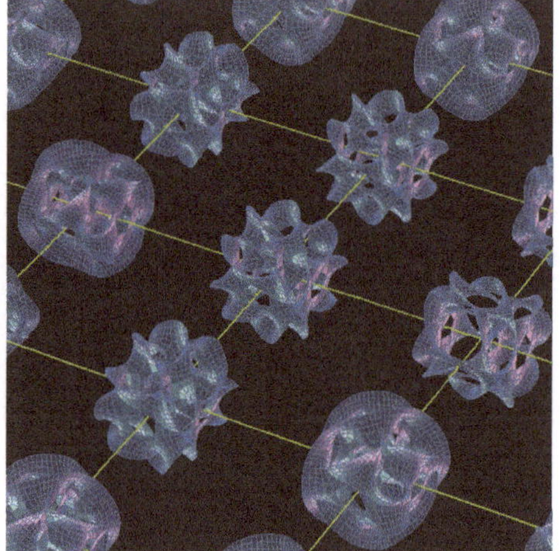

Fig.17.28: *Multiverse – Lacks Empirical Evidence – Curtsy Berkeley Center for Cosmological Physics*

In *The Grand Design* book, Hawking says that M-theory "allows for ***10^{500} different universes***, each with its own laws ... only one of which corresponds to the universe as we know it".

Please observe: 10^{500} is 100,000 billion billion billion ... (and so on to 55 times). So according to Hawking there are 100,000 billion billion billion (55 times) universes and 100,000 billion billion billion (55 times) different laws of nature therein respectively. This would seem to require 100,000 billion billion billion (55 times) different sources, since the various sets of laws of nature postulated are all different, Figure 17.28.

Hawking has given us (his) one source for the laws of our universe—strings. But Prof. Hawking, please tell us what are your 100,000 billion billion billion (55 times; less 1) sources for all the others? Rather than science, this sounds more like presumption times 10^{500}, that is: presumption 100,000 billion billion billion (55 times)!

And even his former collaborator, Sir Roger Penrose, says that Hawking has constructed a huge house of cards resting on an unproven theory: " ... 'M-theory', a popular (but fundamentally incomplete) development of string theory. ... M-theory enjoys no observational support whatever."

HAWKING'S GRAND DESIGN

In case you are wondering 'so what?' to all this, Hawking's purpose is to provide an answer to the problem that frustrated the great Albert Einstein, who sought in vain to construct 'a theory of everything.' That is, a supposed theory of theoretical physics that fully explains and links together all known physical phenomena, and predicts the outcome of any experiment that could be carried out in principle, Figure 19.29.

Fig.17.29: Many Universes in Stephen Hawking's 'Grand Design' – Curtsy - The New York Times

Hawking says:

"Let's return to the idea that the strings of string theory are like the notes played by a string quartet. Each vibration of the strings gives rise to a fundamental particle and to the forces of nature, which between them make up everything in the universe. But of course, the quartet may just as well be playing a different tune with different vibrations. Mathematically, a different tune would produce different particles and different forces of nature, meaning a different universe. Change again, and that's another universe.

"But just as there are an endless number of possible tunes, so our universe must be just one of billions of universes. We can't see them because they are beyond the limits of our own universe, each with their own history and properties. Some are unstable and collapse back to where they came from. Some will produce no stars or planets. and so be dark and cold. Others will expand and go on to produce stars and galaxies, like ours.

"As we ponder this, we should not be surprised to find ourselves in a universe that is perfect for us. Our very presence means our universe must be just right. So, the search for the key to the universe has had one unexpected result; we have found the key to every other universe too.

It seems that M-theory is the system of laws that governs everything—the Grand Design."

Accordingly, there we have it: Hawking's theory of everything, it seems, is M-theory. Yet Hawking, himself belatedly recognized that a theory of everything is logically impossible, because of Gödel's Incompleteness Proof. That is, in any theoretical system as complex as arithmetic or above, there would always be true statements that cannot be proven within the system.

Not so according to Hawking's supporters; Prof. Paul Davies, one of the most eloquent and vociferous promoters of the big bang theory over the years, asks:

"How is the existence of the other universes to be tested? ... more and more must be accepted on faith, and less and less is open to scientific verification. ... Indeed, invoking an infinity of unseen universes to explain the unusual features of the one we do see is just as ad hoc as invoking an unseen Creator. … Appealing to everything in general to explain something in particular is really no explanation at all."

See also the Sir Roger Penrose quote, above. Then too, there is Occam's (or Ockham's) Razor, otherwise known as the law of parsimony, economy or succinctness. It is the principle that if we have competing hypotheses, we should select the one which makes the fewest assumptions.

Finally, there is one other major fallacy with Stephen Hawking's atheistic reasoning—apart from the fact that a simile is not a proof. Did you spot it? For a violin string to produce music, somebody must play the violin. And for a string quartet to produce "an endless number of possible tunes" four intelligent musicians are needed to create the harmonies.

String theory, M-theory, and multi-universe theory are ideological attempts to explain the universe we live in. Unfortunately, many of those postulating these ideas have an additional agenda—to discount the appearance of design in the universe we live in, and hence to explain the universe without God, who for them does not exist.

During the 20th century quantum mechanics and general relativity emerged as twin pillars of modern physics. Both have substantial experimental support. Now, string theory is being promoted in the 21st century as the next great idea in physics. Williams and Hartnett comment:

"To explain something means to describe the unknown in terms of the known. … Cosmologists endeavor to explain the unknown—in their case the origin of the universe. In so doing, string theorists have developed 10- or 26-dimensional models to explain the four-dimensional universe. So, they have *not* explained the unknown in terms of the known. They have appealed to further unknowns (dimensions that we don't know about) to explain the existing unknown (the origin of the universe) so it does not qualify (at this stage) as an explanation. If they turn out to be correct and one day, we discover that there *are* strings and 26 dimensions, that will mean that the universe is unimaginably more complex than it appears to be now." (Emphases in original.)

In the same way, 10^{500} universes are much harder to explain than one—the one we live in. And lest any readers think that we are myopic with respect to anything additional to our own universe, we refer them to The Gospel in time and space. We further suggest that Stephen Hawking *et al*, could find wisdom in pondering the following description by the Apostle John of Jesus Christ, whom we Christians worship as God:

"In the beginning was the Word, and the Word was with God, and the Word was God. He was in the beginning with God. All things were made through Him, and without Him was not anything made that was made. In Him was life, and the life was the light of men. The light shines in the darkness, and the darkness has not overcome it. …

"He was in the world, and the world was made through Him, yet the world did not know Him. He came to His own, and His own people did not receive Him. But to all who did receive Him, who believed in His name, He gave the right to become children of God, who were born, not of blood nor of the will of the flesh nor of the will of man, but of God.

"And the Word became flesh and dwelt among us, and we have seen His glory, glory as of the only Son from the Father, full of grace and truth" (Gospel of John 1:1–5, 10–14).

GOD'S ENORMOUS EXPANSE

Psalm 150:1, **… Praise Him in his mighty expanse.** (NAS), or
… praise him in the firmament of his power. (KJV)

God made the expanse (firmament) on the second day and called it "heavens" (Genesis 1:8, plural from literal Hebrew). Later, on the fourth day, He populated the expanse with the sun, moon and stars (Genesis 1:14-19), Figure 17.30.

Fig.17.30: God's Enormous Expance

Accordingly, the expanse is not the heavenly bodies, but rather the *space* that contains the heavenly bodies. Normally people think of interstellar space, and also the space in which we ourselves exist, as an empty nothingness.

Praise God in His mighty nothing? So, what does God mean here?

Scripture itself gives a clue: it looks like the expanse (firmament, heavens, space) is an actual *material* that we cannot perceive as we move through it and it moves through us.

For example, it can be *stretched out* (Job 9:8 and 16 other Old Testament verses, *torn* (Isaiah 64:1), *worn out* like a garment (Psalm 102:26), *shaken* (Hebrews 12:26, Haggai 2:6, Isaiah 13:13), *burnt up* (2 Peter 3:12), *split apart* like a scroll (Revelation 6:14), and *rolled up* like a mantle (Hebrews 1:12) or a scroll (Isaiah 34:4).

Many physics theories and experiments seem to require that space be a real material:

1. The observed "displacement" electric current of James Clerk Maxwell, the greatest theoretical physicist (and a fine creation scientist) of the 19th century. (Maxwell based his theory on the experimental work of another great creation scientist, Michael Faraday.) With that idea he was able to predict the existence of radio waves, and to lay the foundations of all 20th-century devices using electricity and magnetism.
2. Einstein's theories of special and general relativity not only stem from Maxwell's work, but at bottom they only make sense if space (and time) is some kind of "stuff," as Einstein finally acknowledged in a little-known speech in 1920. The famous limit for the speeds of light and particles, *c*, could only work if there were a real material to enforce the speed limit. (Why should there be a limit if space were completely empty?) Space could be "warped" or "bent" only if it were actual solid matter.
3. The esoteric but well-verified quantum field theory starts with the premise that space is filled with the particles of a non-perceivable material (the "quantum vacuum") that is very dense. According to the theory, this material exists in and around all visible-matter particles and transmits the forces between them, thus, enabling visible matter to exist.

Experiments stemming from quantum field theory show that electrons in atoms influence the space around them and in turn are influenced by it ("vacuum polarization").

In the 1930s, quantum theorist P.A.M. Dirac correctly predicted the existence of antimatter on the basis of his theory that required all space to be filled with a "sea" of electrons. The quantum theory of solids offers a way to understand how space could be very dense but not felt or seen, in the same way that free electrons can move through a perfect crystal without any hindrance.

Biblical Clues

These and other physics clues suggest that the material, which from the biblical clues I call the "fabric" of space, is an elastic solid, like a very rigid and enormously massive crystal. That could be why the Hebrew word for the expanse (*raqia*), and the Greek and Latin translations of it (*stereoma* and *firmamentum*) all have some connection with solidity and firmness, as does the English word, "firmament" used (coined?) by the King James translators.

Strangely, academic materialists have tried to ignore the physics clues that space is a material, probably for religious and philosophical reasons. They even ignored Einstein's 1920 recantation (see second item above) of his 1905 denial of the 19th-century idea of an "ether" (or "aether") meant to propagate light waves. The academics have made the word "ether" politically incorrect.

Now that it looks as if the ether idea merely needed a bit more sophistication, physicists use many code words for it, such as various combinations of: "spacetime," "continuum," "manifold," "quantum vacuum," "the Vacuum," "substratum," "Dirac Sea," "plenum," and "medium,"—all to avoid using the word "ether." This verbal beating-around-the-bush amuses me. It prevents academics from explaining relativity and quantum mechanics in simple, visualizable terms that solve the various paradoxes. I suspect the academic experts on relativity and quantum theories prefer to keep them arcane and perplexing (to other academics as well), because the mystery makes them the high priests of a secular religion for which the public needs interpreters.

Putting aside the foibles of academia, my main point is that the expanse (firmament) is a real material that God made early in Creation Week. It is invisible and very clear, since we can observe through it for cosmic distances. Though we can't perceive it directly, our new knowledge of its massiveness and strength shows forth the glory of its mighty Creator.

Looking for the God Particle at the Large Hadron Collider

The Large Hadron Collider at CERN, Figure 17.31, was scheduled to begin colliding protons in October 2022. Two of its main goals are to look for both the Standard Model's Higgs particle and supersymmetry's sparticles. An objective overview of these theories, along with string theory, reveals the precarious balancing act that high-energy physics is in because of theorists' attempts to explain away fine-tuning which points to an intelligent designer. Ironically, their foundational theory is rooted in a faith in very large numbers that they hope will cancel out impossible odds.

In the autumn of 1993, a vote came before the United States Congress to continue funding construction of the Superconducting Super Collider near Waxahachie, Texas. It was to be a machine that would produce 40 trillion electron-volts (TeV) of energy, twenty times the total energy of the Tevatron near Chicago, which was, at the time the highest-energy particle accelerator in the world.

Although two billion dollars had already been poured into the project, and 22 km of the total 87 km tunnel had been excavated, the demise of the project turned out to fall on the answer to a profoundly misplaced question. A congressman asked a consulting physicist if we would find God with this machine. If so, he would vote for it. The physicist replied in a rather lackluster fashion that we could find the Higgs boson. Knowing the difficulty of justifying $12 billion for another subatomic particle, Congress voted down the great-American particle smasher.

Fig.17.31: Large Hadron Collider (LHC) - The Large Hadron Collider is the world's largest and most powerful particle accelerator CERN

When men refuse to acknowledge God as Creator and believe the lie of evolution, they find themselves looking for God in all the wrong places, even in particle accelerators! The long tentacles of Darwinian philosophy, which attempts to explain the emergence of life without God, reach even into the mathematical world of theoretical physics. The theories of supersymmetry and string theory endeavor to wipe out the fine-tuning of different parameters that make our universe conducive to life as we know it.

An objective survey of these theories, however, reveals that they bring in more problems than they solve for evolutionists in their efforts to conceal the fingerprints of the Designer on his design.

When the Americans dropped the baton of particles physics by nixing the Texas supercollider, the Europeans at CERN quickly picked it up and ran with it by building the Large Hadron Collider. CERN is a French acronym for the *European Organization for Nuclear Research*. It is the international laboratory that houses the Large Hadron Collider, or LHC.

Straddling the Swiss-French border, the tunnel of this $8 billion collider is 27 km in circumference. As 'the world's most powerful hammer', the LHC will catapult particles to an energy of 7 TeV, effectively smashing protons together to see what they are made of and to give new particles a chance to form.

These protons will make 11,000 loops per second around the circle, reaching a speed to within 10 km per hour of the speed of light, making 50 million collisions within a second.

Powerful particle detectors will register the direction and energy of the resultant particle debris and collect a full DVD's worth of data every five seconds. According to current schedule, the first beam of protons was circulated through the entire LHC on September 10, 2008. The first collisions have occur after its official unveiling on October 21, 2008.

Scientists Say the Large Hadron Collider Takes Us Back to the Big Bang

The LHC is being promoted as a machine that will take us back to within a fraction of a second after the big bang. What physicists actually mean by this dramatic claim is that by colliding nuclei, they will create very high temperatures, exceeding 100,000 times the temperature in the core of the sun, in a volume smaller than an atomic nucleus.

Scientists who believe that such high temperatures prevailed during the big bang naturally surmise they will be looking back into our remote origins.

Creation scientists, on the other hand, can separate these subjective conclusions of historical science from the hard facts of empirical science and analyses the information coming out of the LHC without swallowing the evolutionary hype.

Nevertheless, the great energy of the LHC should untie quarks and create the quark-gluon plasma in which quarks and gluons roam freely. The elementary particles called the strange, charm, beauty and top quarks can only exist in these extreme conditions. Physicists will be recreating them in the LHC.

As the temperature cools, these particles quickly decay, leaving the surviving up and down quarks permanently glued together inside protons and neutrons by gluons which transmit the strong nuclear force. This is the purpose of the **ALICE** experiment at the **LHC**: to study how elementary particles are organized under the action of the strong force.

BACKGROUND FOR THE HIGGS PARTICLE

As for the elusive Higgs boson whose mention to Congress derailed the Superconducting Supercollider, it is like the missing piece to an otherwise complete puzzle of particle physics called the Standard Model.

The Standard Model is the current explanation of the known elementary particles and their interactions with electromagnetism and the strong and weak nuclear forces. The Standard Model holds that electromagnetism and the weak force merge into one force around 1013 GeV. The LHC is built to probe physics at this weak interaction scale. The Standard Model says all particles in the universe would be massless at this energy, travelling at the speed of light and unable to settle down into atoms. When the hypothesized Higgs field froze out of the primordial cosmic soup, the symmetry broke, electroweak forces split and the weak force was unable to propagate freely.

	Quarks				Leptons		
Fermions (½ spin)	u Up	c Charm	t Top		e Electron	μ Muon	τ Tau
	d Down	s Strange	b Bottom		ν_e Neutrino	ν_μ Neutrino	ν_τ Neutrino
Hadrons	Baryons Fermionic Hadrons (Three Quarks)		Mesons Bosonic Hadrons (Quark & Anti-Quarks)				
Force		Electromagnetic	Strong	Weak		Gravity	
Vector Bosons (Full spin)		γ Photon	g Gluon	$W^{-/+}$ Boson	Z^0 Boson	Graviton (hypothetical)	
Scalar Boson	Higgs						

Officially Gravity and the Graviton are not part of the Standard Model, but they are included here for completeness.

Fig.17.32: The Standard Model of Particle Physics - Curtsy - Forsyth Astronomical Society

Figure 17.32 - Since all the other elementary particles have been experimentally confirmed, the Higgs particle is the missing piece to an otherwise complete Standard Model of particle physics. Without it, however, the math would be consistent only if all particles are massless, moving at the speed of light. Fermion = particle of half-integral intrinsic spin, and are the building blocks of matter. Boson = particle of integral intrinsic spin; the fundamental forces are mediators by bosons, although some have mass.

The Higgs field is a scalar quantum field. Nobel laureate Leon Lederman referred to the particle that carries the force of this field as the '***god particle***', for this Higgs boson is touted as being the *godlike* source of all mass in the universe.

Apart from being a high-tech version of the ancient animistic error of projecting attributes of the Creator onto the creation, this notion is a scientific misrepresentation as well.

The Higgs field can only account for the mass of elementary particles because composites such as the proton and neutron have binding energy that acts as mass. Nevertheless, the Weinberg–Salem model of the electroweak force in the Standard Model has been very successful. It has led to predictions of three particles that carry the weak force, called

a the W+ ,

b the W– , and

c the Z.

All three of these have been found with the exact properties predicted by the theory. We are left with only one prediction that has not been experimentally verified in this version of the Standard Model, and that is the Higgs particle.

HOW THE HIGGS SAVES THE STANDARD MODEL

In order for electroweak unification to work mathematically, it requires that the force-carrying particles have no mass. Experiments show this is not true. If, however, all particles had no mass until the Higgs field congealed, then any particles that interact with the Higgs field are given a mass via the Higgs boson. The importance of this hypothetical particle to the Standard Model cannot be overstated.

If physicists do not find it with the LHC, they will be forced to develop a completely new theory to explain the origin of mass. In the words of Lee Smolin in his chapter The Trouble with Physics, 'First of all, we want the LHC to see the Higgs particle, the massive boson responsible for carrying the Higgs field. If it doesn't, we will be in big trouble.' The Higgs, you see, is the savior of the Standard Model.

A Fine-Tuned Problem for the Higgs Mass

Trouble comes with the fact that the Higgs mass is undetermined and must be put in by hand, while all we know is that it should be greater than about 120 times the mass of a proton. The large differences between the masses of the particles of the Standard Model (example:

electron mass = 1/1,800 × proton mass) require fine-tuning of their intrinsic masses

(mass minus quantum effects).

The intrinsic masses of bosons (the category for particles of force) and fermions (the category for particles of matter) is proportional to the mass coming from quantum effects. Thus, we say their masses are 'protected'. In other words, if their intrinsic masses are small, so are their total masses. The Higgs boson, it turns out, is the only unprotected particle in the Standard Model. It exhibits an intense self-interaction by emitting particles and reabsorbing them, the energy of which acts as mass.

The only way to keep its mass from being pulled up to the Planck mass where quantum gravity effects come into play, a whopping 1016 times too heavy, is to fine-tune the score of constants in the Standard Model to an incredible precision of 32 decimal places.

There is no room for inaccuracy in any one of these 32 decimal places; otherwise, the Higgs mass becomes much too large. The miraculous fine-tuning required to keep the Higgs mass low is more extreme than the precision needed balance a pencil on its sharpened end. This dilemma is known as the *Higgs hierarchy problem*.

Hand-picking just the right numbers to make them fit in the Standard Model feels like fudging the answers to many physicists because it implies ***intelligent design in our universe***. Since life can only exist in an extremely narrow range of all physical parameters, they feel the need to find a physical explanation as to why nature fell precisely on those fortunate values that give rise to life. Improbable yet fortunate coincidences, in the minds of evolutionary scientists, need an explanation, and the obvious answer that ***the universe has been designed*** to accommodate man is not a viable option for them. Theorists lament that they see fine-tuning as almost certainly a badge of shame which reflects their ignorance. So, they create a new theory to get around the need for fine-tuning.

BACKGROUND FOR SUPER SYMMETRY

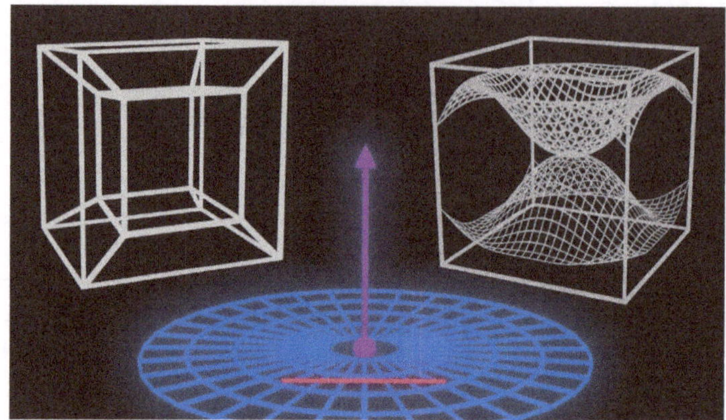

Fig.17.33: *Supersymmetry Theory*

Supersymmetry is a theory introduced to try to explain away the fine-tuning of the Higgs mass, Figure 17.33. Supersymmetry theory relates fermions to bosons by saying that every boson has a fermion partner whose spin differs by one-half and every fermion has a boson partner with the same one-half spin difference. None of these hypothetical supersymmetric partners have been seen in thirty years of looking, but they have been named 'sparticles.' Seeking sparticles is one of the main goals of the LHC.

Besides providing a way to keep the Higgs mass 'naturally' low (i.e., without fine-tuning), there are several other advantages gained with supersymmetry. First, by extending the Standard Model to a supersymmetric quantum field theory, the strong force, the weak force and the electromagnetic force strengths all converge around 2 × 1016 GeV.

In the Standard Model without supersymmetry, these same interaction strengths do not quite come together at the same point. Thus, supersymmetry neatly unifies the three forces of the Standard Model. Another big plus for supersymmetry is that by pushing up the energy at which these three forces converge, supersymmetry slows the predicted decay of the proton to 1035 years, which would explain why no one has seen a single proton decay, but not for lack of trying, in huge underground water tanks like the Super-Kamiokande detector in Japan.

How Supersymmetry Solves the Higgs Hierarchy Problem

Sparticles must be heavier than their particle partners, otherwise we would have detected them in less energetic accelerators by now. To explain why supersymmetric particle pairs need not have the same mass, supersymmetry must be spontaneously broken, but in just the right way to give particles their properties conducive to life as we know it. Lisa Randall, a leading theoretical physicist, discusses the situation in her book Warped Passages,

'We want supersymmetry breaking to be small enough to make the supersymmetry-breaking mass difference between superpartners and Standard Model particles sufficiently small to avoid fudging. It turns out that the quantum contribution to the Higgs particle's mass from a virtual partner and its superpartner, though nonzero, will never have a magnitude much greater than the super-symmetry breaking mass difference between the particle and its superpartner.'

If it were the case that supersymmetry was broken on the scale of the weak interaction, the LHC will see sparticles along with the Higgs particle. Supersymmetry solves the Higgs hierarchy problem because the Higgs field would be emitting a sparticle with each particle. The quantum effect of these sparticles would negate the quantum effect that pulls the Higgs mass up way too high, thus keeping the Higgs mass low, with no fine-tuning required.

PROBLEMS FOR SUPERSYMMETRY

Because the weak-interaction scale is 1013 times smaller than the Grand Unified Theory (or GUT) energy, using the Higgs field to break supersymmetry so far below the GUT scale only substitutes the Higgs hierarchy problem for a new supersymmetry breaking hierarchy problem.

Then only by carefully fine-tuning every term in the perturbation expansion can physicists keep the weak interaction scale from ending up about the same size as the grand unification energy. Considering this, even with all of its explanatory powers, supersymmetry is not a solid foundation for the Higgs mass to be resting on. Not the least of its other problems is the simple fact that no supersymmetric sparticles have been found in 30 years of looking. Another problem is that introducing spontaneous symmetry breaking to supersymmetry makes a very complicated theory called the minimally supersymmetric standard model with 105 more free constants than the standard model's original 20. This leaves theorists to adjust the free constants by hand to get predictions that agree with experiment.

Conveniently, there are many ways to adjust the free constants to make the undetected sparticles too heavy to see. With so many free constants, the supersymmetry theory is difficult to prove or disprove, leaving it open to suspicion. In the words of Lee Smolin, 'The story of supersymmetry is one in which, from the beginning, the game has been to hide the consequences of unification.'

DEVASTATING CONSEQUENCE OF SUPERSYMMETRY

The most devastating consequence of supersymmetry can be alleviated by string theory which requires that spacetime have ten or eleven dimensions, six or seven more spatial dimensions than the three we experience in everyday life. With these extra spatial dimensions, many internal symmetries (symmetries concerning internal properties of particles) could actually be spacetime symmetries (symmetries concerning external properties of particles).

For example, the abstract rotations of the internal symmetries could be real rotations, but in higher dimensions. The following is a good explanation of this important link between internal and spacetime symmetries:

'What makes supersymmetry so super is that it forges a link, not just between the particle categories but also between spacetime symmetries and internal symmetries. Using an abstract version of rotation, you can transform particles into sparticles and then back again. In the process, the particles scoot over a little bit in space. A change in an internal property affects an external one. Before they discovered supersymmetry, physicists had reckoned such a feat impossible.'

Since supersymmetry causes motion in a particle, and motion is described by special relativity, and the local (or 'gauge') symmetry of special relativity is general relativity, then it follows that the force associated with supersymmetry is gravity. Supersymmetry thus unites the elements of quantum theory with gravity into a theory of supergravity. One big perplexity within this apparently happy union is that the particle that transmits gravity, called the graviton, would spiral out of control. String theory provides supersymmetry with a way out of this debilitating dilemma.

Figure 17.34. Imagine the precision needed to balance a pencil on its sharpened end. The extreme fine-tuning required to adjust a score of parameters in the Standard Model to keep the Higgs mass from being pulled up way too high by quantum contributions is much more precise than this.

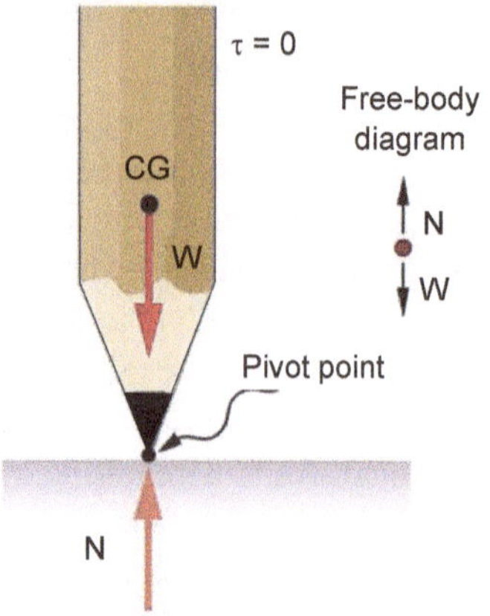

Fig.17.34: Balance a Pencil on its Sharpened End - Unstable Equilibrium, Although Both Conditions for equilibrium are Satisfied.

BACKGROUND FOR STRING THEORY

In order to understand how string theory solves the problem of supersymmetry's spiraling graviton, we first need some background on string theory. Veneziano's formula is an approach that successfully describes the probabilities for the pattern produced when two protons collide at high energy by treating particles as strings that stretch when they gain energy and give up energy when they contract, like a rubber band. The various states of vibration of these strings correspond neatly to the various kinds of particles produced in the proton-smashing experiment.

With enough inventive design, the theory can produce all the particles and forces of the Standard Model. String theory holds that the ends of an open string are charged particles. For example, an electron could be on one end of a string and its antimatter counterpart, the positron, on the other end. The massless vibration of the string separating them describes the photon which carries the electrical force between them. When the two ends of the string come together, the ends go away, the photon is released, and a closed loop of string is left behind. This disappearance of particles corresponds to the annihilation that occurs when an electron and positron meet, creating a photon in their place.

According to string theory, photons come from vibrations of either open or closed strings, while gravitons come only from vibrations of closed loops. The difference between gravity and the other three fundamental forces is naturally explained as the difference between open and closed strings. 'For the first time, gravity plays a central role in the unification of the forces.' Indeed, a unification of force and motion arises from string theory as well, in that the law of motion dictates the laws of the forces because all forces come from the breaking and joining of strings.

In string theory there are only two fundamental constants:

a. string tension (energy per unit length) and

b. the string coupling constant (the probability of a string breaking into two strings, giving rise to a force).

A string's coupling constant is fixed by its multidimensional environment rather than being fixed by the theory. This is an important aspect of string theory, namely that constants migrate from being arbitrarily fixed properties of the theory to being properties of the environment.

Another attraction of strings is that they have one underlying law that unifies all their properties. Strings move so as to minimize the two-dimensional surface area which their one-dimensional line draws out as it moves through time, very similar to the way a soap bubble's shape is the result of its surface taking up the minimal area it can.

Fig.17.35: Balancing Supersymmetry – Curtsy - Illustration by Sandbox Studio, Chicago

Figure 3. The validity of the Standard Model of particle physics is balanced on the existence of the Higgs particle which in turn balances on supersymmetry to avoid extreme fine-tuning to keep the Higgs mass low. Supersymmetry balances on string theory to save it from its spiralling graviton. Likewise, string theory balances on the Landscape concept which balances on faith in exceedingly large numbers of possible universes to make the highly improbable fine-tuning of the observed cosmological constant a mathematical certainty.

Advantageously, string theory unifies the Standard Model and gravity because all particles and forces arise from the vibrations of strings stretched in spacetime following the simple law that the area be minimized.

HOW STRING THEORY SAVES SUPERSYMMETRY

In string theory, particles of matter can change into particles of force. Part of a string can pinch off, creating a new particle, travel through space, and get absorbed by another string. In this way forces are exerted between strings. The strengths of the four fundamental forces are determined by a string's in and out pulsations, called 'dilation'.

A closed string can vibrate by expanding in one direction and contracting in the perpendicular direction, then vice versa. Interestingly, this is the way in which gravitational waves vibrate. Therefore, theorists conclude that a string vibrating like this must be a graviton. These graviton loops can split, and split again, but the splitting has a limit in that it can only continue until the loops are too small to divide further. Being a one-dimensional string keeps the graviton from becoming infinitely small as it would be if it were a dimensionless particle. Thus, string theory shields supersymmetry's graviton from spiraling into infinity. This solution has problems of its own, though.

A Problem for Supersymmetry's String Theory Solution

The way string theory saves supersymmetry from its spiraling graviton is not without an unsettling snag of its own. The difficulty is that supersymmetric string theory does not allow the gravitons to reshape the spacetime around them.

This violates Einstein's essential idea that the geometry of spacetime is dynamical and evolving. Although superstring theory recovers all the solutions to general relativity in which some dimensions are flat and others are curled up, these are very special cases. A quantum theory of gravity should describe how different shapes of spacetime change into each other.

'String theory provides a series of snapshots, but ultimately theorists would like a movie.' Thus, string theory cannot be a theory of gravity since many gravitational phenomena involve time dependence.

LET US RECAP

Let's recap what we have so far. The Higgs field is the theorists' tool that saves the Standard Model by giving the elementary particles their mass. Supersymmetry, in turn, rescues the Higgs from its hierarchy problem by proposing sparticles that cancel out quantum effects on the mass that would otherwise require extreme fine-tuning to keep the Higgs mass low.

Next, we need string theory to salvage supersymmetry by keeping its graviton from spiraling into infinity. So, what saves string theory?

'I didn't know string theory was in trouble', you say. Oh, it is in big trouble, and it takes a theory as large as the vast Landscape to rescue it.

BACKGROUND FOR THE COSMOLOGICAL CONSTANT 'Λ'

The cosmological constant 'Λ'is thought to represent an energy that accelerates the universe's expansion. Quantum theory appears to require a huge cosmological constant. This is because at absolute zero temperature, when a particle is exactly still, it cannot have a definite position and momentum without violating the Heisenberg uncertainty principle. Consequently, there is a small residual energy, called vacuum energy, even at absolute zero. This energy generates virtual particles that pop in and out if existence too fast to be seen individually, but collectively their effect lingers. It turns out that this vacuum energy is synonymous with the cosmological constant.

Fine-Tuning of the Cosmological Constant 'Λ'

Fields have a huge number of modes of vibration. When quantum mechanics is applied to a field, a vacuum energy exists for each of these different modes of vibration, thus quantum mechanics predicts a huge cosmological constant Λ. It cannot be that big in reality because such a large cosmological constant implies an expansion rate of the universe so fast that no structure at all could have formed in the big bang scenario.

The fact that galaxies exist puts a limit on the cosmological constant of some 120 orders of magnitude smaller than predicted by quantum theory! The cosmological constant represents a universal repulsion whose value must coincide with the acceleration rate of the universe.

Observations from type I supernovas have allowed scientists to calculate the acceleration of the universe very precisely. It turns out that the negative and positive contributions to energy density provided by the virtual particles of quantum theory cancel each other out to 119 decimal places.

If the energy density cancellation had been only an order of magnitude or two bigger, no galaxies, stars, or planets could have formed in the presumed big bang. If the cosmological constant were not extremely small, its universal repulsion would have instantly destroyed the universe.

The cosmological constant is so incredibly ***fine-tuned*** that no one could imagine it accidental. It is exactly the value that would make our universe hospitable to life.

How the Landscape Saves String Theory from a Finely-Tuned Cosmological Constant

These observations of type I supernovas came out in 1998. They appear to indicate that the expansion of the universe is accelerating, giving a positive cosmological constant. String theory until this time had concluded that the cosmological constant could only be *zero or negative*. The work of a group of theorists at Stanford University solves the problem of making string theory consistent with a positive cosmological constant, as well as the problem of stabilizing its higher dimensions, but with very bizarre consequences.

They start with a string theory that has a flat four-dimensional spacetime with a small six-dimensional geometry over each point. Wrapping a large number of electric and magnetic fluxes (which can only be wrapped in discrete units) around the compact six-dimensional spaces over each point tends to stabilize the geometry.

Next, they wrap antibranes (which are the antiparticle analogue to the two-dimensional 'branes' that string theory also predicts) around the geometry. In this way energy can be added so as to make the cosmological constant small and positive in accordance with the common interpretation of the astronomical observations.

To get a small cosmological constant, you have to wrap many fluxes, and there are many ways to wrap a flux. There is evidence for 10500 solutions to this string theory, each having different predictions for the elementary particles and the parameters of the Standard Model. The problem comes when we realize that there is no principle that selects a unique string theory, so one can get any outcome he one wants.

The term Landscape was coined by the co-discoverer of string theory, "Leonard Susskind." It denotes a mathematical space representing all the possible environments the theory allows, each one having its own laws of physics, its own elementary particles, and its own constants of nature. He sees us as living in one tiny pocket of a mega-verse where these values happen to be consistent with our kind of life.

The unlikely odds of a small but positive cosmological constant are outweighed by the gargantuan number of other possible universes. A theory with an enormous Landscape causes unfathomably improbable events to be completely inevitable. To evolutionists, the Landscape makes it a mathematical certainty that some parts of space will evolve into a universe like ours where life is possible.

THE PROBLEM WITH THE LANDSCAPE

String Theorists – Claims Many Inteligent Scientists Stupid to Understand!

String theorists claim many intelligent scientists are too stupid to understand, including the author. The Landscape notion of string theory results from a flailing attempt to explain away the fine-tuning of the cosmological constant. It is described by a mathematical solution so complicated that non-specialists often feel they are not qualified to form a responsible judgment of the situation. String theorists seem to perpetuate other scientists' feelings of inadequacy with a widespread intellectual arrogance that thinks only real geniuses are able to work on the theory, and anyone who criticizes their work is probably too simple to understand it.

CHILDLIKE – THE EMPEROR HAS NO CLOTHES

The Theory has No Science

With all of the mathematical sophistication involved in string theory, it takes childlike forthrightness to expose the true state of affairs by exclaiming the equivalent of, '***The emperor has no clothes!***' or rather, in this case, '***The theory has no science!***' Because string theory makes no predictions, it is a theory that cannot be falsified. This makes it debatable whether it can be called science at all, Figure 17.36.

The great Cal Tech physicist Richard Feynman commented on the unscientific nature of string theory with these words:

'I don't like that they're not calculating anything. I don't like that they don't check their ideas, I don't like that for anything that disagrees with an experiment, they cook up an explanation—a fix-up to say, "Well, it still might be true".'

Lee Smolin describes the Landscape's deplorable lack of vindication by commenting thus, 'If an attempt to construct a unique theory of nature leads to **10500 theories**, that approach has been reduced to absurdity.' Most string theorists do not acknowledge this "reductio ad absurdum." They turn an equally deaf ear to recent results that raise questions about whether any of these theories describe stable worlds.

This lamentable lack of objectivity causes many scientists to worry that *string theory is becoming a religion* rather than a science.

'Some physicists have joked that, at least in the United States, string theory may be able to survive by applying to the federal government for funding as a faith-based initiative!'

Fig.17.36: *The Emperor has No Clothes – String Theory Has No Science – Curtsy – Navdanya International*

It seems, then that when the science of high-energy particle physics gets boiled down to its basic ingredients, scientists are left with a simple choice, namely to ***believe in the divine Designer***, or to believe in the Landscape of 10500 possible universes. Both options are a matter of faith.

THE CONCLUSION OF THE MATTER

In their attempts to offset the implications that the fine-tunings of the Higgs mass and the cosmological constant imply, some physicists find themselves doing suspicious science. They propose one theory to bolster the shortcomings of another in a chain that ends in a belief system which falsely perceives the vast Landscape of possible universes as being a more rational explanation for apparent design than an ***intelligent designer***.

These are exciting times to be a physicist! Results coming from the Large Hadron Collider at CERN will give us all a lot to mull over in the next few years. Anything that addresses the fine-tuning problems that evolutionists try to obscure should have measurable experimental consequences in the LHC, such as the absence or presence of sparticles and the Higgs boson.

These are equally exciting times to be a creation scientist! I, too, eagerly await the results to come out of the LHC, for I see the whole endeavor as one way, in the words of Solomon, '***to seek and search out by wisdom concerning all things that are done under heaven***' (Ecclesiastes 1:13a).

The problem comes when men begin to suppress the obvious truths and prefer giving glory to the created things rather than to the Creator, as Romans 1:18–25 outlines. Instead of having a worldview that hangs by the proverbial thread, as most string theorists do, creation scientists have their worldview built on the firm foundation of Jesus Christ and His written Word.

The prophet Job encourages us in the book that bears his name to ask the beasts, the birds and the fish about their origins, '***or speak to the earth, and it will teach you … who among all these does not know that the hand of the Lord has done this?***'

So, be assured that whatever is found at the LHC, when interpreted correctly, will point to God the Creator, even His eternal power and Godhead.'

APPENDIX

a. This was the response Sky & Telescope Senior Editor Alan M. MacRobert gave to my letter-to-the-editor concerning this problem. I wrote back, clarifying, as below, that I was referring to the overall system of both 'universes', an overall system that is still bound by the Second Law. There was no response to this second letter.

b. One might suggest that the Second Law does not necessarily apply to the universe (consisting of two branes) as a whole. But the general methodology of scientists is to assume that known physics is correct until a problem is found with it (whether theoretical or observational), not to assume it is false until proven correct. The known laws of physics are 'assumed innocent until proven guilty'—so to speak. All of astronomy would be impossible if this approach were not taken. Astronomers base their conclusions about distant objects on the assumption (not always provable) that the known laws of physics apply in those distant parts of the universe. The Second Law applies in all known cases, and there is no reason from theory or observation to suppose it does not apply to the universe as a whole.

c. It is also interesting that a cosmology based on unproven physics is taken so seriously. Cosmology is a very speculative enterprise even when it is based on solid, tried-and-true physics, but the ekpyrotic model is a speculative theory based on untested physics. Imagine what the response from evolutionists would be if a creationist used this approach!

d. Supersymmetric version of string theory adds another dimension, i.e. totalling 11 and then other modifications require as many as 26 dimensions.

e. M-theory extends the strings to branes.

f. Creationists and most evolutionary cosmogonists agree that the universe had a beginning.

g. This means that string theory requires the existence of at least six extra dimensions of space, in addition to the three normal spacial dimensions and one of time that we are familiar with.

h. And many centuries before the Wright brothers flew the first airplane, or Sir Isaac Newton wondered why an apple fell down instead of up, Australian Aboriginals were using yet another force that made a thrown boomerang come back to the thrower if it missed its target. Namely the twist imparted by the maker to the blades of his boomerang. But note that not all boomerangs come back; some are designed to fly straight on, again determined by how the maker shapes his weapon.

i. An axiom is an assumption made for the purpose of argument.

REFERENCES

1. Ideas needed—The hunt for a theory of everything is going nowhere fast, *New Scientist* **188**(2529):5, 10 December 2005.
2. Baffled in Brussels, *New Scientist* 188(2529):6, 10 December 2005.

3. Sample, I., Ultimate equation is pie in the sky, says Hawking, *The Guardian*, 23 February 2004, 22 June 2006.
4. Greene, B., *The Elegant Universe: Superstrings, Hidden Dimensions and the Quest for the Ultimate Theory*, Vintage Books, New York, pp. 127–129, 1999. Return to text.
5. Naeye, R., Delving into extra dimensions, *Sky & Telescope* 105(6):43–44, 2003.
6. An Open Letter to the Scientific Community (published in *New Scientist* 22 May 21 ,(2004 February 2007; see CMI commentary, Wieland, C., Secular scientists blast the big bang: What now for naïve apologetics? *Creation* 27(2):23–25, 2005; creation.com/bigbangblast. Return to text.
7. Sarfati, J., *Refuting Compromise*, Master Books, Green Forest, USA, p. 179, 2004; If God created the universe, then who created God? *Journal of Creation* 12(1):20–22, 1998, creation.com/whomadeGod. Return to text.
8. See Hartnett, J., Has 'dark matter' really been proven? Clarifying the clamour of claims from colliding clusters, creation.com/collide, 8 September 2006. Return to text.
9. The last word, 'Strings and M-theory are based on little more than fancy maths and a grab-bag of ideas', *BBC Focus*, p. 98, May 2006. Return to text.
10. For a thorough refutation of Hugh Ross's theological and scientific claims, read *Refuting Compromise* by Jonathan Sarfati, and for a refutation of big bang teaching, see *Dismantling the Big Bang* by Alexander Williams and John Hartnett (both available from CMI).
11. See also Grigg, R., The Gospel in time and space, *Creation* 21(2):50–53, 1999. Also creation.com/timespace.
12. Logically, God must exist outside of our spacetime because He created it in the first place. However, He also interacts in our own spacetime dimension.
13. Even self-described Ross supporter, the philosopher/apologist William Lane Craig, has severely criticized Ross's teachings on this: ' … I find his attempt to construe God as existing in hyper dimensions of time and space and to interpret Christian doctrines in that light to be both philosophically and theologically unacceptable,' Hugh Ross' extra-dimensional deity: a review article, *J. Evang. Theol. Soc.* 42(2):293–304, 1999.
14. Kaku, M., Mini Black Holes and the Large Hadron Collider, Interview on The Circuit Mojo, Youtube, accessed August 12, 2008.
15. Smolin, L., The Trouble with Physics, Spin Networks, Ltd., New York, p. 69, 2007.
16. Randall, L., Warped Passages, Unraveling the Mysteries of the Universe's Hidden Dimensions, Harper Collins, New York, p. 253, 2005.
17. Woit, P., Not Even Wrong, Basic Books, New York, p. 102, 2006.
18. Musser, G., The Complete Idiot's Guide to String Theory, the Penguin Group, New York, p. 272, 2008.
19. 13. Susskind, L., The Cosmic Landscape, Little, Brown and Company, New York, pp. 20–21, 2006.
20. Hawking, S., & Mlodinow, L, *The Grand Design*, Bantam Press, London, 2010.
21. *Hawking* writes: "Quarks … are a model to explain the properties of the protons and neutrons in the nucleus of an atom" (p. 65). And " … according to model-dependent realism, quarks exist in a model that agrees with our observations of how subnuclear particles behave" (p. 66).
22. Note: Williams and Hartnett write concerning the dimensions required by string theory: "10 dimensions if we include both forces and matter, or 26 dimensions if we only deal with forces." Williams, A., & Hartnett, J., *Dismantling the Big Bang*, Master Books, 2005, p. 104.

23. Ideas needed—The hunt for a theory of everything is going nowhere fast, *New Scientist* 188(2529):5, 10 December 2005 (emphasis added).
24. Norman L. Geisler, *Miracles and the Modern Mind* (Grand Rapids, MI: Baker, 1992) p. 111.
25. Sir Roger Penrose and Stephen Hawking were jointly awarded the Wolf Foundation Prize for Physics in 1988 and also the Eddington Medal of the Royal Astronomical Society in 1975; Penrose was awarded the Albert Einstein Medal in 1990.
26. Penrose, R., review of *The Grand Design, Financial Times* (UK), 4 September 2010.
27. Davies, P., *A Brief History of the Multiverse*, The New York Times, April 12, 2003.

CARBON DATING

FUNDAMENTALS OF CARBON DATING

It is imperative to remember that the ***material must have been alive at one point to absorb the carbon,*** meaning that carbon dating of rocks or other inorganic objects is nothing more than inaccurate guesswork. All living things absorb both types of carbon; but once it dies, it will stop absorbing, Figure 18.1.

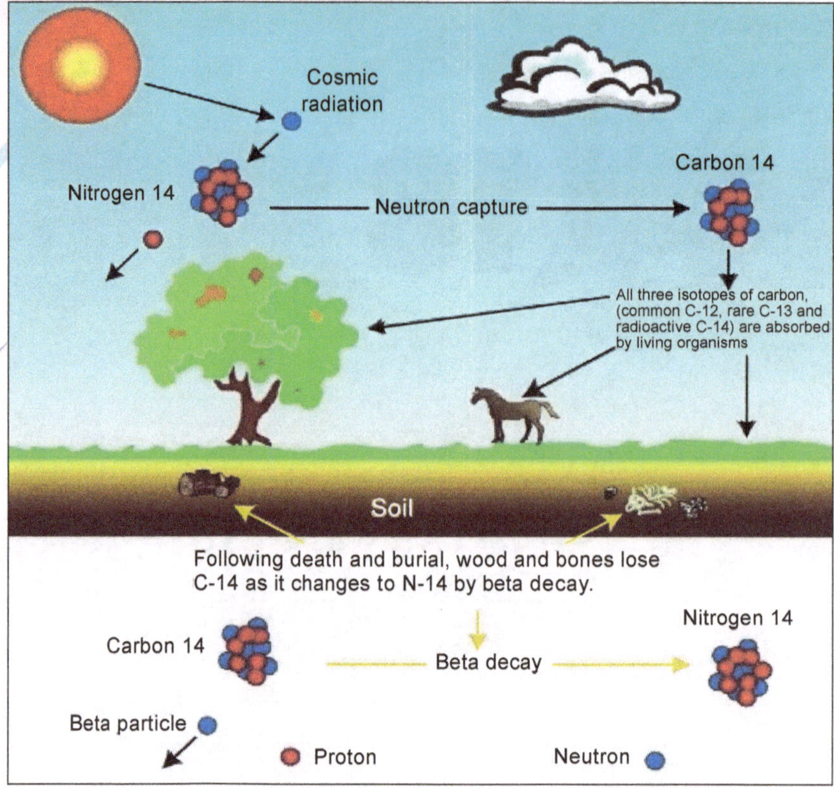

Fig.18.1: Generation of Carbon for Cabon Dating – Curtsy Quora

At least to the uninitiated, carbon dating is generally assumed to be a sure-fire way to predict the age of any organism that once lived on our planet. Without understanding the mechanics of it, we put our blind faith in the words of scientists, who assure us that carbon dating is a reliable method of determining the ages of almost everything around us. However, a little more knowledge about the exact ins and outs of carbon dating reveals that perhaps it is not quite as fool-proof a process as we may have been led to believe, Figure 18.2.

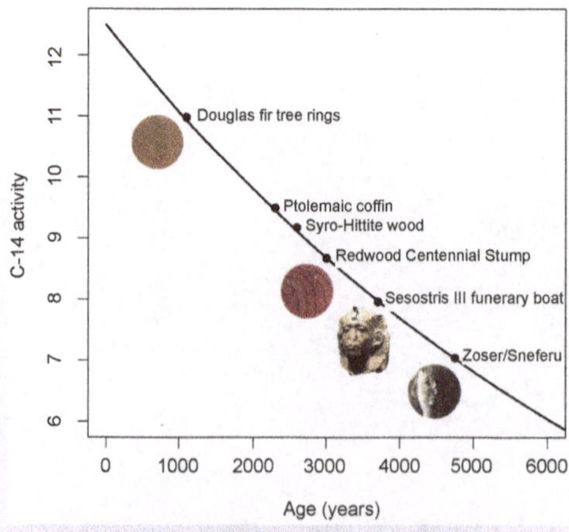

Fig.18.2: Carbon Dating: Science in the service of History – Curtsy Scientific Gems

WHAT IS CARBON DATING?

At its most basic level, carbon dating is the method of determining the age of organic material by measuring the levels of carbon found in it. Specifically, there are two types of carbon found in organic materials: carbon 12 (C-12) and carbon 14 (C-14). It is imperative to remember that the material must have been alive at one point to absorb the carbon, meaning that carbon dating of rocks or other inorganic objects is nothing more than inaccurate guesswork.

All living things absorb both types of carbon; but once it dies, it will stop absorbing. The C-12 is a very stable element and will not change form after being absorbed; however, C-14 is highly unstable and in fact will immediately begin changing after absorption. Specifically, each nucleus will lose an electron, a process which is referred to as decay. This rate of decay, thankfully, is constant, and can be easily measured in terms of 'half-life.' Figure 18.3.

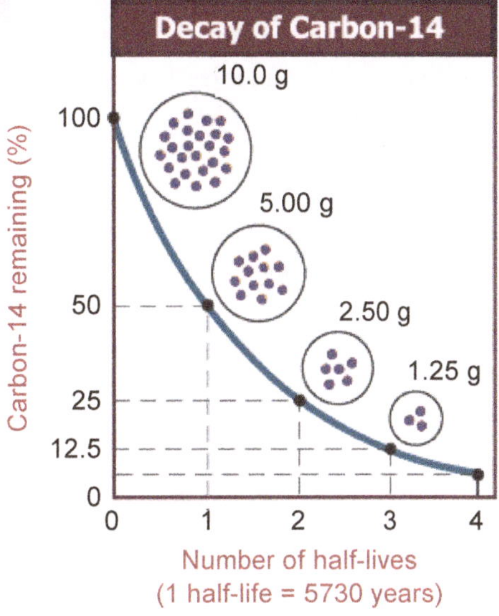

Fig.18.3: Carbon Half-Life

CARBON DECAY - HALF LIFE

Half-life refers to the amount of time it takes for an object to lose exactly half of the amount of carbon (or another element) stored in it. This half-life is very constant and will continue at the same rate forever. The half-life of carbon is **5,730 years**, which means that it will take this amount of time for it to reduce from 100g of carbon to 50g – exactly half its original amount. Similarly, it will take another 5,730 years for the amount of carbon to drop to 25g, and so on and so forth. By testing the amount of carbon stored in an object, and comparing to the original amount of carbon *believed* to have been stored at the time of death, scientists can estimate its age.

INACCURACY IN CARBON DATING

Unfortunately, the believed amount of carbon present at the time of expiration is exactly that: a belief, an assumption, an estimate. It is very difficult for scientists to know how much carbon would have originally been present; one of the ways in which they have tried to overcome this difficulty was through using carbon equilibrium.

Carbon Equilibrium

Equilibrium is the name given to the point when the rate of carbon production and carbon decay are equal. By measuring the rate of production and of decay (both eminently quantifiable), scientists were able to estimate that carbon in the atmosphere would go from zero to equilibrium in **30,000 – 50,000 years**. Since the universe is hypothetically, estimated to be millions of years old, it was assumed that this equilibrium had already been reached.

However, in the 1960s, the growth rate was found to be significantly higher than the decay rate; almost a third in fact. This indicated that equilibrium had not in fact been reached, throwing off scientists' assumptions about carbon dating. They attempted to account for this by setting 1950 as a standard year for the ratio of C-12 to C-14, and measuring subsequent findings against that.

Not Reliable

In short, the answer is… sometimes. Sometimes carbon dating will agree with other evolutionary methods of age estimation, which is great. Other times, the findings will differ slightly, at which point scientists apply so-called **'correction tables'** to amend the results and eliminate discrepancies.

Most concerning, though, is when the carbon dating directly opposes or contradicts other estimates. **At this point, the carbon dating data is simply disregarded**. It has been summed up most succinctly in the words of American neuroscience Professor Bruce Brew:

"If a C-14 date supports our theories, we put it in the main text. If it does not entirely contradict them, we put it in a footnote. And if it is completely out of date, we just drop it."

Contemporary Carbon Dating?

Essentially, this means that carbon dating, though a useful tool, **is not 100% reliable**. For example, recently science teams at the British Antarctic Survey and Reading University **unearthed the discovery** that samples of moss could be brought back to life after being frozen in ice. The kicker? That carbon dating deemed the moss to have been frozen for over 1,500 years. Now, if this carbon dating agrees with other evolutionary methods of determining age, the team could have a real discovery on their hands. Taken alone, however, the carbon dating is unreliable at best, and at worst, downright inaccurate.

CARBON DATING NOT ALWAYS ACCURATE

Scientists wish that was true, but in reality, only 50% of corpses can be dated using this method because in some skeletons there isn't enough organic material or it is contaminated. Many exciting finds have been inaccurately dated or not dated at all, meaning the skeletons' clues from the past are still locked away.

Carbon Dating is Accurate – 50%

The method's true-positive rate ranges from 20–90%, with the most realistic rates being between 30 and 50%. The method's false positive error rate is approximately 10%. Increasing the number of radiocarbon dates used to date the time-series above five had no noticeable effect on the true- or false-positive rates.

Earth is Hypothetically 4.5 Billion Years-old!

Fig.18.4: Inaccurate Carbon Dating of Earth Age

By dating the rocks in Earth's ever-changing crust, as well as the rocks in Earth's neighbors, such as the moon and visiting meteorites, scientists have calculated that Earth is 4.54 billion years old, with an error range of 50 million years. Rocks are often misleading indication to the actual age of the object. Billions of years can be a mere few thousands, Figure 18.4.

Carbon Dating Inaccurate After Far-Less than 50,000 Years

By about 58,000 years (ten half-lives) after an organism has died, there's so little radioactive carbon left (less than 1/1000) that calculations of age are no longer accurate. That's why radiocarbon dating is only reliable for samples far less than 50,000 years old. The best data if it were less than 10,000 years.

Limitations of Carbon Dating - 5,730 Years

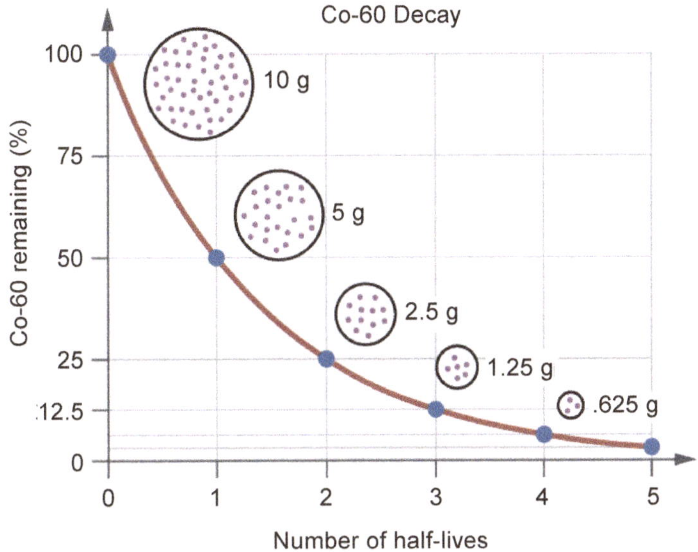

Fig.18.5: Carbon Decays by Time – Half Life 5,730 Years

The hypothetical practical upper limit is about 50,000 years, because so little C-14 remains after almost 9 half-lives (9x5,730 years) that it may be hard to detect and obtain an accurate reading, regardless of the size of the sample, Figure 18.5. Additionally, the instrument± conducting the data has an inherit design allowable inaccuracy ± %5. Accordingly, carbon dating is best if the same is around 5,730 years old. The main limitations of Radiometric Carbon Dating is that it only works on certain igneous rocks as most rocks have insufficient Re and Os or lack evolution of the isotopes. This technique is good for iron meteorites and the mineral molybdenite.

Fossil fuels can shift the radiocarbon age of new organic materials today, making them hard to distinguish from ancient ones.

Accurate Calibrated Carbon Dating – 5,000-Year-Old Tree

Since the 1960s, researchers have mainly done this recalibration with trees, counting annual rings to get calendar dates and matching those with measured radiocarbon dates. The oldest single tree for which this has been done, a bristlecone pine from California, was about 5,000 years old.

Carbon-14 Eventually Disappear

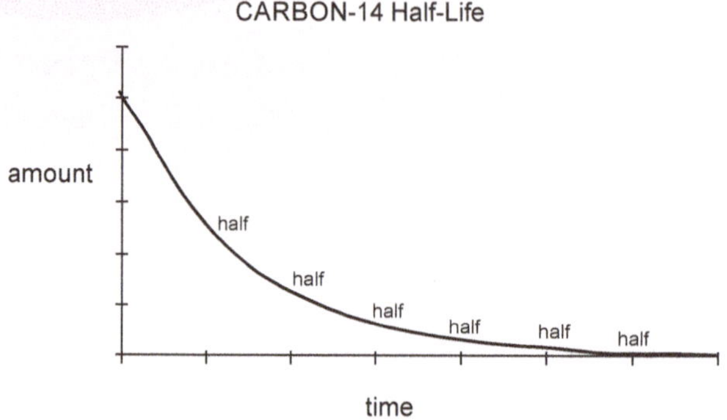

Fig.18.6: Carbon Eventually Disappear

When a plant stops assimilating carbon dioxide or when an animal or human being stops eating, the ingestion of carbon-14 also stops and the equilibrium is disrupted. From that time forward, the only process at work in the body is radioactive decay. Eventually, all the carbon-14 in the remains will disappear, Figure 18.6.

Is Carbon Dating ever Wrong?

Radiocarbon dating is a key tool archaeologists use to determine the age of plants and objects made with organic material. Nevertheless, new research shows that commonly accepted radiocarbon dating standards can miss the mark -- calling into question historical timelines.

CARBON RADIOACTIVE DATING UNRELIABLE IN MOST SITUATIONS

Radioactive dating is unreliable in most situations because most objects have multiple sources of radioactive material. For example, sedimentary rocks are built using sediments that are derived from many different rocks.

Carbon 14 Dating

Carbon-14 (^{14}C) or radiocarbon as it is often called, is a substance manufactured in the upper atmosphere by the action of cosmic rays. Ordinary nitrogen (^{14}N) is converted into ^{14}C as shown in Figure 18.7.

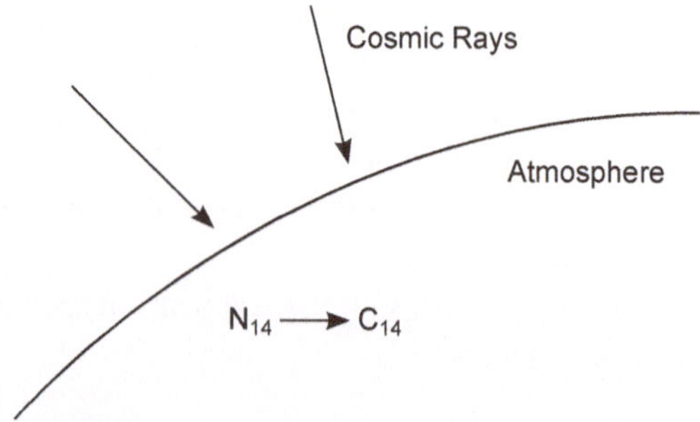

Fig.18.7: Nitrogen (^{14}N) is converted into Carbon ^{14}C

Ordinary carbon is carbon-12 (^{12}C). We find it in carbon dioxide in the air we breathe (CO_2), which of course is cycled by plants and animals throughout nature, so that your body, or the leaf of a tree, or even a piece of wooden furniture, contains carbon. When ^{14}C has been formed, it behaves just like ordinary carbon (^{12}C), combining with oxygen to give carbon dioxide ($^{14}CO_2$), and also gets freely cycled through the cells of all plants and animals.

The difference is this: once ^{14}C has been formed, it begins to decay radioactively back to ^{14}N, at a rate of change which can be measured. If we take a sample of air, and measure how many ^{12}C atoms there are for every ^{14}C atom, this is called the $^{14}C/^{12}C$ ratio. Because ^{14}C is so well 'mixed up' with the ^{12}C, we find that this ratio is the same if we sample a leaf from a tree, or a part of your body. Think of it like a teaspoon of cocoa mixed into a cake dough—after a while, the 'ratio' of cocoa to flour particles would be roughly the same no matter which part of the cake you sampled.

The fact that the ^{14}C atoms are changing back to ^{14}N doesn't matter in a living thing—because it is constantly exchanging carbon with its surroundings, the 'mixture' will be the same as in the atmosphere and in all living things.

As soon as it dies, however, the ^{14}C atoms which decay are no longer replaced by new ones from outside, so the amount of ^{14}C in that living thing gets smaller and smaller as time goes on. Another way of saying it is that the $^{14}C/^{12}C$ ratio gets smaller. In other words, we have a 'clock' which starts ticking at the moment something dies.

Obviously, this only works for things which once contained carbon—it can't be used to date rocks and minerals, for example. We know how quickly ^{14}C decays, and so it becomes possible to measure how long it has been since the plant or animal died.

However, how do we know what the $^{14}C/^{12}C$ ratio was to start with? We obviously need to know this to be able to work out at what point the 'clock' began to tick. We've seen that it would have been the same as in the atmosphere at the time the specimen died. So how do we know what that was?

Do scientists assume that it was the same as it is now? Well, not exactly. It is well known that the industrial revolution, with its burning of huge masses of coal, etc. has upset the natural carbon balance by releasing huge quantities of ^{12}C into the air, for example. Tree-ring studies can tell us what the $^{14}C/^{12}C$ ratio was like before the industrial revolution, and all radiocarbon dating is made with this in mind. How do we know what the ratio was before then, though, say thousands of years ago?

It is assumed that the ratio has been constant for a very long time before the industrial revolution. Is this assumption correct? (For on it hangs the whole validity of the system.)

Why did "W.F. Libby," the brilliant discoverer of this system, assume this?

Libby knew that ^{14}C was entering and leaving the atmosphere (and hence the carbon cycle). Because Libby believed that the Earth was *millions of years old*, he assumed that there had been plenty of time for the system to be in *equilibrium*. This means that he thought that ^{14}C was entering the atmosphere as fast as it was leaving—calculations show that this should take place in about 30,000 years, and of course the Earth was much older than that, said the geologists.

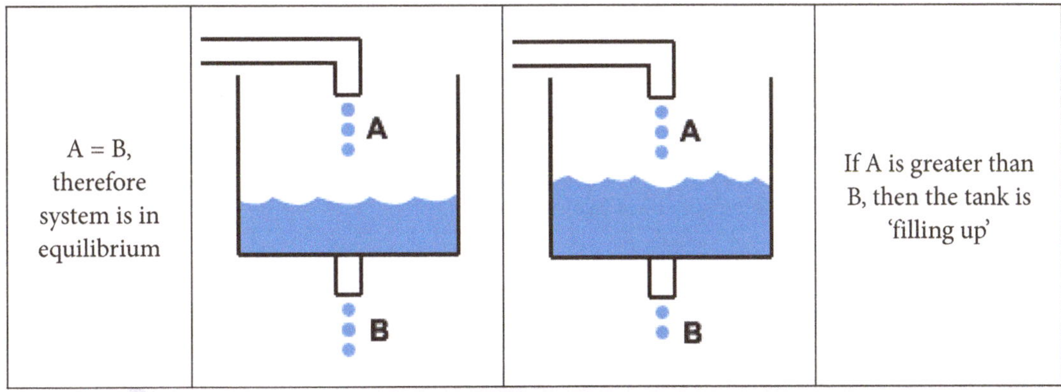

Fig.18.8: Carbon Equilibrium

Imagine a tank with water, Figure 18.8, flowing in at a certain rate, and flowing out again at the same rate. This system is in equilibrium. If you saw it for the first time, you wouldn't be able to work out how old it was—how long it had been since it was 'switched on'.

Was Libby right in this assumption? Was the ^{14}C entering and leaving the system at the same rate?

In his day, the measurements and calculations, which he knew about, showed that ^{14}C was entering the system some 12–20% *faster* than it was leaving. Imagine the same tank, this time it is not yet full and the top tap is flowing more quickly than the bottom one is leaking out—this gives you a way of measuring how long ago the whole system was 'switched on' and it also tells you that that can't have been too long ago, Figure 18.8.

Libby knew that if these figures were correct, it would mean that the atmosphere was young, so he dismissed the results as being due to experimental error! (We are not implying dishonesty here, merely showing how powerfully the evolutionary/uniformitarian concepts of Earth history influence great scientists to mold or discard evidence which appears to contradict that viewpoint.)

What about modern measurements, using advanced technology such as satellites? Unfortunately for the 'old-Earth' advocates, the studies of such renowned atmospheric physicists as "Suess and Lingenfelter" show that ^{14}C is entering the system some 30–32% *faster* than it is leaving it. The model of radiocarbon dating which Libby developed, using his incorrect 'uniform' assumption, must therefore be corrected to fit the *facts* about ^{14}C—let us call the new, corrected model the 'non-uniform' model.

What does this Mean?

It implies that if the ^{14}C is still 'building up', we can calculate how old the whole system is—this puts an upper limit on the age of the atmosphere of some 7 to 10,000 years. Also, it means that a thousand years ago, the $^{14}C/^{12}C$ ratio in the atmosphere was less than today (because the ^{14}C was still building up).

Therefore, a specimen which died a thousand years ago will show an older age than its true age. Two thousand years ago, specimens would have still less ^{14}C to start with, so they have an even greater error. In other words, the further you go back, the more you have to shrink the radiocarbon dates to make them fit the facts. Remember that this correction is based on *measurable scientific data*, not on any creation scientist preconceptions.

We need to consider two other effects:

1. If, as many creation scientists propose, there was a vast water vapor canopy around the Earth before the Flood, then this would have shielded the atmosphere from much of the cosmic radiation. Therefore, the amount of ^{14}C in the pre-flood world would have been very small, perhaps even negligible. So, a specimen from before the flood would appear to be 'very old' or even on 'infinite' age because it had so little ^{14}C in it, making it look as if it had been decaying for tens of thousands of years - Noah' Flood:
2. The measured exponential decay of the Earth's magnetic field, as described by Dr Thomas Barnes suggests that as you go back in history, the strength of the field increases rapidly. A stronger magnetic field would mean more protection against cosmic rays, therefore again much less ^{14}C produced and again this gives artificially 'old' ages the more you go back in time.

In summary then:

1. The ^{14}C in the atmosphere is not in equilibrium, but is building up rapidly.
2. This seems to put an upper limit on the age of the whole system in the order of 10,000 years.
3. All radiocarbon dates have to be adjusted from the obviously incorrect 'uniform' model which is still in use today, and when this is done there is a shrinking in these dates. The older the date, the greater the reduction.

4. The protective water vapor canopy and the greater magnetic field before the flood would decrease ^{14}C levels in the past, causing greatly exaggerated ^{14}C 'ages.'

In any case, even the incorrect 'uniform' model has given, in many cases, serious embarrassment to evolutionists by giving ages which are much younger than those he expects in terms of his model of earth history. Consider this—if a specimen is older than 50,000 years, it has been calculated, it would have such a small amount of ^{14}C that for practical purposes it would show an 'infinite' radiocarbon age. So, it was expected that most deposits such as coal, gas, petrified trees, etc. would be un-dateable by this method. In fact, of 15,000 dates in the journal *Radiocarbon* to 1968, only three were classed 'un-dateable'—most were of the sort which should have been in this category. This is especially remarkable with samples of coal and gas supposedly produced in the carboniferous 100 million years ago! Some examples of dates which contradict orthodox (evolutionary) views:

- Coal from Russia from the 'Pennsylvanian', supposedly 330 million years old, dated at 1,680 years. [Incorrect—this particular sample was ***charcoal*** from Kyrgyzstan, not coal from Russia. But numerous instances of carbon-14 in coal have been reliably recorded.
- Natural gas from Alabama and Mississippi (Cretaceous and Eocene, respectively)—should have been 50 to 135 million years old. ^{14}C gave dates of 30,000 and 34,000, respectively.
- Bones of a sabretooth tiger from the LaBrea tar pits, supposedly 100,000 years old, gave a date of 28,000, Figure 18.9.
- A block of wood from the Cretaceous (supposedly more than 70 million years old) found encased in a block of Cambrian rock (hundreds of millions of years earlier), gave a date of 4,000 years.

Fig.18.9: A Block of Wood from the Cretaceous

Please remember that all these dates are using the incorrect 'uniform' model.

A question which could be asked after all this is: does radio-carbon, adjusted to fit the 'non-uniform' model, give any independent evidence of a worldwide catastrophe such as the Flood?

Certainly, if there was such a Flood, as we maintain from several other lines of evidence and reasoning, most living things would have perished, and so we would expect a 'cut-off' point at this time.

In other words, going into the past, we should reach a period of time in which there is a sharp reduction in the number of specimens compared to the period just older than that, and as we went forward in time, we would expect a gradual buildup, as plant and animal populations recovered their numbers, Figure 18.10.

Such a study has been done by Dr "Robert Whitelaw." Using the 15,000 published dates previously mentioned after adjusting them as described, he grouped them into 500-year 'blocks' and found a dramatic drop-off about 5,000 years ago, with a worldwide distribution. Note that the data presented does not necessarily endorse a particular age for the Earth, but reveals a pattern *consistent* with a recent creation and global flood model.

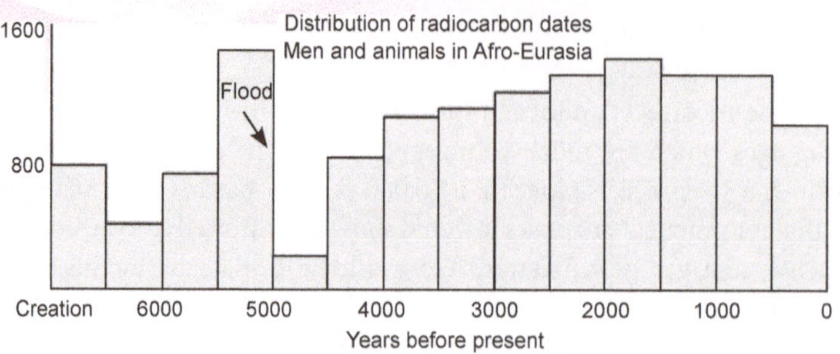

Fig.18.10: *Distribution of Radio Carbon Before and After Moah's Flood*

We see, then, that far from being an embarrassment to the creation scientists who believes in a young Earth, the radiocarbon method of dating—when fully understood in accordance with modern atmospheric data—gives powerful support to his position.

STRANGE DATA OF CARBON DATING

Historical Examples

King Richard III

Fig.18.11: *Skeleton of King Richard III.*

Note the extreme curvature in the mid-portion of his spine, Figure 18.11. Writing over a century later, Shakespeare incorrectly describes him as a hunchback, and he adds that he had a withered arm and walked with a limp, for which there is no evidence. After his mount in killed in battle, Shakespeare has him crying, "A horse, a horse, my kingdom for a horse!" (*Richard III*, Act V, Scene iv, line 13), but Richard would have been very uncomfortable on horseback.

Buried under a carpark not far from CMI's UK office, the bones of the last Plantagenet King, Richard III (1452–1485) lay in peace. The monastery church where he was hurriedly laid to rest after his death in the Battle of Bosworth Field has long since been forgotten as the growing city of Leicester engulfed the land. Some enterprising historical sleuths realized where his grave should be, and there he was, not near the altar in the Grey Friar's Priory, but under parking spot 'R'.

He had an extreme case of scoliosis, mirroring Shakespeare's claim that he was a hunchback, and they found parasite eggs in the soil around where his innards should have been—parasites that are contracted from eating undercooked beef, and most commoners could not afford that. He has also been stabbed, multiple times, and his skull had been perforated. Clearly, this person died in battle. Every datum was suggesting that this was indeed Richard, except for the carbon date. It was off by decades.

Upon reflection, they realized that, since he was a meat-eating king, they needed to apply a different historical model that they would to a chick-pea–eating peasant. Voilà! His re-dated remains fell right where they should.

This is the state of the art in carbon dating. Yes, it works, but this illustration clearly illustrates that a 'date' is not independent from the many assumptions that are used to obtain it.

Syphilis in Europe

Fig.18.12: Manuscripts and art support archaeological evidence that syphilis was in Europe long before explorers – Curtsy The Cleveland Museum of Art

A debate has raged for centuries. Did syphilis originate in Europe or in the New World? In other words, did the rapacious Europeans give it to Native Americans or did the Europeans get it from the "Noble Savage" (to use the racist terminology of the era), Figure 18.12.

It should be easy to resolve the debate, for syphilis is a horrible disease that leaves a person disfigured. In fact, the bones of syphilitic skeletons often have characteristic lesions, as if the bacterium (*Treponema pallidum*) was turning the person's skeleton into Swiss cheese, and/or thickened bones. Different strains of *T. pallidum* cause yaws, bejel, and other diseases of varying severity. It is not always sexually transmitted.

When some syphilitic skeletal remains were discovered in Portugal several years ago, they thought the debate could be settled. The carbon dates placed them prior to 1492. In other words, syphilis was in Europe before the New World was discovered. Some scientists cried foul, however. These people lived on the coast, in a fishing village. They would have eaten a great deal of seafood and, due to the marine reservoir effect, their carbon dates would be skewed downward. They re-calibrated the remains based on a seafood diet and they came out *post*-1492. Syphilis came from the New World after all?

A few years later, skeletal remains from the 1300s were found under the floor of church in England. They had the characteristic lesions of syphilitic patients. Maybe it was a world-wide phenomenon? Or perhaps the disease was brought to England via the Vikings, who had already made contact with the Americas. Further studies have yielded a diverse assemblage of ancient *T. pallidum* genomes, including ones associated with syphilis, yaws, and bejel from across northern Europe. The carbon dates for these samples are not necessarily post-Columbian, but the diversity of bacterial strains discovered certainly suggests it had been in Europe for a long time. In the end, carbon dating did not help resolve the issue and the question has yet to be answered.

The Shroud of Turin

The Shroud of Turin is a famous medieval artifact that displays a ghostly image of the face and body of a man. This is a controversial subject and not every question about it can be answered here, Figure 18.13. However, the Shroud was carbon dated in 1988. In fact, four samples were removed and sent to three AMS labs (in the US, UK, and Switzerland) for independent testing. Three control samples were also included:

1. a piece of linen from a Nubian tomb dating to the 11th–12th centuries BC,
2. a linen cloth from a mummy of Cleopatra of Thebes from the early second century BC, and
3. threads removed from the cope (a type of coat) of St. Louis d'Anjou from the Basilica of Saint-Maximin, France from the turn of the 13th century ad.

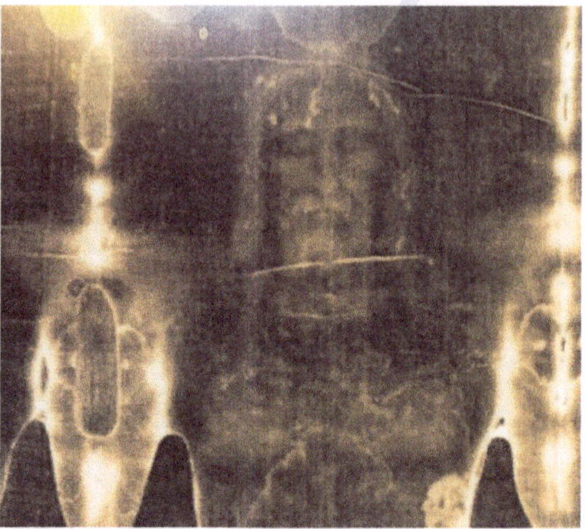

Fig.18.13: What is the Shroud of Turin and is it evidence of Jesus

After applying strict calibration methods, the radiocarbon date of the Shroud ranged from 1260 to 1390 AD.

The main counterclaim is that the cloth had been contaminated by modern handling, by soot from medieval candles, by water, and by fire. However, consider what it would take to bring a carbon date forward by *thirteen centuries*. Datable objects from the time of Christ should have about 78 pMC (percent modern carbon). Items from 1300 AD would have about 92 pMC. At least 1/3 of the sample would have to be 'contamination', yet the cloth, even if it is dirty and stained, is clearly the original cloth without massive amounts of clotted wax, soot, or dirt.

A second major claim is that the samples were taken from a section of the Shroud that had been repaired. In other words, the material was not original. However, this would require something called 'invisible reweaving' with threads that had the same color and texture as material that was more than 1,000 years older. Instead, the 'repair', no matter how expertly done, should have been performed by sewing on a patch, and this should be obvious to any careful observer. The sample site has no appearance of being anything but original threads.

We cannot just wish away carbon dating. It is good science. The techniques are sound. Errors can happen, yes, but most of those are easily explainable. Also, the possible dates of the Shroud (e.g., ~33 vs ~1300 AD) are well within the range where we have *thousands* of other samples that have been dated. The -13century difference is too much to simply claim 'error.'

THE DESTRUCTION OF JERICHO

Fig.18.14: *The ancient stone retaining wall at Jericho that Joshua marched around. Archaeologists found mudbricks piled up at the base of these stones, which fell from the former free-standing walls above. This allowed the Israelites to attack "every man straight before him" (Joshua 6:20) – Curtsy Deror avi, Wikimedia commons*

According to the archaeological record, the ancient city of Jericho was destroyed in an unusual way. The walls of the city fell outward, creating a slope that an invading army could easily climb Figure 18.14. One section of the wall, however, did not fall. Strangely, the city was not looted (at least of food and other non-precious goods) afterward, and it was deliberately burned by fire. With the exception of one small edifice, it was another 500 years before any rebuilding happened at the site. Those are the archaeological details.

Compare the evidence above to the biblical details in the book of Joshua chapters 2 through 6. The Bible claims the walls fell outward, but that Rahab's house on the wall was still standing (Joshua 2:15–21), that they deliberately burned the city, and that the perishables were left in place, with only gold and silver objects being taken for use in the Tabernacle. Note the curse Joshua laid, "Cursed before the Lord be the man who rises up and rebuilds this city, Jericho. At the cost of his firstborn shall he lay its foundation, and at the cost of his youngest son shall he set up its gates." (Joshua 6:26). With the exception of a brief occupation by the Moabite king Eglon, which was ended in gruesome fashion by Ehud (Judges 3:12–30), it was not until the time of the idolatrous Israelite King Ahab that Jericho was rebuilt, and Joshua's curse was fulfilled (1 Kings 16:34).

Everything about ancient Jericho supports the Bible story. Yet, the carbon dates are off by several centuries. Is the Bible in error? No! The carbon dates need to be adjusted.

EGYPTIAN HISTORY

Contrary to public opinion, there is no consistent timeline of Egyptian history. They left behind a lot of amazing things (e.g., the pyramids of Giza), but they did not give us many easily datable historical records. The best records for Egyptian history are the king lists of Manetho, who was writing after the time of Alexander the Great. He included many reign lengths and total dynasty lengths, but we don't have his original writings, only fragments that were recorded by later writers, and these disagree with each other in major ways.

There is also a list of kings on the walls of a temple dedicated to Seti I (19th Dynasty) at Abydos, Figure 18.15. This is an incomplete list, but there are 61 kings listed. Several kings that are left off other lists (e.g., that of Manetho) are included and other kings that we know existed (e.g., Akhenaten, Tutankhamen, Queen Hatshepsut, the Hyksos kings, and others) are skipped. There are also no dates associated with the list of names.

When you examine Egyptian history carefully, you will see well-attested periods interspersed with periods of complete confusion. Some "dynasties" have but a single pharaoh listed; others have an impossible number of pharaohs in a short amount of time. The late periods (e.g., the Ptolemaic and New Kingdom periods) have the most consistent and detailed histories.

They also have the greatest number of artefacts that can be carbon dated. The dates are probably off, but not by centuries. Instead, it is the earliest periods where most of the problems lie.

Fig.18.15: *King Seti I Facts & Names - Pharaoh Seti I Family Tree - Seti I Mummy & Death – Curtsy - Copyright: https://www.tripsinegypt.com*

Because of these and other difficulties, Egyptologists turned to carbon dating. This gave them a general scheme where the major pieces could be placed in order, but a clear pattern emerged where the carbon dates were consistently older than the dates derived from archaeology. This has caused much contention. Carbon dating the most recent periods of Egypt has shown to be relatively accurate to within tens of years to a couple of hundred years from known dates.

However, carbon dating the older pre-dynastic and Old Kingdom artefacts have shown the greatest discrepancies. As these are from the earliest post-Flood periods, it is expected that they would vary greatly compared to the biblical timeline.

Consider all that was written above. The effects of the Flood and the decaying magnetic field of the earth would combine to magnify carbon dates as one goes back in time. Egypt was clearly settled soon after the Flood, so the dates of the earliest remains would be magnified the most. Currently, a biblical re-calibration curve for carbon dates does not exist so we can only talk in generalities. However, we can accept the *general order* of the major events in Egyptian history. The earliest events need to be brought forward in time, the middle dates need to be adjusted a little, and the latter dates do not need to change much at all. In fact, carbon dating is very accurate over the last 2,000 years.

Is carbon dating a good argument for biblical creation scientists? Yes! Carbon dating is a great confirmation that the earth is young. It is an indisputable fact carbon-14 can be found in carbon-containing objects that are claimed to be millions, even billions, of years old. Yet, carbon-14 cannot last that long. Period.

There are challenges, however. Some of those deal with unknown calibration points deep in history, which will be strongly influenced by the after-effects of Noah's Flood. Other challenges deal with claims of contamination (which is not possible in a diamond), machine error (which would then question *all* carbon dates), or improbable physical settings (e.g., a diamond that is being bombarded with neutrons from uranium, but no uranium is present in the sample).

We don't have to reject all carbon-14 dates out of hand, but we do have to carefully weigh the strengths and weaknesses of the method. Carbon dating is still a challenge to biblical archaeology and a biblical recalibration curve is desperately needed. On the other hand, we can rest on the fact that ^{14}C found in diamonds, coal, and dinosaur bones is loudly proclaiming the earth is young and we have good reason to believe the 'oldest' carbon dates are artificially inflated due to the events surrounding Noah's Flood, Figure 18.16.

Fig.18.16: Carbon Dates are Artificially Inflated Due to the Events Surrounding Noah's Flood – Curtsy Gregory/Shutterstock

RESTRICTIONS OF CARBON DATING

How Carbon Dating Works

"What about carbon dating?" is a question many people are often asked. In fact, it is such a common question among scholars as well as the public. We have written about how carbon dating of 'ancient' things like diamonds is a strong indication that earth is young. Yet, people are still confused about carbon dating, so we have decided to pour every pertinent knowledge of Carbon Dating in this chapter.

This chapter, will provide you an explanation of what carbon dating is, how the tests are done, and how it impacts our understanding of the Bible.

Many people use 'carbon dating' as shorthand for all radiometric dating techniques. One of my favorite questions to ask after getting the inevitable carbon dating question is, "Do you mean 'carbon dating' or radiometric dating in general, like uranium to lead or potassium-40 to argon-40?"

The look on their face is usually one of confusion as the questioner suddenly realizes they did not even realize what they were asking. Each of the other techniques has its own answers, but as far as carbon dating is concerned, it is a wonderful tool for biblical creation scientist. It gives us solid evidence that the earth is *young* (if we can really call such a great age as ~6,000 years "young"). Before we draw that conclusion, however, we have a few things to learn.

RADIOACTIVE DECAY RATES AND HALF-LIFE

Carbon-14 cannot be used to date very ancient things. As Dr Jonathan Sarfati pointed out in a documentary '*Evolution's Achilles' Heels*,' if the entire earth were composed of carbon-14, it would completely vanish in less than a million years.

To understand why, one needs to understand the concept of a *half-life*. This is simply the amount of time it takes for ½ of a radioactive material to break down. If the earth were a giant ball of pure carbon-14, after only one half-life (5,700 ± 30 years) fully one-half of all the atoms would have already decayed (into nitrogen-14 atoms). After less than 12,000 years, only ¼ of the earth would remain. There are about 1×10^{50} atoms within our planet. If, instead of being mostly iron, these were all carbon-14, and if half of those disappear every 5,700 years, there would not be a single carbon-14 atom left before one million years had passed. For this reason, carbon dating can only be used on *recent* material, and it cannot be used for estimating ages in the millions- or billions-of-years range. Therefore, carbon dating does not prove the earth is old.

THE INVENTION OF CARBON DATING

The carbon dating technique was invented by the American scientist "Willard Libby" in the 1940s. He received the 1960 Nobel Prize in Chemistry for his pioneering work. Modern carbon dating does not use his original method, but it is worth describing. Essentially, by placing a carbon sample in a radiation-shielded box, you can use a scintillation counter to measure the decay of carbon-14 atoms in the sample.

Most people are familiar with the hand-held Geiger counter and its characteristic 'clicking' sound. This is a scintillation counter. Libby's method involved counting the beta particles (high energy electrons) emanating from a sample of carbon as the carbon-14 in it decayed into nitrogen-14.

Since carbon-14 has a short half-life, the amount of carbon-14 in a sample should diminish rapidly. For example, a sample from 3,700 BC (one half-life ago) should have about half the carbon14- as a sample from the modern era. At least in theory, for as we shall see, there are multiple issues with a date like this. For starters, that would place the sample before Noah's Flood!

As expected, Libby found that younger samples produced more beta particles than older samples, and by correlating the numbers with samples of known age, he could make an educated guess about the age of undated samples. Today, most carbon dating laboratories use an accelerator mass spectrometer (AMS), which will be described below. These came into popular use in the 1980s, Figure 18.17.

Fig.18.17: An accelerator Mass Spectrometer at Lawrence Livermore National Laboratory.

BIBLICAL IMPLICATIONS

This discovery had immediate and strong implications for Bible believers. Libby published the first carbon-14 dates for historical artifacts in 1949, and they seemed to invalidate the biblical timeline. He dated sarcophagus lids from the tombs of Djoser (the first pharaoh of Egypt's 3rd Dynasty) and Sneferu, Figure 18.18 - (the first pharaoh of Egypt's 4th Dynasty). They were 'carbon dated' to 2,800 BC ± 250 years. Clearly, ***the pyramids could not have been built before the Flood***. Something was wrong.

Fig.18.18: *King Seneferu - Egypt the 4th dynasty*

Later work re-dated the Egyptian pre-dynastic period, using carbon dating to move it forward by over 400 years, but this was not nearly enough.

In the 1950s, Kathleen Kenyon used the newly invented technique to date the destruction layer at Jericho. She concluded that Jericho was *not* destroyed by the Israelites. The carbon dates placed the destruction several centuries prior to the Exodus. Of course, academics argue strenuously about the date of the Exodus, and these early carbon dates would be skewed by the lingering effects of the biblical Flood and by earth's weakening magnetic field, so perhaps Kenyon went too far out on a limb. Yet, the seeds for doubting biblical history were sown. They would germinate quickly as the last vestiges of biblical thought were stripped from the field of archaeology.

If you were to stop here, and many do, you would conclude that carbon dating invalidates the biblical timeline of history. However, this would be a gross error. Understood correctly, and when a biblical timeline is applied, carbon dating is great evidence that the earth is not millions of years old. Can we now use it to show that the earth is as young as the Bible claims? Absolutely!

Is Carbon Dating Accurate?

Generally, carbon dating is relatively accurate. We get consistent results and can often place an artifact into its proper historical context, plus or minus a few years, a few decades, or a few centuries, depending on the age of the sample. Thus, the question is not about the accuracy of the machines but about the ***assumptions*** behind the techniques. *Carbon dating is dependent on the environmental reference where the condition of the samples existed prior to its existence or death.* Inaccurate environment can lead to substantial error in tens of thousands, hundreds, or even in millions, as presented in the historical data stated previously. Our questions concern the oldest samples, not the more recent ones. In fact, carbon dating works very well on objects from the past 2,000 years.

How are the Dates Obtained?

The science of carbon dating depends on a standardized calibration curve. This is often made by comparing carbon dates for tree rings. Yet, there are several time periods where the curve is essentially flat, meaning the 'date' could fall within a range of values. In other places, the calibration line wiggles up and down, meaning a 'date' could be any of several possible values. This does not affect the overall conclusion that older samples have less carbon-14 than younger samples, but it does let us know the state of the art.

There are no historical artifacts that can be precisely dated beyond a few thousand years before Christ, and even then, archaeologists argue strenuously about the exact date for many of these. There are, however, physical and biological records that can be used. Specifically, even if no actual tree has lived for tens of thousands of years, tree rings, Figure 18.19, can be correlated from one dead tree to another. Using this statistical approach, scientists have built a putative historical record going back about ~13,000 years. commons.wikimedia.org

Fig.18.19: Because the required amount of material for carbon dating is now so small, we can sometimes 'date' individual tree rings. This has led to one of the most significant advances in carbon dating historical artefacts.

The science of deducing history from tree rings is called dendrochronology, but this is not as straightforward as many believe. First, bristlecone pines can lay down more than one ring per year. Living in the marginal environment of the White Mountains of California, the trees tend to grow whenever the weather conditions are optimal. Also, there is not one, single tree that gives us a complete date range. Instead, the rings from living and dead trees are compared with statistical measures to give 'best fit' approximations.

This is even true of the highly esteemed Irish oak tree ring record, which was created from trees found in bogs, archaeological wood samples, and beams found in old buildings. The Dendrochronology Laboratory at Queen's University Belfast has kindly made all their data publicly available so that anyone could build their own dendrochronology. On the data portal page, they wrote:

"Doing so involves making decisions about which ring patterns to include, and is thus an intellectual exercise. Other solutions may exist from the same basic collection of ring patterns, depending on choices made by the analyst, which can then be interpreted as seen fit."

Being that the tree ring record is such a critical tool for carbon dating ancient artefacts, this is an important consideration. Another problem is that tree ring alignments from locations as close as Northern Ireland, southwest England, and southern Germany disagree strongly.

Likewise, they can use the density bands in dead coral skeletons Figure 18.19, to model historical changes in carbon-14, but due to the marine reservoir effect and isotopic fractionation, both phenomena are described below, this is problematic.

Additional historical and physical models must be applied to the results of any carbon-14 measurement to obtain an estimated date. Also, coral growth can be quite irregular. If you take a core out of any particular colony, you might encounter periods of die back and re-growth or times when one lobe of the colony had overgrown another.

In the end, to build a long-term record of coral banding, scientists are forced to correlate the bands from multiple colonies, many of which died a long time ago. This is identical to the situation with tree rings, Figure 18.20. One cannot simply 'date' a coral skeleton.

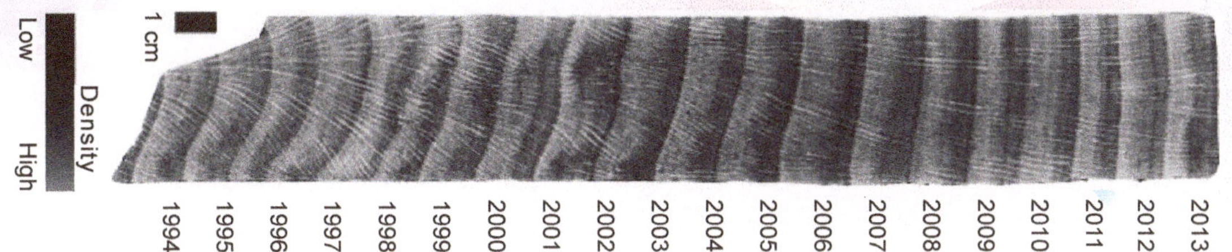

Fig.18.20: Banding in coral skeletons can also be used to estimate historic levels of carbon14- in the atmosphere, but since the oceans and atmosphere do not mix freely, additional historical and physical models must be applied to any dates obtained.

A third measure involves the thin mud layers (called varves) that form at the bottom of certain lakes. We can question if these are truly annual bands, but there is no question that there are many thousands of varves in places like Lake Suigetsu in Japan, Figure 18.21). This is a shallow lake with an outlet that leads to another lake that is then connected to the ocean. Only very fine silty particles make it into the lake, which then settle out seasonally. And even if the annual bands are not always possible to distinguish, multiple cores have been taken and the banding patterns matched across the cores. This has allowed scientists to (theoretically) piece together a timeline of about 70,000 annual layers. By carbon dating organic material (e.g., leaves) trapped between the layers, they can estimate how much carbon-14 was in the air at specific points in time. They can also see evidence of historic earthquakes (which disturb and mix the surface layers and bring in more sediment than average), periodic floods (which deposit thicker bands), volcanic eruptions (which deposit ash), and westerly winds (which blow in yellow dust from the Gobi Desert of China in the fall and winter months).

Fig.18.21: The ~70,000 varves in Lake Suigetsu in Japan have been used to create a carbon clock spanning the past 70,000 years, or so it is claimed. Curtsy - commons.wikimedia.org, Phonon.b

Please observe, however, that this depends on a very specific set of parameters. The modern lake is only 34 m (111 ft) deep. There are 45 m of well-developed varves and an additional 30 m of mostly non-varved mud beneath that. Even at an estimated deposition rate of 0.7 mm per year, the lake floor must have subsided in reference to the surrounding hills to keep apace of sediment buildup, all while not affecting water influx from the shallow channels that connect the five neighboring lakes. The bottom of the lake also must have remained anoxic for all these thousands of years (this prevents the organic material [leaves] from being consumed by decomposing organisms).

Please observe also that a sediment core does not reveal the horizontal extent of any specific layer. Not only do we not know if these are annual layers, we also do not actually know if each varve spans the whole lake.

Using tree rings, corals banding, and lake varves, the new IntCal20 calibration currently stretches carbon dating back to about 70,000 theoretical years.

Once one has a standardized calibration curve, you can take the results from a carbon-dating test and compare it to known values Figure 18.22. There are some places where these calibration curves produce nice, clean, unambiguous carbon dates. However, there are other places that are much more difficult to interpret, including some areas where the line wiggles up and down, meaning any object could be assigned any of several possible dates.

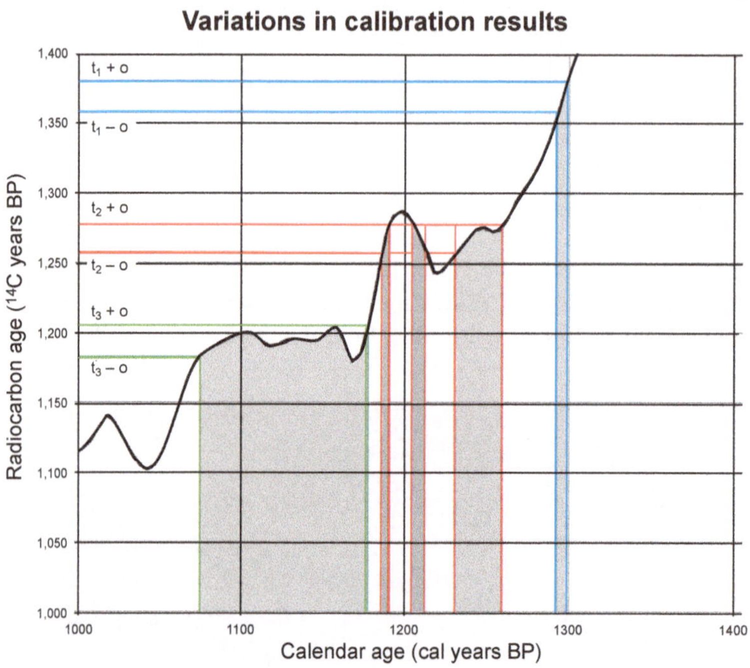

Fig.18.22: *This portion of the INTCAL13 calibration curve covers the medieval period. The X axis is in calendar years before present (BP). The Y axis shows the 'Radiocarbon Age', which is a little confusing. This is an uncalibrated measurement of the amount of ^{14}C in the samples. To obtain the radiocarbon "age", you start with the measurement on the y-axis, trace across until you run into the measurement curve, then trace down until you hit the known age on the X axis.*

For example, at t_1 (blue), the date is clear and unambiguous. Given a little bit of uncertainty (σ) in the initial measurement, we can be confident that a sample with that amount of carbon-14 can be dated to within a specific decade. Interestingly, this is the approximate timespan for the carbon dates obtained from the **Shroud of Turin**. At t_2 (red), however, there are three possible dates for the sample, and t_3 gives a wide range of possibilities that span a century.

In other places, the curve is flat for decades, sometimes centuries. For example, the timespan from 800 to 400 BC has been called the "Hallstatt plateau" or the "1st millennium radiocarbon disaster area". It is a flat spot on the calibration curve that spans a *very* important time in history. In biblical archaeology, this includes the end of the Divided Monarchy, the fall of Jerusalem, and the Babylonian captivity. The Assyrian, neo-Babylonian, and Persian empires and some of the most important events in Greek antiquity are a carbon-dating cypher. None of these events can be currently resolved with carbon dating! Archaeologists are working hard to fix these issues, but this has been a difficult issue to resolve.

CARBON DATING IS AFFECTED BY THE STRENGTH OF EARTH'S MAGNETIC FIELD

There are two historical problems with carbon dating that are generally ignored by secularist scientists and archaeologists. The first is that the earth's magnetic field is measurably weakening over time. In fact, it is weakening exponentially, by about 5% per century. This gives an upper limit to the age of the earth because the earth does not have an internal power source.

The earth itself is not a magnet, because the internal temperature is above the Curie point (the temperature at which magnetism breaks down in a solid metal). Instead, the planet's magnetism is thought to emanate from circulating electric currents, and all physical models would expect them to decay exponentially over time, even if the field 'flips' occasionally.

The decaying magnetic field is partially accounted for when they validate a carbon date with a sample from a known point in history. That way, it does not matter how much carbon-14 was in the world at the time the sample was made. All you must do is correlate an unknown to a known, Figure 18.23.

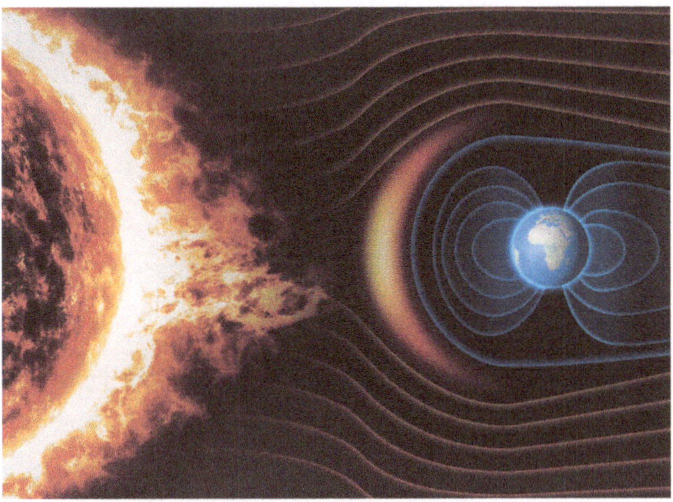

Fig.18.23: Earth Magnetic Field – Curtsy Night Magazine

The problem arises when there is a sample with no known historical matches.

The reason this is a problem is that the amount of carbon-14 in the atmosphere is controlled by the earth's magnetic field. Carbon-14 is formed in the upper atmosphere when the fast-moving protons and atomic nuclei we call **'cosmic rays'** strike nuclei of gases in the upper atmosphere. Among other things, they can sometimes knock neutrons out of the nuclei. These neutrons can then hit the nuclei of **nitrogen-14** atoms and kick out a proton. This produces **carbon-14** via the nuclear reaction:

$$ {}^{14}_{7}N + {}^{1}_{0}n \longrightarrow {}^{14}_{6}C + {}^{1}_{1}p $$

Here, the top numbers are the mass number (defined below). Neutrons and protons have a mass of "one" unit. Carbon-14 and nitrogen-14 have the same mass number, but nitrogen-14 has an additional proton while carbon-14 has an additional neutron. The lower number represents the atomic number. Each element has a unique atomic number, which is equivalent to the number of protons in the nucleus. Note that a hydrogen atom has one proton, so protons and hydrogen both have the atomic number "1".

But a strong magnetic field will deflect more cosmic rays than a weaker one. If the earth's magnetic field was stronger in the past, there should have been less carbon-14 in the atmosphere in ancient times.

If ancient samples started out with less carbon-14 than we expect, they will *appear* to be older than they really are. Having less carbon14- will be *interpreted* as the result of more time to decay, so the sample will be assigned an older age. This age inflation should also increase as you sample older and older objects. Thus, we might expect samples from ancient Egypt or Jericho to be inflated by a few centuries.

The carbon date of any sample that was formed immediately after the Flood would be even more inflated, perhaps by 50,000 years or more.

We do not know how much carbon-14 was created by God initially (it could have been 'zero') and we don't know how strong the earth's magnetic field was prior to the Flood. It is entirely possible that 'ancient' carbon dates are simply due to a general lack of antediluvian carbon-14. Yet, it does not appear that there was zero carbon-14 at the outset of the Flood, for it has been detected in coal, oil, natural gas, dinosaur bones, and diamonds. Given the short half-life of carbon-14, none of these should contain any carbon-14 at all if they are as old as claimed. Yet, it is there, and in similar concentrations within these diverse forms of carbon. This is excellent support for a young earth.

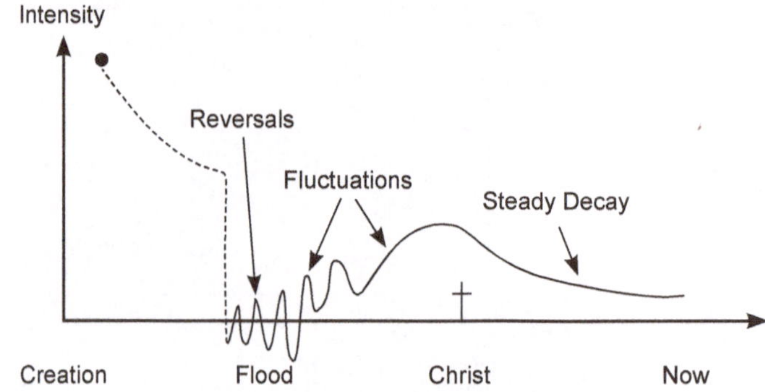

Fig.18.24: *A very Rough Model of the Earth's Changing Magnetic Field.*

A theoretical model of the earth's changing magnetic field appeared in our article The earth's magnetic field: evidence that the earth is young Figure 18.24. This will no doubt be revised over time, but the main point is that we cannot assume the earth's magnetic field has always been constant, and the strength of the field controls how much carbon-14 is being formed at any given time. Therefore, we must be careful about drawing grand historical conclusions from carbon dating.

CARBON DATING IS AFFECTED BY CARBON BURIAL DURING THE FLOOD

There is another reason why pre-Flood remains have old carbon dates. It has to do with the dramatic reduction from the high antediluvian atmospheric carbon dioxide levels to the modern value. This was due to

1. the burial of masses of organic matter (i.e., most of the pre-Flood biota),
2. the formation of chalk ($CaCO_3$) deposits from coccolithophore and foraminifera blooms during and after the Flood year, and
3. the revegetation of the earth after the Flood.

All these drew down the atmospheric CO_2 to the modern level, which happens to be suboptimal for plant growth. Secular sources tell us that atmospheric CO_2 levels in the early Paleozoic (e.g., pre-Flood) were at least 15 times higher than they are today.

Given a constant influx of cosmic rays and the assumption that the amount of carbon-14 is proportional to cosmic ray influx and not the availability of carbon in the atmosphere, this alone is equal to about four half-lives, accounting for over 20,000 years of the inflated pre-Flood carbon dates.

A reduction in atmospheric CO_2 levels would result in a relatively rapid and large increase in the $^{14}C/^{12}C$ ratio following the Flood, even without the effect of a declining magnetic field. To this we can add volcanic activity, which adds ^{14}C-depleted CO_2, but we have no idea how much of this there was. Many volcanoes don't produce much CO_2. Clearly, there was a huge drop in atmospheric CO_2 from the pre-Flood to the post-Flood world, so the amount of volcanic CO_2 contributed during the Flood was either not large or the volcanic CO_2 was absorbed by the formation of limestone.

Either way, the effects of the reduced atmospheric CO_2 is going to be large and will occur in a much shorter time frame than the increase in ^{14}C over time due to a decaying magnetic field.

How far Back can You Go?

At present, estimated ages of about 50,000 'years' are easily obtainable. This amounts to about 9 half-lives of carbon-14 ($\frac{1}{2}^9$ or $^1/_{512}$ of the original amount of ^{14}C). It is possible to measure older dates, but there are no fixed historical events that far back from which we can obtain samples for calibrating the machines.

Also, being that so little carbon-14 is left after that much time, any errors in the measurement have a much greater potential effect. For these reasons, archaeologists have shied away from reporting older dates, even though the machines can certainly produce results for these samples.

Interestingly, we have done a lot to throw off the carbon clock in modern times. For example, since coal, oil, and natural gas have less carbon-14 than wood, the burning of fossil fuels during the Industrial Revolution dumped a lot of 'old' carbon into the atmosphere. This creates a calibration plateau from 1700 to 1950 that is a bit of an annoyance to carbon dating specialists. Yet, that was nothing compared to what happened in the nuclear age.

The atmospheric testing of nuclear weapons added a massive amount of carbon14- to the atmosphere. This was one of the best things to happen to the carbon dating community, however. The detonation of hundreds of atomic weapons in a short period of time, followed by a sudden cessation when the Partial Nuclear Test Ban treaty went into effect in 1963, produced a "bomb peak" of carbon-14, which has been decaying ever since Figure 18.25. Samples from 1955 through about 1990 are easily dated with best-ever precision because of the "bomb peak" calibration curve.

Oceanographers have even picked up a plume of modern carbon spreading across the sea floor. Atlantic bottom water is formed at the ice-ocean interface. As ice is formed, salts are excluded. Thus, dense, cold, salty "thermohaline" waters sink to the bottom in the far north. This creates a bottom current of seawater that is slowing flowing southward. You can tell how far the current has travelled since 1950 by simply measuring the carbon-14 content of the water!

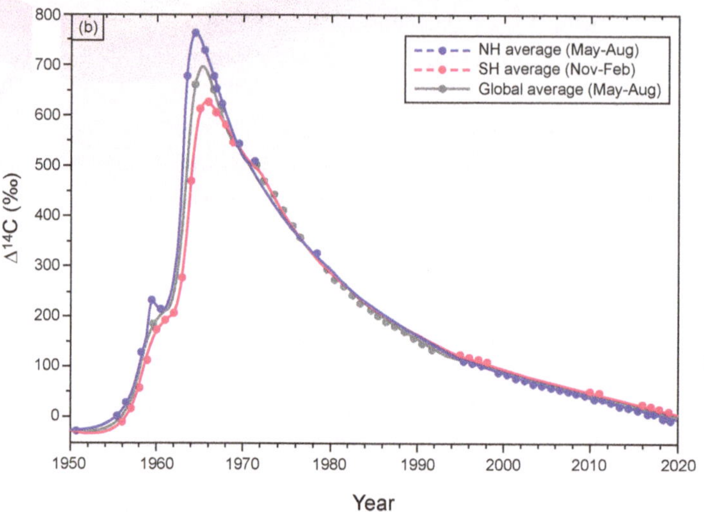

Fig.18.25: The amount of carbon-14 in the atmosphere changed radically at the advent of the atomic age (after Hua et al. 2021). These data were obtained from the rings of modern trees growing across the world. NH = Northern Hemisphere. SH = Southern Hemisphere. $\Delta^{14}C$ refers to a ratio between a standard and a measured carbon-14 value. The peak in carbon-14 levels is due to the atmospheric testing of nuclear weapons. Note that carbon-14 does not decay as fast as this. Instead, the reduction in carbon-14 levels is partially due to decay and partially due to removal from the atmosphere by plants, rainfall, and in chemical reactions with certain rocks. Atmospheric equilibrium appears to take about a century. Credit: Hua, Q. et al., 2021

What is an Isotope?

Each of the elements on the periodic table has a specified number of protons and electrons. An uncharged atom has the same number of each. If an atom gains or loses an electron, it is called an *ion*. It is still the same element, but it now has a charge. Elements are defined by the number of protons, also called the *atomic number*, sometimes designated by the symbol Z. By this definition, carbon has the atomic number 6 (Z = 6). Z is sometimes written as a subscript to the left of the chemical symbol of the element.

Atoms also contain neutrons. These have the same mass as the positively charged protons, but they are neutral. Unlike the number of protons, the number of neutrons in an atom of a given element can vary. This gives rise to the various *isotopes* of certain elements. Hydrogen, uranium, plutonium, and many other elements have several different isotopes, but some isotopes are much more common than others.

To differentiate them, isotope names are usually combined with their *mass number* (the number of protons + number of neutrons). The name can be written in long form (e.g., carbon14-) but often you will see a superscript number next to the atomic symbol (e.g., ^{14}C). Some isotopes also have a common name, e.g., deuterium (^{2}H), and tritium (^{3}H), while ordinary hydrogen is sometimes called protium (^{1}H). The reason they are all hydrogen is that they all have only one proton, or Z = 1.

While mass number is just a number, there is also a quantity called *atomic mass*. How do you calculate the mass of an atom? Protons and neutrons are assigned 1 atomic mass unit (AMU), where 1 AMU = 1.660×10^{-27} kg. Electrons have essentially no mass (about $\frac{1}{1836}$ of a proton) and can be ignored. Carbon has 6 protons and 6 neutrons, so it should have 12 AMUs. But no, its mass is 12.0107.

The reason for this is twofold.

1. The first reason explains why nuclear reactions are so energetic. For example, the atomic mass of ^{14}C is not 14, but 14.003241989. The tiny difference is caused by the binding energy, which is released when the isotope breaks down.
2. The second reason is that ^{13}C and ^{14}C will be mixed in with any natural source of carbon. They are not as common as ^{12}C, but they are abundant enough that we must account for them when studying chemistry. In the atmosphere, 98.98% of the carbon is ^{12}C. About 1.11% is ^{13}C. Only a tiny fraction, about one in a trillion atoms of carbon, is ^{14}C.

Many isotopes are stable—including ^{12}C and ^{13}C—but some are not. These are the radioactive isotopes. When they break down, they release various subatomic particles. If they lose a proton, or more often an alpha particle (a helium nucleus with two protons plus two neutrons), they change into another element entirely. Thus, uranium famously turns into lead while emitting multiple alpha particles, and carbon-14 changes into nitrogen-14 when one of the neutrons turns into a proton and the atom releases an electron (beta particle).

How does Carbon-14 Reach into a Bone from the Air?

Carbon-14 is made in the upper atmosphere. It then combines with oxygen to form CO_2. Winds slowly circulate it everywhere, and it is eventually taken up by plants during photosynthesis. While the plant is alive, the proportion of carbon-14 in the leaves, bark, sugars, and other plant parts essentially (but not exactly) matches the amount in the atmosphere.

Animals then eat the plants and the carbon-14 becomes part of the animal. Carnivores eat the herbivores, decomposers consume what the animals leave behind, etc. In this way, Carbon-14 is slowly distributed throughout the biosphere.

Once an animal or plant dies, it is no longer exchanging carbon with the atmosphere. Whatever carbon-14 is included in the wood, bone, or fibers left behind will begin to slowly decay.

HOW CARBON DATING IS ACHIEVED

The ^{14}C dating technique involves multiple, complex steps, but it is a fascinating scientific process.

Samples to be analyzed are sent to one of the many carbon-dating laboratories in the world. Technicians visually examine the material under a microscope and manually pick out any odd particles that don't match the material type. While they are doing this, they confirm that the sample is correctly labeled as bone, cloth, wood, etc.

The second step is to perform a thorough chemical cleaning. This often involves washing with inorganic solvents, acidic solutions, and/or alkaline solutions to remove contaminants like fingerprints, oils, etc. This is very important, considering the controversy about the **Shroud of Turin** and supposed 'contamination.'

In recent years, many laboratories started burning the cleaned sample in an elemental analyzer, which is coupled to the graphitization system. The sample is gasified under high heat in the elemental analyzer. In this way the lab can gain other information about the molecular composition of the sample from the analyzer. The CO_2 is then turned into graphite by reacting with hydrogen at high temperature in the presence of iron powder, which acts as a catalyst.

An older method involved placing the sample in a quartz tube with some copper II oxide (CuO) and silver wire.

The air in the tube was evacuated, the tube was sealed, and it was then heated to 900°C overnight. At this point, all the carbon in the sample had been converted into CO_2, using the oxygen from the copper oxide. The silver removed any nitrogen oxides and sulfur oxides produced by the nitrogen and sulfur in the sample. The tube was opened in a closed system of pipes and the CO_2 was collected by cooling it in liquid nitrogen. This was then used to create graphite, as described above.

The cleaned, purified, burned, sometimes liquified, and re-solidified carbon is now ready to be tested. A small amount (typically 0.5–2 mg, about 1,000 times less than older methods required) of the graphite is placed in a sample holder, along with several calibrated carbon standards, and inserted into the front end of a long machine. The air is evacuated from the sample chamber. The graphite is then atomized by hitting it with a high-energy beam of cesium ions.

Cesium is famous for being one of the few metals that are liquid at room temperature. It is also the easiest metal to ionize. When the heavy ions strike the carbon source, the surface is blasted into carbon atoms and ions, which are then sent into the detection part of machine.

ELECTROMAGNETISM

One of the most basic facts of physics is that electricity and magnetism are linked. For example, an electric current can be produced by moving a magnet across a conducting wire. It works better when the wire is formed into a coil, but the principle is that changing magnetic fields induce electric currents in conductive materials. This is called Faraday's Law of Electromagnetic Induction and is one of the most essential principles of electromagnetism.

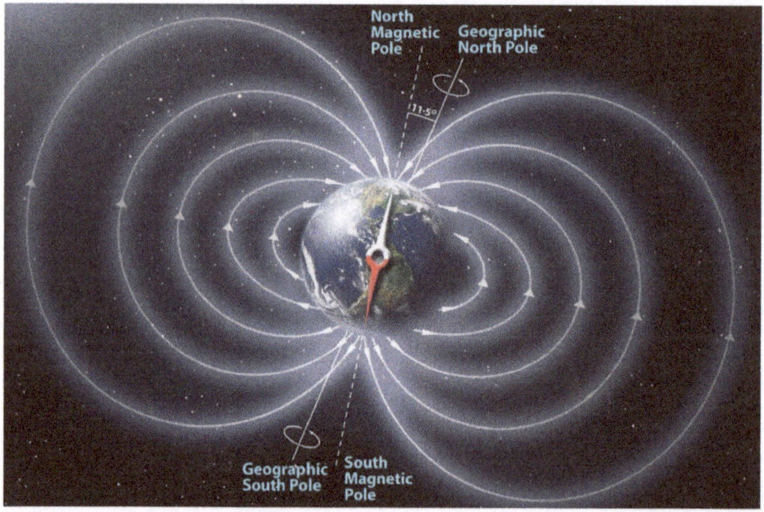

Fig.18.26: Earth Electromagnetic Field

Michael Faraday, by the way, was also a Bible-believing Christian and creation scientist. His law predicts how a magnetic field will interact with an electric circuit to produce an electromotive force (EMF). Cell phones, FM radios, your home WiFi, and Bluetooth all operate using this principle, Figure 18.26. In each of these cases a 'radio' signal is picked up by a small wire called an antenna. The signal 'induces' an electric current in the wire that is then picked up by the electronics in your device.

But, since electricity and magnetism are linked, the opposite is also true: moving electrical charges produce magnetic fields. This is the principle behind the old-fashioned 'fat' TV screens that everybody used until just a few years ago. A beam of electrons was created at the back of the TV and shot toward the phosphorescent glass at the front. The beam, however, was deflected left and right and up and down by magnetic fields. By scanning the beam across the entire glass surface, an image could be formed. And, since the phosphorescence was so fleeting, a series of images could be shown. Flickering at 30 frames per second, your brain would think that it was looking at a moving object instead of a series of still pictures. The screen on which you are reading this probably does not operate like this, but I can assure you that Faraday's Law of Electromagnetic Induction is at work somewhere behind the scene.

Detecting Carbon-14

There are several different configurations used in radiocarbon labs, and the techniques are constantly being refined. Thus, there is no single explanation of the measurement techniques. However, there is a general approach that most AMS labs share Figure 18.27.

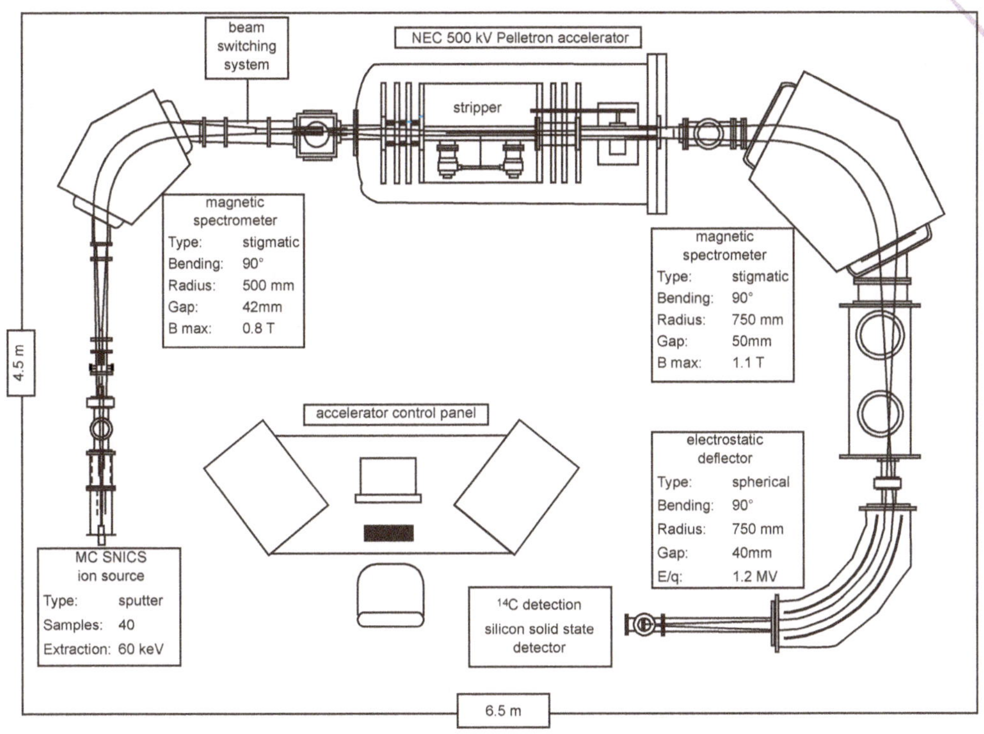

Fig.18.27: The Basic Setup of a Modern AMS – Curtsy Synal, Jacob, and Suter 2000

The cesium ions donate electrons to some of the carbon atoms they are striking to form negatively charged carbon ions. A strong (~40 kV) electric field is applied to the chamber, which causes the ions to shoot past a magnet and into an acceleration tube full of argon gas. The argon is a stripper gas, turning the former negative ions into positive ones. Thus, instead of being pulled back toward the magnet, they are now pushed away.

The stream of positively charged carbon ions are then sent down a long tube toward a series of detectors and more strong magnets.

These fast-moving, positively charged ions will produce localized magnetic fields as they move. Thus, if you send one of these ions past a magnet, its path will bend as the two magnetic fields interact. By passing the ions past a magnet with just the right strength and angled in just the right direction, you can get the ion beam to turn in a specific direction, but the heavier ions will not bend as much. Thus, the ion beam is separated into distinct ion streams inside the machine.

"Atom smashers" operate on the same principles. For example, at the **Large Hadron Supercollider** in Europe, propulsion is caused by electric fields. Strong magnets are used to get the charged particles to go in a circle. Linear accelerators also exist. They use oscillating magnetic fields to accelerate charged particles in a specific direction. Accelerator mass spectrometers are a hybrid of these.

The streaming ions are detected with a Faraday cup. This is a simple device that is hooked up to an electric circuit. When an ion strikes the inside of the cup, it creates a small electric current. Importantly, the current is solely based on the charge, and has nothing to do with mass. The circuit must be extremely sensitive and extremely accurate, for even if several billions of ions are striking the detector per second, the current will only be about one nanoamp. An ampere is defined as a current of 1 coulomb of electric charge per second, and a coulomb is 6.241×10^{18} electric charges.

In other words, the current in the circuit can be directly converted into the number of atoms (each with a charge of -1) striking the detector per second.

Here's where the magic comes in. The carbon ions have the same charge and the same magnetic field strength, so they have the same Lorentz force acting on them. But the ions do not all have the same mass. Since they have different numbers of neutrons, the masses of the three isotopes of carbon are significantly different. ^{13}C is 8% heavier and ^{14}C is 17% heavier than ^{12}C.

Some AMSs can even differentiate the tiny mass difference between ^{14}C and ^{14}N.

As the ions strike each detector, the measurements are sent to a computer. This machine is highly accurate but note that it is not counting individual atoms for the most abundant isotopes. For ^{14}C, the oldest samples will only generate a few to a few hundred detections in a typical three-minute run, so atoms can absolutely be counted individually.

However, the machine is just comparing the amount of current generated in the detectors. It is the ratio of carbon isotopes that is important. This way, it does not matter how much of the sample made it into the beam. One microgram of a sample will have the same relative isotope ratios as one gram of that same sample.

The AMS is orders of magnitude more sensitive and more accurate than the old scintillation counters. Also, the scintillation counters can only measure the atoms that decay. The AMS is measuring all the atoms present.

How Much does it Cost?

At present, a 'date' can be obtained from a radiocarbon lab for about $500 per sample.

INTERPRETING THE DATA

Carbon dates are usually presented in 'years before present' (with the 'present' being defined as the year 1950). Sometimes, a date will be converted to AD or BC. Occasionally, you might see a mention of 'percent modern carbon' (pmc) or you might also see a phrase like "del ^{13}C". This refers to a complicated ratio of ratios between the ^{13}C and ^{12}C values of an unknown sample and a standard reference.

Yet, you cannot stick a sample into an AMS and get a 'date.' Instead, all results must be compared to a known standard, usually several. The goal is to have a standard that is close enough in age that it yields similar values. Operators will generally load sample references that bracket the expected date for the sample. There are reasons why a carbon-14 lab will ask you the expected age of a sample before they run the experiment. It is not like they are deliberately giving you the 'answer you expect.'

ISOTOPIC FRACTIONATION

The results will almost always be corrected for 'isotopic fractionation.' There are many processes that discriminate among the three carbon isotopes. Since carbon-12 is lighter, it is preferentially used in most biological and chemical reactions.

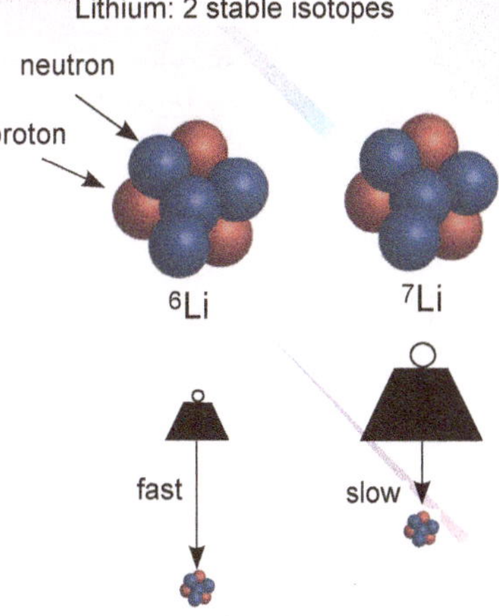

Fig.18.28: Principles of Isotope Fractionation

Isotopic fractionation affects how much carbon-14 is absorbed by plants from the atmosphere, by herbivores when they eat the plants, and by carnivores when they eat the herbivores. Thus, a tree branch, a cow bone, and a lion bone from the same date in history will have measurably different carbon ages. We must also account for the differences between land and sea. It takes time for carbon-14 to get into the ocean, and the oceans have a long (~1,000 year) cycle time. Fractionation also affects the rate of carbon-14 penetration into the oceans and how fast it leaves (because it is lighter, $^{12}CO_2$ evaporates faster), Figure 18.28. The net effect is that ocean water appears to be about 400 years 'older' than the atmosphere. This is known as the marine reservoir effect (MRE), but do not make the mistake of thinking that all ocean water has the same carbon 'age.' There are significant differences from one place to another, even at the same place at different times.

Isotopic fractionation also affects the carbon dating methods used in the laboratory. Each step (vaporization, liquefaction, solidification, and ionization) involves some degree of isotopic fractionation.

Note, this does not mean that carbon dates are made up. However, it does mean that there is a lot of art to this science. Carbon dating experts must be dedicated to their craft.

After this, the measurement must be calibrated using the standards included in the run and a historical model will be applied to the final results. This is one of the most difficult steps.

DELTA VALUES

When measuring a difference between two values, scientists often use the Greek letter delta (Δ). In the case of isotope differences, however, they often use the less familiar lowercase form (δ) to represent a ratio. In fact, this is a ratio of ratios.

This comes up very often in discussions about global warming. $\delta^{18}O$ is a proxy for seawater temperature. Oxygen has three stable isotopes, ^{16}O, ^{17}O, and ^{18}O. When looking for evidence of ancient temperatures, scientists will measure the ratio of ^{16}O and ^{18}O in a sample, for example, in an ice core. By comparing this to a standard, they can estimate the temperature of the water as it evaporated and then turned into snow. Yet, since snowflakes that contain ^{18}O are heavier, they tend to precipitate out faster. Thus, the ratio is also affected by the distance between the sample area and the edge of the ancient ice shelf, which is often unknown.

Note also that you cannot "count" layers in these ice cores, like you can count rings on a tree. The reason for this is that the great weight of the glacier presses down on the ice, flattening the layers. Instead, scientists use a historical model to estimate the date of the ice at each depth.

In discussions about carbon dating, $\delta^{13}C$ often comes up. Since the process of photosynthesis preferentially takes up ^{12}C, you can use $\delta^{13}C$ as an estimate of global primary productivity (e.g., photosynthesis) in the past. For ^{13}C, the standard is the PDB, or a more recent substitute that has been calibrated to the PDB.

If you know, for example, the ratio of ^{13}C to ^{12}C in the PDB, you can measure the ratio in a piece of wood and compare the two. But the difference is going to be very small. It is not worth reporting in percent (%, parts per hundred), so they report the results in parts per thousand (‰). They also report the value as a negative number, because the value in the sample is usually less than the value in the standard.

This is the basic formula, where s = sample and r = reference:

$$\delta^{13}C = ((^{13}Cs/^{12}Cs)/\ (^{13}C_r/^{12}C_r) - 1) \times 1000‰$$

You can do this for ^{14}C, ^{18}O, or any other isotope that you can compare to a known standard.

APPLYING HISTORICAL MODELS

As stated above, you cannot obtain a carbon date without referencing a known standard, but it is a lot more complicated than simply comparing a known date to an unknown date. The amount of carbon-14 in the atmosphere varies seasonally and latitudinally. It even changes depending on the direction of the wind (i.e., is it coming off the ocean or off the land?).

When building a calibration curve, data might come from tree rings, coral skeletons, cave formations, or laminated lake sediments, but note that these data are associated with many historical assumptions of their own. Even the newest calibration curve, IntCal20, is only a general model and does not include regional differences:

The conclusion drawn there is that there is as yet insufficient information to quantify such effects or indeed to fully understand the contribution from different underlying mechanisms. The potential issues are: different growing seasons for the material dated compared to the calibration datasets, localized addition of CO_2 from different reservoirs (such as ocean upwelling, anthropogenic sources, and local volcanic vents), mixture of air masses from both hemispheres in the tropics [reference], and overall trends with latitude or altitude.

With the exception of the potential mixing of carbon from different reservoirs, they expect that the seasonal and regional differences will be small, approaching the measurement precision of the machine.

Can a global calibration curve really be applied to material from, say, ancient Egypt? Seasonal winds have a strong effect here, sometimes blowing steadily north, sometimes blowing steadily south, and sometimes not blowing at all. Long-term changes in weather patterns should mean that Southern Europe and Northern Africa need to be treated as separate carbon reservoirs, but we simply do not have the data to do this (yet). This also affects the dating of artifacts in Israel, so these calibration curves are of utmost importance to get right.

Historically, there have been multiple events that should have major effects on carbon-14, but we are unable to measure them at present. For example, at the end of the Ice Age (about 10,000 years before present in the evolutionary dating scheme; about 4,000 years ago in the biblical model) the Mediterranean was a freshwater lake. So much snowmelt was pouring down the tributary rivers in Europe and Asia that it formed a freshwater lens across the entire surface, similar to the situation in the Black Sea today. The deeper saltwater layer was cut off from ready sources of carbon-14. Yet, the Mediterranean is not fresh water today. Sometime in the historical period, well within the abilities of carbon-dating techniques to detect, the Mediterranean turned over.

In the evolutionary scenario, this would have dumped a massive amount of 'old' carbon into the air and would affect the carbon dates of anything downwind (e.g., Israel and Egypt). This illustrates the tenuous nature of carbon dates. Yes, we can accurately measure the amount of carbon-14 in a sample, but how well this reflects a specific historical time period is debatable.

KNOWN ISSUES WITH CARBON DATING

There are multiple issues that complicate the art of carbon dating. Many of these are not usually addressed in public. For example, it is known that trees growing in the vicinity of volcanic areas can often 'date' much older than they really are. The reason for this is that any CO_2 coming up from under the ground will have little to no carbon-14. This 'infinitely old' carbon will mix with the carbon in the atmosphere, meaning the carbon taken in by the plant will have a false appearance of age. This was first noticed in Italy, but examples of this phenomenon have been documented worldwide.

Carbon dating has also been performed on anomalously old material. This includes wood from the Hawkesbury Sandstone near Sydney, Australia. It was assigned an age of 33,720 ± 430 years bp, with a $\delta^{13}C_{PDB}$ value of –24.0%. That sandstone is supposedly 225–230 million years old. Another piece of wood, from central Queensland, was carbon dated to 37,500 years bp, with a $\delta^{13}C_{PDB}$ value of –25.69‰. An adjacent basalt layer, formed from lava that had also charred the wood, was 'dated' to 47.5 MA by potassium argon dating. Not only did the $\delta^{13}C$ results rule out contamination for these wood samples, but the carbon dates are impossible for the evolutionary model. The same can be said of carbon-dates on dinosaur bones (multiple dates, generally ranging from 30–40 thousand years old).

Carbon dating has also been applied to coal, from '20-million-year-old' Miocene coal to '300-million-year-old' Pennsylvanian coal. Every coal sample contained carbon-14. They even contained approximately the same amount of carbon-14. There is no way to physically contaminate all the coal in the world with carbon-14. The evidence is that coal is young, because there should not be any carbon-14 left in anything that that is claimed to be millions of years in age.

"OLD" CARBON DATES DO NOT INVALIDATE THE BIBLE

But do these carbon dates invalidate the biblical timeline? The earth, after all, is only a few thousand years old. Nothing can even be tens of thousands of years old, yet these ancient samples consistently 'date' older than is allowed. This is not a problem, because the dates should be strongly constrained by the amount of carbon-14 in the antediluvian world (i.e., little to none). In other words, even if there was zero carbon-14 in this material, it would fit with biblical expectations. Yet, the opposite is not true. Even a little bit of carbon-14 in extremely 'old' material is fatal to evolutionary expectations.

RECALIBRATING CARBON DATES

What is needed is a way to take the raw carbon dates and recalibrate them according to the biblical timeline. Figure 18.29 shows a very rough approximation. I pegged the "antediluvian" era to 30,000 carbon years before present (an average of the values seen in dinosaur bones). The date for Abraham is in the range of carbon dates for the Chalcolithic (6,000 YBP), because we know that Abraham, Isaac, and Jacob lived in Beersheba and the latest archaeological evidence for intensive grain cultivation (Genesis 26:12) in the region is from the Chalcolithic (aka, the Copper Age).

The red dashed line represents the real historical age. The "present" is defined as the year 1950 (the final data point), but "zero" on the X axis is the year 2022. I used straight lines to connect the blue dots to indicate that this is not a curve that can be represented by a formula. It is a conceptual model only and will be subject to much debate and refinement.

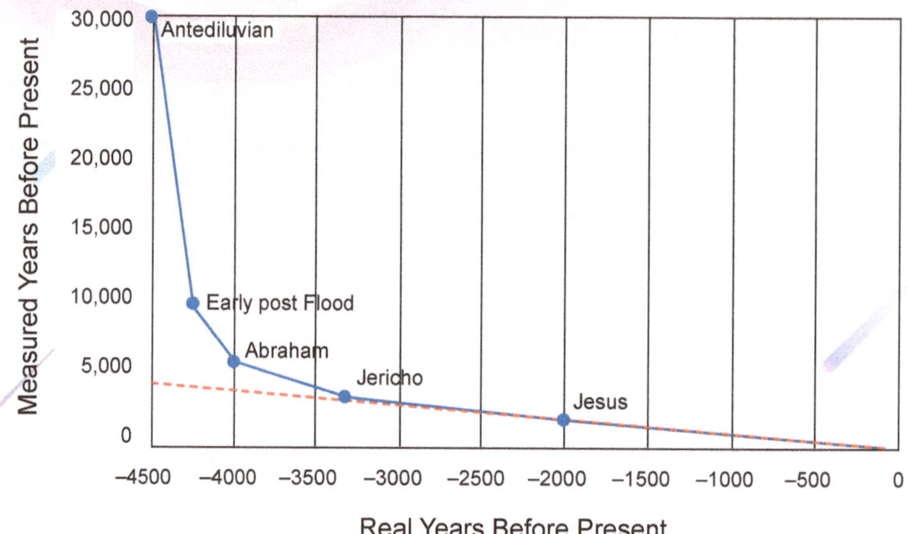

Fig.18.29: A rough carbon14- recalibration curve that can be used to bring carbon dates in line with biblical history.

If, as stated above, the earth's magnetic field strength has been declining by 5% per century, and if the decline has been steady (which is not at all guaranteed), the magnetic field would have been about nine times higher at the time of the Flood, meaning there should have been a lot less carbon-14 in the atmosphere before the Flood. The bomb peak took about 100 years to reach the modern equilibrium, so I assume the concomitant rise in carbon-14 will lag field decay by about a century.

On the other hand, carbon-14 does not have to explicitly follow a simple model of field decay. The magnetic field is only one of several important parameters (e.g., I have discounted the effects of volcanism), and the pre- and post-Flood levels of carbon-14 should not be expected to be the same just because only one year separates them. In fact, the 30,000 YBP measurement for dinosaurs might not be a good measurement for the early post-Flood period at all. It is possible that early post-Flood material could date older than antediluvian samples because of the huge amount of 'old' carbon that was dumped into the atmosphere via volcanism.

COUNTER ARGUMENTS

Of course, evolutionists have not taken this lying down. They have come up with multiple ways to refute the claims.

"Contamination" is something we consistently hear, but rarely is this contamination documented. "Instrumental errors" are another common refrain, but do they really want to throw their own scientists under the bus? As far as the carbon-14 found in diamonds, we often hear that, yes, there is carbon-14 present, but "uranium" is the cause.

There are two proposed ways to explain the 'anomalous' ^{14}C in diamonds using uranium. One is cluster decay, where some of the isotopes in the uranium decay chain emit a ^{14}C nucleus. But this is so rare that the majority of the material in and around the carbon would need to be uranium. We can tell the difference between uranium ore and a diamond, after all. Another way is for the neutrons produced by uranium fission to turn ^{14}N in the sample into ^{14}C, in the same reaction that produces radiocarbon in the atmosphere. But in this case, we would expect a strong correlation of ^{14}C with the nitrogen content of the sample. In fact, it would make the dating unworkable. These arguments go nowhere.

CARBON DATING PREDICTING THE FUTURE

Testing of atmospheric nuclear bombs in the 1950s and 60s dramatically increased levels of radiocarbon (carbon-14, ^{14}C) in the atmosphere—Figure 30. below. Measurements have always shown fluctuations in the level of atmospheric radiocarbon, but since the peak in the mid-1960s, the level has been steadily decreasing overall. A carbon age of thousands of years in the future like the one you described could indicate that the wood in question was cut/separated from the tree in the 1960s, when atmospheric radiocarbon levels were at their highest. (There are other phenomena besides nuclear bombs that also affect carbon isotope levels.)

Several examples of future carbon dates appear in some research done on silk cloth. In this case fake antique silks were easily distinguished from authentic antique silks by their negative carbon dating ages, ranging from 900 to 1,600 years into the future.

Isotope levels can be measured extremely accurately, much more accurately than when carbon-dating methods were first devised. However, this doesn't mean that carbon-dating is any more reliable. Radioisotope dating relies on unprovable assumptions about the past, and unless we have reliable historical dates for objects, we cannot be certain that their calculated ages are accurate. The only reliable way to date anything is by the historical method—relying on trustworthy records such as the Bible.

Archaeologists will readily dismiss carbon-dating ages if they do not agree with the date they want or expect:

'If a C14 date supports our theories, we put it in the main text. If it does not entirely contradict them, we put it in a footnote. And if it is completely 'out of date', we just drop it.'

For example Australian archaeologist Dr Josephine Flood rejected the 30,000 year carbon-dates for 'Mungo Man' because she believes, based on other dating methods such as electron spin resonance, that Mungo Man is 60,000 years old. Both methods can't be right, but both can be wrong, and we know from the accurate history recorded in the Bible that in this case both are wrong.

There is evidence of changes in carbon isotope ratios from era to era, region to region, species to species, and even from one part of an individual organism to another part. For examples, 'What about carbon dating?' Consider the green and red lines on the graph below, Figure 18.29. Volcanism also effects carbon ratios, as does industrialization, and these effects can be regional, such as variation between plant types. And consider too that non-air-breathing creatures, e.g., fish, will tend to ingest less ^{14}C.

Carbon-dating relies on knowing the ratio of carbon-14 to carbon-12 (normal non-radioactive carbon) in the atmosphere throughout history. But the carbon system is not always in **equilibrium**; there is not always a balance between the amount of new carbon-14 forming from nitrogen in the air due to solar radiation, and the amount of ^{14}C leaving the atmosphere due to being absorbed by plants. As a recent Science article noted:

'[T]he amount of ^{14}C produced in the atmosphere varies with the sun's solar activity and fluctuations in Earth's magnetic field. This means that the radiocarbon clock can race ahead or seemingly stop for up to 5 centuries. As a result, raw radiocarbon dates sometimes diverge from real calendar years by hundreds or even thousands of years. Thus, researchers must calibrate the clock to account for these fluctuations, and that can be a challenge.'

Besides these regular changes, which secular scientists try to take into account, there was a unique event in earth's history that also greatly upset the carbon balance, but which the scientific establishment refuses to take into account when calibrating the carbon-dating method. That event was the Flood.

There was massive worldwide volcanism during the Flood, with massive volumes of dust, CO_2, and water vapor being ejected into the atmosphere. Volcanic CO_2 tends not to contain ^{14}C, just normal carbon. The ratio of ^{14}C to ^{12}C in the atmosphere would thus have become greatly diluted during the Flood, and would have remained low for many centuries, as the level gradually built up again towards equilibrium. Animals and plants in the immediate post-Flood centuries would have absorbed and ingested less ^{14}C than modern (i.e., pre-nuclear bomb) era organisms. Hence artefacts from the first post-Flood millennium will generate ages older than they really are, due to starting with less ^{14}C than would be expected based on the incorrect assumption that the ratio then was the same as in modern times. Carbon 'dates' older than 3000 years are likely to be grossly inflated.

Image Wikipedia.org

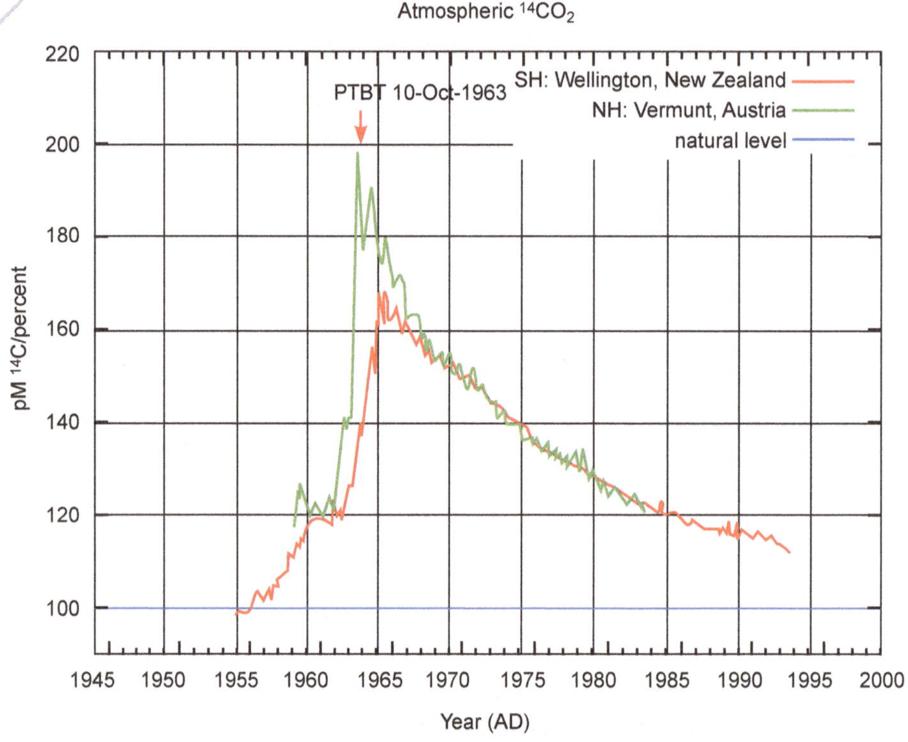

Fig.18.30: Atmospheric carbon-14 levels, showing effect of nuclear bomb testing

The strength of the earth's magnetic field has been steadily decreasing, meaning more solar radiation is reaching the atmosphere, and thus more ^{14}C is being produced in our atmosphere. So even without atomic bomb testing, the atmospheric ratio of ^{14}C to ^{12}C may have been increasing, albeit very gradually. Note that the 'natural level' in the graph above is assumed. In fact, the atmospheric carbon system may never have been in equilibrium. The blue line on that graph represents a mythical uniformitarian ideal that has not existed in the short period that records of atmospheric isotope levels have been kept. And yet the edifice of carbon dating basically relies on extrapolating this romanticized straight blue line 50,000 years into the past! This is not good science.

Carbon dating began in the late 1940s and some of the first things to be carbon dated were **Egyptian artefacts**, including the coffin of **Pharaoh Djoser**. The date of this pharaoh's death was already considered to be accurately established, at **2750 BC**. We know this date is incorrect because Egypt is a post-Flood civilization and this date precedes the Flood by several hundred years. Tests like these on ancient Egyptian artefacts formed the basis for the calibration of the carbon dating method. But scholars are now coming to the realization that the traditionally accepted *Egyptian chronology* was greatly in error, and needs shortening by hundreds of years. A revolution in ancient chronology has been steadily building over the last decade, with a number of scholars publishing hard-to-refute evidence that traditional Egyptian chronologies are overextended by many hundreds of years. More and more archaeologists are beginning to accept that revision is inevitable. On the issue of an accurate chronology of ancient history.

Carbon dating actually tends to be the creation scientists' friend because carbon-14 is found in many samples that are supposed to be millions of years old, when there should be no ^{14}C left according to uniformitarian assumptions.

For those with the inclination, abundant evidence is available with which they can verify to themselves the unreliability of radiometric dating methods, including the carbon method.

RADIOACTIVE DATING TECHNIQUES

Conflicting Results

Fig.18.31: Different Radioactive Dating Methods

With respect to measuring the ages of things, we are told that there are a dozen different radioactive dating methods and that they all give the same answer, Figure 18.31. ***Do they***?

Fossil wood from a quarry near the town of Banbury, England, some 80 miles north-west of London, was dated using the carbon-14 method. The ages calculated ranged from 20.7 to 28.8 thousand years old. However, the limestone in which the wood was found was of Jurassic age, of 183 *million* years. Clearly the dating methods are in conflict.

Diamonds analyzed from mines in South Africa and Botswana, and from alluvial deposits in Guinea, West Africa, found measurable carbon-14 over ten times the detection limit of the laboratory equipment. The average 'age' calculated for the samples was 55,700 years. Yet the rocks that contained the diamonds ranged from 1,000 to 3,000 million years old. Dating methods are in conflict again.

Rock samples from a lava dome within the Mount St Helens crater, USA, were dated using the potassium-argon method. Whole-rock samples gave an age of 350,000 years. When some of the amphibole minerals in the rock sample were extracted and analyzed separately, their age was more than double at 900,000 years. Two mineral samples of a different mineral, pyroxene, gave an age of 1,700,000 and 2,800,000 years. Which age is right? ***None, actually***.

The lava dome formed after Mount St Helens exploded in 1980 and the samples were just 10 years old. Here are more conflicting results between dating methods.

Creation scientists have uncovered dozens of anomalies and conflicts like this. Surprisingly, these conflicting results do not unsettle mainstream geologists. They genuinely believe the world is billions of years old, and the conflicting results do not cause them to question their belief. In their minds, these conflicts are a little mystery that will be resolved with creative thinking and more research.

In his well-known textbook on isotope geology, Gunter Faure explains the various radioactive dating methods, including the so-called isochron method. When the results for a number of rock samples are plotted on a graph and form a straight line, the researcher can calculate an age for the samples. But Faure warns his readers not to accept the calculated age without question.

He gives an example of volcanic lava along the border of Uganda, Zaire and Rwanda, East Africa. That lava is known to be relatively young, possibly erupted within historical times, yet a rubidium-strontium straight-line isochron gave an age of 773 million years. Does this worry these scientists? ***No***. They have total faith in the method. In their minds, the key is the way the results are *interpreted*. Faure says that in this case we should interpret the line, not as an isochron, but a "mixing line." So how can we tell the difference? ***We can't***. The only way we can know it is a mixing line as if the calculated age is wrong—and the only way one can 'know' if an age is right or wrong is to have a pre-existing belief about what the age should be.

In another example, Okudaira *et al.* measured isochron ages of a rock called amphibolite sampled from south-east India. With the rubidium-strontium method they obtained an age of 481 million years but with samarium-neodymium the age was almost double at 824 million years. Did the disagreement cause the researchers to doubt the dating methods? ***Not at all***. They removed the disagreement by the way they 'interpreted' the results. They said the older age was the age the rocks underwent metamorphism, while the younger age was when the rocks were later heated. How did they know? No matter what the numbers are, a plausible story can always be invented *after* the results are obtained.

Another example involves a volcanic region in Southern India, a pluton. Using the lead-lead method, a whole-rock sample gave an age of 508 million years. With the potassium-argon method, samples of mica gave an age of 450 million years. Zircons using the uranium-lead method gave an age of 572 million years. Three different samples; three different methods; three different results. Did this cause the researchers to doubt the radioactive dating methods? ***No***. They just applied some creative interpretation. They said the different ages are because the huge pluton cooled slowly over millions of years and the different minerals were affected in different ways. Instead of a problem, the conflict became a new discovery.

Conflicting radioactive dating results are reported all the time and, on their own, there is no way of knowing what they mean. So, geologists research how other geologists have interpreted the other rocks in the area in order to find out what sort of dates they would expect. Then they invent a story to explain the numbers as part of the geological history of the area. Creation geologists consider that the Bible records the true history of the earth and that the rocks are less than 6,000 years old. Because the Bible is reliable and historically verifiable, we consider it scientifically valid to interpret the radioactive dating results within the biblical scenario.

THE TRUTH ABOUT GENUINE DIAMONDS

The Value of Accurate Carbon

What do hard sparkling diamonds and dull soft pencil 'lead' have in common? They are both forms (allotropes) of carbon. Most carbon atoms are 12 times heavier than hydrogen (^{12}C), about one in 100 is 13 times heavier

(^{13}C), and one in a trillion (10^{12}) is 14 times heavier (^{14}C). Of these different types (isotopes) of carbon, ^{14}C is called radiocarbon, because it is radioactive—it breaks down over time, Figure 18.32.

Radiocarbon Dating

Fig.18.32: The Famous Hope Diamond which was Found about Four Centuries Ago – Curtsy - wikimedia commons

Some try to measure age by how much ^{14}C has decayed. Many people think that radiocarbon dating proves billions of years. But evolutionists know it can't, because ^{14}C decays too fast. Its half-life ($t_{½}$) is only 5,730 years—that is, every 5,730 years, half of it decays away. After two half-lives, a quarter is left; after three half-lives, only an eighth; after 10 half-lives, less than a thousandth is left. In fact, a lump of ^{14}C as massive as the earth would have all decayed in less than a million years.

So, if samples were really over a million years old, there would be no radiocarbon left. But this is not what we find, even with very sensitive ^{14}C detectors.

HARD DIAMONDS

Diamond is the hardest substance known, so its interior should be very resistant to contamination. Diamond requires very high pressure to form—pressure found naturally on earth only deep below the surface. Thus, they probably formed at a depth of 100–200 km. Geologists believe that the ones we find must have been transported supersonically to the surface, in extremely violent eruptions through volcanic pipes. Some are found in these pipes, such as kimberlites, while other diamonds were liberated by water erosion and deposited elsewhere (called alluvial diamonds). According to evolutionists, the diamonds formed about 1-3 billion years ago.

DATING DIAMONDS

Geophysicist Dr John Baumgardner, part of the RATE research group, investigated ^{14}C in a number of diamonds. There should be no ^{14}C at all if they really were over a billion years old, yet the radiocarbon lab reported that there was over 10 times the detection limit. Thus, they had a radiocarbon 'age' far less than a million years! Dr Baumgardner repeated this with six more alluvial diamonds from Namibia, and these had even more radiocarbon.

The presence of radiocarbon in these diamonds where there should be none is thus sparkling evidence for a 'young' world, as the Bible records.

Objections

1. The ^{14}C readings in the diamonds are the result of background radiation in the detector. This shows that the objector doesn't even understand the method. AMS doesn't measure radiation but counts atoms. It was the obsolete scintillation method that counted only decaying atoms, so was far less sensitive. In any case, the mean of the $^{14}C/C$ ratios in Dr Baumgardner's diamonds was close to 0.12±0.01 pMC, well above that of the lab's background of purified natural gas (0.08 pMC).
2. The ^{14}C was produced by U-fission (actually it's *cluster decay* of radium isotopes that are in the uranium decay chain). This was an excuse proposed for ^{14}C in coal, also analyzed in Dr Baumgardner's paper, but not possible for diamonds. But to explain the observed ^{14}C, then the coal would have to contain 99% uranium, so colloquial parlance would term the sample 'uranium' rather than 'coal'.
3. The ^{14}C was produced by neutron capture by ^{14}N impurities in the diamonds. But this would generate less than one ten-thousandth of the measured amount even in best case scenarios of normal decay. And as Dr Paul Giem points out:
4. 'One can hypothesize that neutrons were once much more plentiful than they are now, and that is why there is so much carbon-14 in our experimental samples. But the number of neutrons required must be over a million times more than those found today, for at least 6,000 years; and every 5,730 years that we put the neutron shower back doubles the number of neutrons required. Every time we halve the duration of the neutron shower, we roughly double its required intensity. Eventually the problem becomes insurmountable. In addition, since nitrogen creates carbon-14 from neutrons 110,000 times more easily than does carbon-13, a sample with 0.000 0091% nitrogen should have twice the carbon-14 content of a sample without any nitrogen. If neutron capture is a significant source of carbon-14 in a given sample, radiocarbon dates should vary wildly with the nitrogen content of the sample. I know of no such data. Perhaps this effect should be looked for by anyone seriously proposing that significant quantities of carbon-14 were produced by nuclear synthesis *in situ*.'
5. Also, if atmospheric contamination were responsible, the entire carbon content would have to be exchanged every million years or so. But if this were occurring, we would expect huge variations in radiocarbon dates with porosity and thickness, which would also render the method useless. Dr Baumgardner thus first thought that the ^{14}C must have been there right from the beginning. But if nuclear decay were accelerated, say a recent episode of 500 million years' worth, it could explain some of the observed amounts. Indeed, his RATE colleagues have shown good evidence for accelerated decay in the past, which would invalidate radiometric dating.
6. The ^{14}C 'dates' for the diamonds of 55,700 years were still much older than the biblical timescale. This misses the point: we are not claiming that this 'date' is the actual age; rather, if the earth were just a million years old, let alone 4.6 billion years old, there should be no ^{14}C at all! Another point is that the 55,700 years is based on an assumed ^{14}C level in the atmosphere. Since no one, creationist or evolutionist, thinks there has been an exchange of carbon in the diamond with the atmosphere, using the standard formula for ^{14}C dating to work out the age of a diamond is meaningless. Also, ^{14}C dating assumes that the $^{14}C/C$ ratio has been constant. But the Flood must have buried huge numbers of carbon-containing living creatures, and some of them likely formed today's coal, oil, natural gas and some of today's fossil-

containing limestone. Studies of the ancient biosphere indicate that there was several hundred times as much carbon in the past, so the $^{14}C/C$ ratio would have been several hundred times smaller. This would explain the observed small amounts of ^{14}C found in 'old' samples that were likely buried in the Flood.

THE EARTH'S MAGNETIC FIELD

The Age of the Earth

As early as 1971, creation scientist Dr Thomas Barnes, then Professor of Physics at The University of Texas at El Paso, drew fresh attention to the fact that the strength of the earth's magnetic field was decreasing. He noted that between 1835 and 1965 geophysicists had made some 26 measurements of the magnetic dipole moment of the earth's magnetic field. When plotted against time (that is, the year of measurement) these data points fitted a decay curve which Barnes calculated had a 'half-life' (halving period) of only 1,400 years. On this basis he concluded that *the earth's magnetic field was less than 10,000 years old, and so the earth must likewise be that young*, Figure 18.33.

Was the Earth Evolved or Created?

Needless to say, because of the powerful implications of this evidence Barnes received much opposition from the evolutionary community. Evolutionist geophysicists simply poured scorn on Barnes' conclusions because they argued that any 'decay' of the earth's magnetic field merely represented the latest phase in the ongoing history of waxing and waning field strengths as the field repeatedly reverses during multiplied millions of years. But Barnes stoutly repulsed this objection by denying the validity of the 'fossil' magnetism (paleomagnetism) measurements of reversed magnetic polarity (direction) in rock strata.

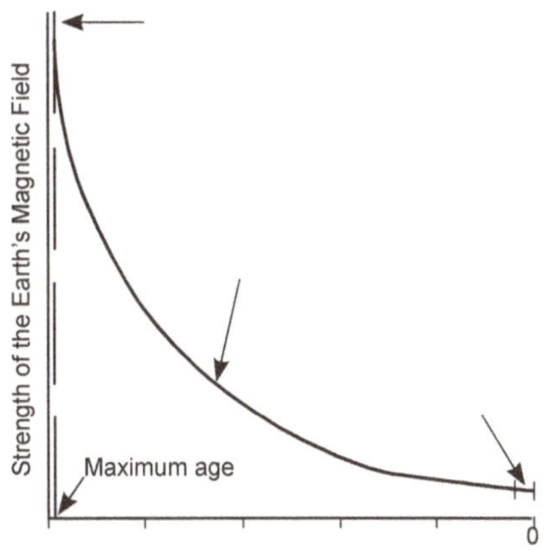

Fig.18.33: Humphrey's model for the history of the earth 's magnetic field.

Evolutionary geophysicists were already 'locked into' their multi-million-year timescale not only because of the radioactive dating of the rocks in which the palaeomagnetic reversals were measured, but because of their presumed 'dynamo' mechanism for operation of the earth's magnetic field. It is generally believed that the earth's magnetic field is generated by electric currents in the earth' innermost region, the core, which is presumed to consist of a metallic iron-nickel mixture. However, according to the 'dynamo' hypothesis, these electric currents are believed to be produced by the slow circulation of molten material that carries unequal amounts of positive and negative electric charge. The energy for this is thought to come from the earth's rotation and/or its internal heat.

So, the generating mechanism is presumed to operate like a dynamo, (similar to an electricity generator) causing the field and maintaining it over large periods of time. Consequently, a reversal of the earth's magnetic field (which is difficult for them to explain anyway) could be expected to be a slow process. Thus, the evolutionary view has been that a transition from one magnetic polarity (direction) to the other generally took millions of years, or several thousand years at the very least. However, this so-called dynamo hypothesis, the operational mechanism preferred by most geophysicists, has many problems associated with it which have been well documented.

EARTH MAGNETIC FIELD REVERSALS

More recently, creation scientist Dr. D. Russell Humphreys (Physicist at the Sandia National Laboratories, Albuquerque, and Adjunct Professor of Physics at the Institute for Creation Research, San Diego) has reviewed the evidence for the validity of these 'fossil' magnetism studies and has found that fully half of all the 200,000 plus geological samples tested have a measurable magnetization whose direction ('polarity') is reversed with respect to the earth's present magnetic field. He was forced to conclude that the variety, extent, continuity and consistency of the reversal data all strongly suggest that most of the data are valid, so that there is no option but to accept that reversals of the earth's magnetic field must have occurred.

Accordingly, how can these reversals of the earth's magnetic field be reconciled with Barnes' evidence that the earth and its magnetic field are less than 10,000 years old? Barnes and Humphreys have both argued convincingly for a viable alternative hypothesis to the evolutionists' dynamo idea. They propose that the earth's magnetic field comes from freely circulating electric currents created initially with a high built-in energy. As these currents subsequently lose their energy due to friction, the magnetic field will decay as the current strength decays.

However, a collapsing magnetic field cutting across a conductor (the earth's iron-nickel core) will generate more current, which helps to retard the rate of decay, otherwise the field would vanish more quickly. In fact, if we were to calculate how much current is being generated from the measured rate of collapse of today's magnetic field, this current is sufficient to account for the actual known strength of the field as it is today! Besides being a good confirmation of the model, this means that the evolutionist's 'dynamo', if it ever existed, must now be switched off.

Now we have already seen that this decay is real, having been measured for over 160 years, as pointed out by Barnes and documented by McDonald and Gunst. So, this Barnes-Humphreys mechanism can account for this Realtime decay of the earth's magnetic field over the past 160 years, the current generated from such field decay correlating well with calculations of the amount of current actually present within the core. Furthermore, Humphreys maintains that it can also account for the magnetic reversals recorded in the rocks having taken place in only days to weeks!

The Flood and Rapid Magnetic Reversals

Now measurements of this 'fossil' magnetism in rock strata (being the local field direction and strength) are different to the global measurements of the strength of the earth's total magnetic field as reported by Barnes, yet the 'fossil' magnetism (paleomagnetism) does record the behavior of the field during the earth's history. Geophysicists have now recognized a continuous sequence of roughly 50 magnetic polarity (field direction) reversals in the magnetism 'fossilized' in rock strata that span the last 600 million years of the evolutionists' timescale, from the so-called Cambrian period when the first metazoan (multi-celled) fossils 'appear' in the rock record to the present. However, since some fossiliferous strata also have reversed polarities preserved in them, the magnetic field must have been reversely polarized when those sediments were being laid down.

Many creation scientists argue that Noah's Flood produced most of these fossiliferous rock layers in a single year. Thus, these reversals of the earth's magnetic field have to be envisaged as occurring on average every week or two during the Flood year. If this were the case, we should then be able to find field evidence of the reversal process having occurred this rapidly, otherwise the Barnes-Humphreys freely decaying electric currents mechanism for the generation of the earth's magnetic field in less than 10,000 years is also in trouble.

But the field evidence has now been found. As already reported, palaeomagnetic measurements of a lava flow at Steens Mountain in Oregon have shown that one of these magnetic polarity transitions (part of a complete reversal) took place in about two weeks, the time period over which the lava would have cooled. As would be expected, the investigators, both evolutionists, were astonished by these results and had difficulty accepting them, but finally had to admit:

'...even this conservative figure of 15 days corresponds to an astonishingly rapid rate of variation of the geomagnetic field direction of 3° per day. ...The rapidity and large amplitude of geomagnetic variation that we infer from the remanence directions in flow B51, even when regarded as an impulse during a polarity transition, truly strains the imagination....We think that the most probable explanation of the anomalous remanence directions of flow B51 is the occurrence of a large and extremely rapid change in the geomagnetic field during cooling of the flow, and that this change likely originated in the (earth's) core.'

THE SUN'S MAGNETIC FIELD

In order to further bolster his case for rapid magnetic polarity reversals, Humphreys has also pointed to a natural object, namely the sun, which demonstrates that a large body can rapidly reverse its magnetic field. Observations show that the sun reverses the polarity of its general magnetic field every 11 years, in synchronism with its sunspot cycle. When the number of sunspots is at a minimum, the observed field on a large scale has its lines of force going mainly north and south.

As the number of sunspots begins to increase, the strength of the north-south part of the field diminishes. In about 5.5 years the north-south component has diminished to zero and the number of sunspots is at a maximum. Then things begin to happen in reverse. A south-north part of the field appears in the opposite direction from its predecessor and the number of sunspots starts to diminish. After another 5.5 years, the number of sunspots is at a minimum again, and the field is back to its original shape, but with the north and south poles of the field having switched places, that is, the sun's magnetic field has reversed its polarity.

Physicists and astronomers do not yet have a theory that completely explains this complex reversal phenomenon. One probable reason why they have had difficulty explaining the sun's reversals is that, because they believe the sun's magnetic field is also generated by a dynamo, they have been looking for a mechanism which would not only reverse the sun's field, but also regenerate and maintain it for billions of years. But if the sun is relatively young (only thousands of years), there is no need for the regeneration requirement. The sun would merely be winding up and unwinding whatever magnetic field it had at creation, losing magnetic energy each solar cycle. Its long-term behavior would thus be a steady decay modulated by the solar cycle of reversals.

Earth's Physical Mechanism

Dr. Humphreys has now proposed a physical mechanism for reversals of the earth's magnetic field during the Flood. We have already seen there is agreement that the earth's magnetic field is generated in the earth's metallic iron-nickel core, most evolutionary scientists preferring a dynamo, as opposed to the free-decay model of Barnes and Humphreys, Figure 18.34. The latter involves an initial endowment of energy in the core at the time of creation and that energy has dissipated and decayed freely as electrical currents in the core since then, the currents generating a magnetic field at the earth's surface and beyond, which has decayed in step with the decay of electrical currents.

However, because the metallic iron-nickel in the earth's outer core is in a fluid state, internal motions occur in this region due to convection flow, for which there is evidence even at present. Humphreys suggests that a powerful event in the earth's core at the beginning of the Flood produced this convection, perhaps by the heating of the core due to a sudden increase of radioactive decay or cooling of the mantle above the core, but these are as yet tentative suggestions that need further analysis. However, once convection flows were initiated in the earth's core, such flows moving upwards in the core towards the mantle would produce a magnetic flux up into the mantle, which would then be conducted to the earth' surface as a magnetic excursion.

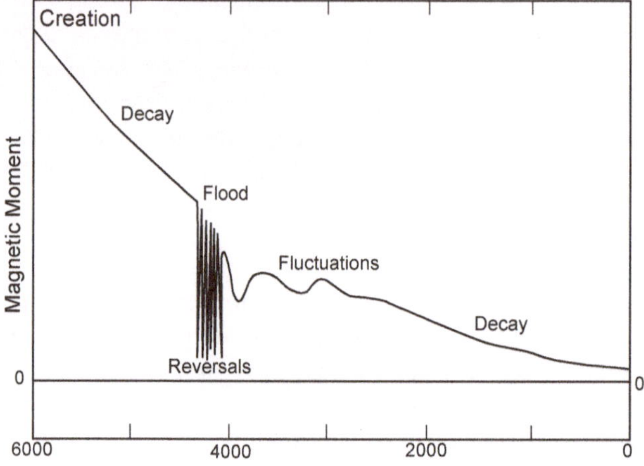

Fig.18.34: Humphrey's Model for the History of the Earth's Magnetic Field

These convective 'updraughts' in the earth's core would have carried more magnetic flux to the surface than 'downdraughts' would have carried away from the surface, so the convective updraughts would have rapidly cancelled out any previous flux above it. The work done by these convection flows in pushing a magnetic flux upward would generate new electric currents, which in turn would generate new flux in the opposite direction. Thus, a cycle of magnetic reversals is set up due to these recurring convection currents, which are maintained as long as there is a strong heat source within the core. In support of his model, Humphreys draws comparisons with convection currents within the sun, which we have already seen are responsible for a rapid magnetic reversal cycle there.

EARTH'S MAGNETIC FIELD HISTORY

Humphreys' model for the history of the earth's magnetic field is more complex than Barnes' original picture of a steady state decay from creation to now, but it does not differ on the essential hypothesis that the earth's magnetic field has freely decayed since creation. Figure 18.34 shows Humphreys' model for the history of the earth's magnetic field, which he divides into five episodes:

1. Creation of the magnetic field along with the earth.
2. Steady decay for nearly 2,000 years until the Flood.
3. Rapid reversals during the year of the Flood.
4. Large fluctuations continuing for up to two thousand years after the Flood.
5. Resumption of steady decay from about the time of Christ until now.

The last period includes the historical measurements which show decay, as reported originally by Barnes.

The model for the reversal process as proposed by Humphreys is simple compared to the evolutionary 'dynamo' theories. It differs fundamentally from the dynamo theories in that it is not intended to maintain the earth's magnetic field for billions of years. Rather, it inverts a previously existing field over and over again. Far from maintaining a field indefinitely, this process accelerates the decay of a planetary magnetic field. The field strength at the peak of each cycle is less than the peak of the previous cycle, because the inverting process does not completely reproduce the flux. This means that the energy contained in the post-Flood magnetic field would be considerably less than that of the pre- Flood field.

According to Humphreys, even though creation scientists explanations of planetary magnetic fields are still in their infancy, they appear to be more complete and successful than the 40-year-old dynamo theories. Indeed, recent magnetic measurements by the Voyager spacecraft as it flew past Uranus and Neptune have confirmed Humphreys' predictions on the origin of the planetary magnetic fields. Furthermore, recent measurements have cast doubt on whether a dynamo really does operate in the earth's core at present. In addition, there is no dynamo theory which accounts for the extremely rapid variations reported at Steens Mountain by Coe and Prevot, but Humphreys' model accounts for this data particularly well. Dynamo theorists even acknowledge that their theories are incomplete, very complex, and not very successful at making predictions. As one such theorist has said:

'...you would have thought we would have given up guessing about planetary magnetic fields after being wrong at nearly every planet in the solar system...'

Age of the Earth

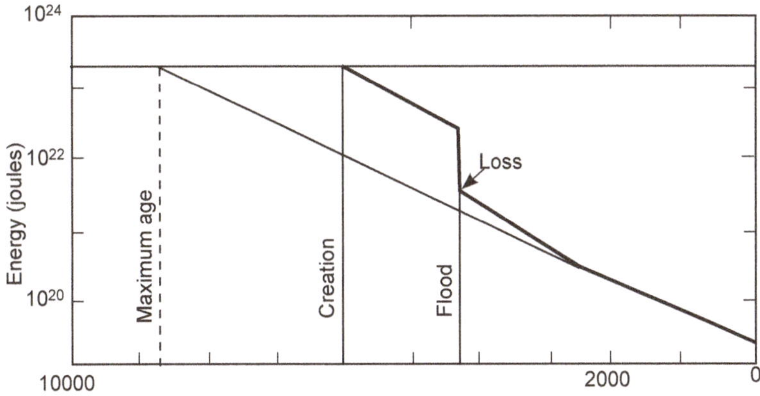

Fig.18.35: Humphrey's model for the history of the earth 's magnetic field

If the creation scientist Humphreys is correct, and seeing that his predictions about planetary magnetic fields in the solar system have been verified, and his model for magnetic reversals here on earth does fit well to the geophysical and rock palaeomagnetic data compared to the woeful state of the dynamo model, then such a decrease in the energy of the earth's magnetic field implies that it is not eternal but relatively recent. Consequently, Humphreys has extrapolated today's energy decay rate back to a theoretical maximum energy, and so has derived an upper limit for the age of the Earth's magnetic field at 8,700 years.

However, he concludes that the rate of energy loss would have been greater during and just after the Flood, due to the postulated powerful heating event in the core at the time of the Flood which set in motion the convection flow, that in turn produced the magnetic reversals and rapid dissipation of the field energy. Figure 18.35 shows one scenario suggested by Humphreys in which about 90% of the earth's magnetic field energy was lost during the Flood or shortly thereafter. He suggests, thereby, that this would make the age of the field about 6,000 years, thus again providing powerful evidence that the earth is as young as the text of Scripture clearly implies.

APPENDIX

a. Yet, something strange happened when these machines were first put into use. As a cross-check for contamination AMS labs measure the amount of ^{13}C in something called the Pee Dee belemnite (PDB). Belemnites like the famous *Belemnitella americana* were squid-like animals with 10 tentacles and a hard internal shell. This one came from the Late Cretaceous Peedee formation in North and South Carolina (USA), which evolutionists 'date' to at least 66 million years old. That shell, the PDB, was the standard used for calibrating $^{13}C/^{12}C$ isotopic ratio measurements across the world for decades. I am still trying to track down the source for this claim, so it is being placed here in a footnote and not in the main text. They were not using the PDB to test for ^{14}C, yet it has a small but measurable amount of ^{14}C. In fact, the amount of ^{14}C is 80 to 100 times above the detection level of the machine. Perhaps it is not as old as they thought!

b. You can do this calculation yourself on a spreadsheet or even the calculator function on your phone. Start with 1×10^{50} and divide that number by 2 until you get to a number less than 1. Then, multiply 5,700 years by the number of times you divided by 2. I get 957,600 years. See also the calculation in Diamonds: a creationist's best friend, Ref. 3.

c. A magnetic field deflects a moving electric charge in a direction perpendicular both to the field direction and the charge's direction of movement. The strength is proportional to charge, field strength, and the sine of the angle between them. The magnetic field causes a deflection. An electric field exerts a force on any charge, proportional to charge and to field strength. The combination, the electromagnetic or Lorentz force, is given by $\mathbf{F} = q(\mathbf{E} + \mathbf{v} \times \mathbf{B})$.Acceleration is inversely proportional to mass, as per Newton's Second Law of Motion.

d. In one cesium ion gun, cesium metal is heated to about 150°C. This is well below the boiling point 671°C (1240°F), but enough for some cesium vapor to rise. The vapor rises through a tube, until it hits the ionizer, a hollow tube of molybdenum heated to about 1,300°C. At this temperature, the cesium atom readily gives up an electron to form Cs^+ ions. They are accelerated by an 8kV voltage towards the graphite. See: Cesium sputter ion source for carbon accelerator mass spectrometry, whoi.edu, accessed 28 Feb 2022.

e. C14 dating was being discussed at a symposium on the prehistory of the Nile Valley. A famous American colleague, Professor Brew, briefly summarized a common attitude among archaeologists towards it, as follows:

f. "If a C14 date supports our theories, we put it in the main text. If it does not entirely contradict them, we put it in a foot-note. And if it is completely 'out of date', we just drop it."

g. Few archaeologists who have concerned themselves with absolute chronology are innocent of having sometimes applied this method, and many are still hesitant to accept C14 dates without reservations.

h. For example, the 'Rev.' Barry Lynn, leader of the anti-Christian group Americans United for the Separation of Church and State, proclaimed in a nationally televised debate, 'Carbon dating, that shows the earth is billions of years old!' (*Firing Line*, PBS, 19 December 1997).

i. The time t since radioactive decay commenced can be given by $\mathbf{N/N_0 = e^{-\lambda t}}$, where N is the number of atoms measured in the present; N_0 is the initial number; λ, the decay constant, which is related to the half-life $t_{½}$ by $\boldsymbol{\lambda} = \mathbf{ln2/t_{½}}$. This presupposes that the system is closed, so that the loss of atoms is solely by decay, and that the decay rate is constant. See also Sarfati, J., *Refuting Compromise*, ch. 12, Master Books, Arkansas, USA, 2004.

j. The earth's mass is 6 $\times 10^{27}$ g; equivalent to 10 $\times$ 4.3^{26} moles of ^{14}C. Each mole contains Avogadro's number (N_A = 10 $\times$ 6.022^{23}) of atoms. It takes only 167 halving to get down to a single atom ($log_2(4.3 \times 10^{26}$ mol $\times$ 10 $\times$ 6.022^{23} mol^{-1}) = $log_{10}(2.58 \times 10^{50})$ / $log_{10}2$), and 167 half-lives is well under a million years.

k. AMS (accelerator mass spectrometry) counts the atoms themselves, and can detect one ^{14}C in more than 10^{16} atoms, or measure a $^{14}C/C$ ratio of $<10^{-16}$ or %0.01 of the modern ratio (0.01 pMC, percent modern carbon).

l. Otherwise, the diamond would anneal into graphite, so-called pencil 'lead'. See Snelling, A., Diamonds—evidence of explosive geological processes, *Creation* **16**(1): 42–45, 1993; cf. Diamond Science, diamondwholesalecorporation.com, accessed 22 May 2006.

m. Vardiman, L., Snelling, A. and Chaffin, E., *Radioisotopes and the Age of the Earth*, Vol. II, ch. 8, Institute for Creation Research, California, USA, 2005. Dr Baumgardner also investigated many coal samples, and they also turned out to have ^{14}C.

n. Baumgardner, J., ^{14}C evidence for a recent global flood and a young earth; in ref. 6, ch. 8. See also his paper at globalflood.org: Measurable ^{14}C in fossilized organic materials: confirming the young earth creation-flood model, 5th International Conference on Creationism, 2003.

REFERENCES

1. Vinogradov, Alexander Pavlovich *et al.*, Radiocarbon dating in the Vernadsky Institute I–IV, *Radiocarbon* 8(1):292–323, 1966; p. 319.
2. Giem, Paul, Carbon-14 content of fossil carbon, *Origins* 51:6–30, 2001.
3. Lowe, D.C., Problems associated with the use of coal as a source of ^{14}C-free background material, *Radiocarbon* 31(2):117–120, 1989.
4. Linick, T.W. *et al.*, Accelerator mass spectrometry: the new revolution in radiocarbon dating, *Quaternary International* 1:1–6, 1989.
5. Arnold, J.R. and Libby, W.F. Age determinations by radiocarbon content: checks with samples of known age, *Science* **110**(2869):678–680, 1949.
6. Morelle, R., New timeline for origin of ancient Egypt, bbc.co.uk/news/science-environment-23947820, 4 Sep 2013. See also Bates, G., Egyptian chronology and the Bible—framing the issues, 2 Sep 2014.
7. The raw data for the Irish oak tree ring can be found at chrono.qub.ac.uk/bennett/dendro_data/dendro.html, last accessed 14 Mar 2022.
8. Hua, Q. *et al.*, Atmospheric radiocarbon for the period 1950–2019, *Radiocarbon* 1–23, 2021.
9. Carbon also weighs 12.0107 kg/mol, where 1 mol equals the number of atoms in 12 g of carbon-12, or Avogadro's Number 6.02214 $\times$ 10^{23} atoms. An AMU is also called a Dalton (Da), after the great creationist chemist John Dalton (1766–1844) who introduced atomic theory into chemistry.
10. If the room is quite warm—the melting point of caesium is 28.5°C (83.3°F).
11. Synal, H.-A., Jacob, S.A.W., and Suter, M., The PSI/ETH small radiocarbon dating system, *Nuclear Instruments & Methods in Physics Research Section B-beam Interactions with Materials and Atoms* **172**:1–7, 2000.
12. For ^{18}O, the isotopic standard is called the Vienna Standard Mean Ocean Water (VSMOW).
13. van der Plicht, J. *et al.*, Recent developments in calibration for archaeological and environmental samples, *Radiocarbon* 62(4):1095–1117, 2020.

14. Snelling, A., Radioactive 'dating' in conflict! Fossil wood in 'ancient' lava flow yields radiocarbon, *Creation* 20(1):24–27, 1997.
15. Thomas, B. and Nelson, V., Radiocarbon in dinosaur and other fossils, *CRSQ* 51:299–311, 2015.
16. Giem, P., Carbon-14 content of fossil carbon, *Origins* 51:6–30, 2001; grisda.org/origins-51006.
17. Baumgardner, J.R. *et al.*, Measurable ^{14}C in fossilized organic materials: confirming the young earth creation-flood model, *Proc. 5th ICC*, pp. 127–142, 2003.
18. Thomas, B., Contamination claims can't cancel radiocarbon results, *Acts & Facts* 49(4), 2020, icr.org.
19. Cupps, V.R. and Thomas, B., Deep time philosophy impacts radiocarbon measurements, *CRSQ* 55(4):212–222, 2019.
20. Woodmorappe, J., The anti-biblical noble savage hypothesis refuted: Do peoples free of biblical influence *actually* live in harmony with nature and each other? rae.org, accessed 28 Feb 2022.
21. Hagmann, M., Columbus didn't do it, science.org/content/article/columbus-didnt-do-it, 31 Jul 2020.
22. Majander, K. *et al.* Ancient bacterial genomes reveal a high diversity of *Treponema pallidum* strains in early modern Europe, *Current Biology* 30(19):P3788–3803.E10, 2020.
23. Dee, M. *et al.*, An absolute chronology for early Egypt using radiocarbon dating and Bayesian statistical modelling, *Proc. R. Soc. A.* 469:20130395, 2013.
24. Ramsey, C.B. *et al.*, Radiocarbon-based chronology for dynastic Egypt, *Science* 328(5985):1554–1557, 2010.
25. Kutschera, W. *et al.*, The chronology of Tell el-Daba: a crucial meeting point of ^{14}C dating, archaeology, and Egyptology in the 2nd millennium BC, *Radiocarbon* 54(3–4):407–422, 2013.
26. Turnbull, J., Sparks, R., and Prior, C., Testing the effectiveness of AMS radiocarbon pretreatment and preparation on archaeological textiles, *Nuclear Instruments and Methods in Physics Research B* 172: 469–472, 2000.
27. Brew, J.O., cited in T. Säve-Söderbergh and I. U. Olsson (Institute of Egyptology and Institute of Physics respectively, University of Uppsala, Sweden), 'C14 dating and Egyptian chronology', in *Radiocarbon Variations and Absolute Chronology*, Proceedings of the Twelfth Nobel Symposium, Ingrid U. Olsson (editor), Almqvist & Wiksell, Stockholm, and John Wiley & Sons, Inc., New York, p. 35, 1970.
28. Christopher Bantick, Extending the dreamtime, *West Australian*, 10 June 2000.
29. Michael Balter, Radiocarbon dating's final frontier, *Science* 313(5793):1560–1563, 15 September 2006.
30. For example, David Rohl in his book *Pharaohs and Kings*. (The US version of his book is titled *A Test of Time.)* CMI published a review of this book in *Journal of Creation* 11(1):33–35. Another such call for drastic shortening of the Egyptian chronology came from Peter James, in his book *Centuries of Darkness*, Pimlico, London, 1992. In the introduction to this book the highly regarded Cambridge Professor Colin Renfrew wrote:
31. 'The revolutionary suggestion is made here that the existing chronologies for that crucial phase in human history are in error by several centuries, and that, in consequence, history will have to be rewritten.' (page xiv)
32. 'The authors of this book . . . show the frailty of the links by which the whole ramshackle chronological structure is held together. I feel that their critical analysis is right, and that a chronological revolution is on its way.' (page xvi)
33. Another worthwhile resource on the need to drastically shorten Egyptian chronology is James Jordan, The Egyptian Problem, *Biblical Chronology* 6(1), January 1994. See also: Ian Taylor, Willard Libby and carbon dating Egyptian artifacts, *Journal of Creation* 17(3):58–59, 2003.

34. Snelling, A., Geological conflict: Young radiocarbon date for ancient fossil wood challenges fossil dating, *Creation* 22(2):44–47, 2000; creation.com/geological-conflict.
35. Baumgardner, J., ^{14}C evidence for a recent global flood and a young earth; in: Vardiman, L., Snelling, A. and Chaffin, E., *Radioisotopes and the Age of the Earth*, Vol. II, Institute for Creation Research, California, USA, pp. 609–614, 2005. See also Diamonds: a creationist's best friend, *Creation* 28(4):26–27, 2006; creation.com/diamonds.
36. Austin, S.A., Excess argon within mineral concentrates from the new dacite lava dome at Mount St Helens Volcano, *Journal of Creation* 10(3):335–343, 1996; creation.com/lavadome.
37. Faure, G., *Principles of Isotope Geology*, 2nd ed., John Wiley & Sons, New York, pp. 1986 ,147–145.
38. Okudaira, T., Hamamoto, T., Prasad, B.H. and Kumar, R., Sm-Nd and Rb-Sr dating of amphibolite from the Nellore-Khammam schist belt, S.E. India: constraints on the collision of the Eastern Ghats terrane and Dharwar-Bastarcraton, *Geological Magazine* 138(4):495–498, 2001; geolmag.geoscienceworld.org/cgi/content/abstract/138/4/495.
39. Miyazaki, T. and Santosh, M., Cooling history of the Puttetti alkali syenite pluton, Southern India, *Gondwana Research* 8(4):576–574, 2005.
40. Cupps, V.R. and Thomas, B., Deep time philosophy impacts radiocarbon measurements, *CRSQ* 55(4):212–222, Spring 2019. See also the summary article, Thomas, B., Contamination claims can't cancel radiocarbon results, *Acts & Facts* 49(4), Apr 2020.
41. Rotta, R.B., Evolutionary explanations for anomalous radiocarbon in coal? *CRSQ* 41(2):104–112, September 2004. ^{14}C in coal was reported by: Baumgardner, J., Humphreys, D., Snelling, A. and Austin, S., The Enigma of the Ubiquity of ^{14}C in Organic Samples Older Than 100 ka, *Eos Transactions of the American Geophysical Union* 84(46), Fall Meeting Suppl., Abstract V32C-1045, 2003. And also: Lowe, D., Problems associated with the use of coal as a source of ^{14}C free background material, *Radiocarbon* 31:117–120, 1989.
42. Giem, P., Carbon-14 content of fossil carbon, *Origins* 51:6–30 (2001), grisda.org.
43. Barnes, T.G., 1971. Decay of the earth's magnetic moment and the geochronological implications. *Creation Research Society Quarterly* 8(1):24–29.
44. For example, Brush, S.G., 1982. Finding the age of the earth by physics or by faith? *Journal of Geological Education* 30:34–58.
45. Barnes, T.G., 1983. Origin and destiny of the earth's magnetic field. *Institute for Creation Research Technical Monograph No. 4,* Institute for Creation Research, San Diego, California.
46. Jacobs, J.A., 1984. *Reversals of the Earth' Magnetic Field*, Adam Hilger Ltd, Bristol, pp. 13–18.
47. Merrill, R.T. and McEilhinny, M.W., 1983. *The Earth's Magnetic Field*, Academic Press, London, pp. 209–263.
48. Opdyke, M.D., Kent, D.V. and Lowrie, W., 1973. Details of magnetic polarity transitions recorded in a high deposition rate deep sea core. *Earth and Planetary Science Letters* 20:315–324.
49. Barnes, T.G., 1972. Young age vs. geologic age for the earth's magnetic field. *Creation Research Society Quarterly* 9(1), pp. 47–50. Return to text.
50. James, R.W., Roberts, P.H. and Winch, P.E., 1980. The Cowling anti-dynamo theorem. *Geophysical and Astrophysical Fluid Dynamics* 15:149–160. Return to text.
51. Inglis, D.R., 1981. Dynamo theory of the earth's varying magnetic field. *Reviews of Modern Physics* 53(3):81–496.

52. Humphreys, D.R., 1986. Reversals of the earth's magnetic field during the Genesis Flood. *Proceedings of the First International Conference on Creationism*, Creation Science Fellowship, Pittsburgh **2**:113–126.
53. Humphreys, D.R., 1988. Has the earth's magnetic field ever flipped? Creation Research Society Quarterly, vol. 25(3), pp. 130–137.
54. Barnes, R.G., 1973. Electromagnetics of the earth's field and evaluation of electric conductivity, current, and joule heating in the earth's core. *Creation Research Society Quarterly* 9(4), pp. 222–230.
55. Humphreys, D.R., 1983. The creation of the earth's magnetic field. *Creation Research Society Quarterly* 20(2), pp. 89–94.
56. Humphreys, D.R., 1984. The creation of planetary magnetic fields, *Creation Research Society Quarterly* 21(3), pp. 140–149.
57. McDonald, K.L. and Gunst, R.H., 1967. An analysis of the earth's magnetic field from 1835 to 1965. ESSA Technical Report IER 46–1ES 1, US Government Printing Office, Washington.
58. Humphreys, D.R., 1990. Physical mechanism for reversals of the earth's magnetic field during the Flood. *Proceedings of the Second International Conference on Creationism*, Creation Science Fellowship, Pittsburgh, Vol. 2.
59. Snelling, A.A., 1991. Fossil magnetism reveals rapid reversals of the earth' magnetic field . *Creation* 13(3):46–50.
60. Coe, R.S. and Prevot, M., 1989. Evidence suggesting extremely rapid field variation during a geomagnetic reversal. *Earth and Planetary Science Letters* 92:296–297.
61. Newkirk, G. Jr. and Frazier, J., 1982. The solar cycle. *Physics Today* 35(4):25–34.
62. Humphreys, D.R., 1990. Beyond Neptune: Voyager II supports creation. *Institute for Creation Research.* Impact No. 203.
63. Lanzerotti, L.J., et al. 1985. Measurements of the large-scale direct-current earth potential and possible implications for geomagnetic dynamo. *Science* 229:47–49.
64. Dessler, A.J., 1986. Does Uranus have a magnetic field? *Nature* 319:174–175.
65. Bagenal, F., 1989. The emptiest magnetosphere. *Physics World*, October 1989, pp. 18–19.

CARBON DATING HISTORICAL MODELS

ARTHUR HOLMES (1890– 1965)

The fascinating biography of Arthur Holmes (1890-1965), Figure 19.1, was written by Dr. Cherry Lewis's, the English geologist famous for his work on radioactive dating and the age of the earth. It traces how ideas about the age of the earth changed over one man's lifetime. Dr Lewis is a geologist/geochemist, currently working as Research Communications Manager at Bristol University, and is Secretary of the History of Geology Group (HOGG) there on the United Kingdom.

Fig.19.1: *Arthur Holmes (1890– 1965)*

Holmes, Figure 19.1, was 'the only child of staunchly Methodist parents' and he 'well remembered his parents' Bible, and the magic fascination of the date of Creation, 4004 BC.' In later years Holmes reminisced 'that the Earth has grown older much more rapidly than I have—from about six thousand years when I was ten, to four or five billion years by the time I reached sixty.' I would like to learn more about Holmes' own reasons for his apostasy from his Christian heritage, because his story sadly is all too common.

A short, interesting book, The Dating Game is filled with photos and human interest. Dr. Cherry Lewis has researched her topic thoroughly, and quotes widely from diaries and letters. Clearly, Holmes was a perceptive and independent thinker, a dynamic lecturer and diligent worker.

WESTERN CULTURE AND THE AGE OF THE EARTH

Is it Really a Game?

Those who think that science has proved the earth is billions of years old should find this book disturbing. Lewis clearly shows that the quest for the age of the earth is not objective science but a subjective, arbitrary and erratic pursuit.

Even her name for the book illustrates that point. Some reviewers must have urged her to use a different title, but what could be more fitting than **The Dating Game**? Lewis refused to change it, but apologized to any readers who found the book 'on the "Romance" shelves.'

My dictionary defines 'game' as 'a contest for amusement in the form of a trial of chance, skill or endurance according to a set of rules.' She vividly paints the characters of the players in the 'dating game', and tracks the progress of the score for a hundred years. To the uninitiated, a history like hers is one of the best ways to understand a subject.

In a game, the score is determined, not by impersonal scientific measurement, but by the strength, skill and creativity of the players. The rules of play are not laws of nature, but arbitrarily agreed by the players, and sometimes changed during play. We see this acutely demonstrated in the events she describes.

Holmes' interest in the game was aroused in his teens one summer holiday. This is when the great physicist William Thomson (Lord Kelvin, 1824- 1907) instigated the dating game in The Times. Lewis described how Arthur and his friend 'were on the edge of their seats with the excitement of it all, for not only did they become familiar with all the arguments, they also got to know all the big names in science at that time—William Ramsay, Ernest Rutherford, Frederick Soddy and Robert Strutt.' Watching The Times exchange influenced Holmes to take up the sport.

THE KELVIN AFFAIR

Holmes began his career at the most interesting time, as Lewis describes. For forty years Lord Kelvin, Figure 19.2, had completely demolished all opposition. But by the early 1900s, Kelvin was gradually losing his dominance. The upcoming generation had a new weapon and were about to dislodge him. Holmes would soon be a key player.

Fig.19.2: Lord Kelvin was a major player in the dating game.

By the end of the 1800s, Lord Kelvin was saying that the earth was between 20 million and 40 million years old,, with a 'personal preference' for the lower value. Of course, Kelvin had not measured the age (otherwise he would not talk about 'personal preference'), but he had calculated it (or rather, logically his methods could calculate at best an upper limit for the age). And before he could start his calculations, he had to make assumptions about the past. In particular, he had to assume a history for the earth.

In fact, every age calculation is based on an assumed history—assumed because, without an eyewitness report, we cannot travel back in time to observe what happened. This reality is not generally recognized—that it is impossible to measure the age of something scientifically— impossible. The numbers quoted for the age of the earth (or the age of the dinosaurs or the age of a volcano) are the outworking of personal beliefs, made to look authoritative by much technical equipment and complicated calculations. It is not until we understand this fundamental fact that the antics of the players in the dating game make sense.

In Kelvin's case, he assumed the earth was initially a ***molten blob*** and calculated how long it would take for the blob to cool (assuming numbers for all the relevant parameters such as its initial temperature, conductivity and reflectivity).

WAS KELVIN'S ANSWER, RIGHT?

Scientifically, it is impossible to say. How could anyone check? What would you compare it with? You could only say whether it seemed reasonable, and Kelvin certainly thought so. For a start, it agreed with similar calculations of the age of the sun by Hermann von Helmholtz (1821–1894). But for the geologists (and evolutionary biologists as shown below), 20 Ma was far too short because they envisaged the earth was unimaginably old. Kelvin knew their long ages flowed from their assumption that geological processes have always operated slowly, like they do at present. They were not based on experimentally established laws such as the laws of heat transfer. Kelvin was blunt, 'A great reform in geological speculation seems now to have become necessary.' In other words, Kelvin challenged them to assume a different history. He knew the geologists could easily harmonize their age for the earth with his result by imagining a bit of catastrophe.

The geologists did not take kindly to Kelvin's suggestion. 'Although incensed,' Lewis explained, 'most geologists were clearly intimidated by Kelvin's authority and felt obliged to heed his arguments.'

The problem was not his 'authority.' The problem was that they agreed with his assumed history for the planet, and could find no mistake with his chosen parameters. After that it was just a matter of cranking the mathematical handle and the age popped out.

But T.C. Chamberlin (1843–1928), head of the Department of Geology at the University of Chicago was not prepared to concede defeat. He speculated that there might yet be discovered new sources of energy within the particles of matter (unknown to him then) that would allow more time than Kelvin had calculated. In geological circles this response is regarded as heroic but it really shows that the age issue cannot be resolved scientifically. Clearly, he was simply defending his belief in long-ages as a matter of faith, without any observational basis whatever.

The evolutionary biologists such as Charles Darwin and T.H. Huxley were not happy with Kelvin either. 20 Ma was nowhere near enough time for evolution to occur. Darwin was particularly dissatisfied, describing Kelvin as his 'sorest trouble.' That suited Kelvin because he opposed evolution. Loren Eiseley writes:

'It can be observed from Darwin's letters that this development in physics gravely troubled him. He refers to Lord Kelvin as an "odious specter," and in a letter [of 1869] ... he writes: "Notwithstanding your excellent remarks on the work which can be effected within a million years, I am greatly troubled at the short duration of the world according to Sir W. Thomson [Lord Kelvin] for I require for my theoretical views a very long period before the Cambrian formation." ... Painfully and doubtfully he [Darwin] wrote to Wallace in 1871, "I have not as yet been able to digest the fundamental notion of the shortened age of the sun and earth."'

Christians often regard Kelvin as a great creationist apologist because of his opposition to evolution. They credit him with keeping Darwin at bay for 40 years. But Kelvin did serious damage to the credibility of the Christian worldview because he publicly promoted an earth history that contradicted the Bible. The Bible says the earth was originally covered with water; Kelvin said it was a molten blob. The Bible indicates the earth is 6,000 years old; Kelvin said 20 million was acceptable. If Christians won't stand on the plain teaching of their own book, why should anyone else accept the Bible as authoritative?

RADIOACTIVITY CHANGED THE GAME

So, the dating game was set for a fascinating turn when Holmes began his career. The dynamics changed dramatically when (Antoine-) Henri Becquerel (1852–1908) discovered radioactivity in 1896. Heat generated by the radioactive decay of elements within the earth was quickly invoked to explain cooling of the earth over long ages. Thus, radioactivity allowed a different history for the earth to be proposed, one that could extend 'for as long as geologists and biologists might need'—notice the word 'need'! It is widely paraded that the discovery of radioactivity solved the heat problem but that is not so. Empirical evidence still favors the view that the earth is much younger than presently believed.

Most importantly, radioactivity allowed age calculations to be applied to individual rocks and minerals, and this is where Holmes became famous. It was Ernest Rutherford (1871–1937), Lewis explains, who was the 'very first person ever to date the true age of a rock.' Her word true is curious because her explanations demonstrate that ages are not found but assumed. (Remember, every age calculation is based on assumptions about the past.) Two pages later, Lewis says that Robert Strutt 'recognized the flaw in the method.' So much for Rutherford's true age.

Apart from the 'flaw', Lewis reveals that in those early days they did not know there were two different uranium decay chains or different isotopes of uranium and lead. In fact, they did not know isotopes existed—this had to wait till Frederick Soddy (1877–1956) in 1913. This illustrates another vital fact about the dating game: assumptions are always made in ignorance. That's why every age result is always tentative, just waiting for a new finding to knock it over, as Lewis illustrates again and again.

Holmes was just 21 when he published his first uranium-lead result for a rock from Norway (still before the discovery of isotopes). He also recalculated ages from data previously published by Boltwood, the oldest result being 1,640 million years. That was a vast increase on any numbers previously published. The response of the scientific community was stunned incredulity. Geologists 'had been given vast time scales to fill with sediments of which there was no evidence.'

METHODS AND DATES ARE SELECTED

One interesting episode Lewis describes involved 'dating' two rocks from northern England using the helium method. The 'dates' Holmes obtained were 182 million years for the igneous sill, and 26 million years for the dyke. Holmes was delighted. He considered the results 'to be in excellent agreement with the geological evidence'. But then, how would anyone know?

But today the sill is considered to be 295 million years old, not 182 million, and the dyke 60 million years compared with 26 million. The helium method has been blamed for the discrepancy because it is held that helium leaks from the rocks, giving too low an age.

But why did Holmes so quickly accept a faulty method? Lewis explains,

'So strong was the desire to find a successful dating technique, he convinced himself that although the helium results were "slightly low", they concurred "quite satisfactorily with the scanty results based on lead ages."'

SO MUCH FOR OBJECTIVE SCIENTIFIC METHODS.

Interestingly, the helium method continued to be used on meteorites and some very ancient dates were obtained. It was argued that, unlike terrestrial samples, meteorites did not lose helium (again, how would you know?). However, when ages were obtained that were anomalously high (i.e., did not agree with what was expected), it was suggested that meteorites gained helium by being bombarded by cosmic radiation as they cruised the heavens. This is another example of an *ad hoc* assumption invoked to dismiss anomalous results.

These stories illustrate how in the dating game, the methods used, and dates obtained, are selected after the event according to whether the results agree with what is already believed and desired to be correct. Recent creationist work is particularly relevant to the helium method. Helium retained in zircons from granite actually provides strong evidence against the idea of millions of years, and for the idea that accelerated nuclear decay occurred in the past. A main pillar of the argument is precisely the rapid leakage of helium noted above—yet much helium still remains in the zircons!

THE HUBBLE AFFAIR

Lewis describes a curious complication that emerged in the late 1940s. As the age of the earth gradually crept up towards 3000 million years, and beyond, the earth eventually became twice as old as the universe. As with Kelvin, the issue became a battle of wills across scientific disciplines, Figure 19.3.

The age of the universe was calculated from the Hubble constant, assuming the big bang history for the universe. Edwin Hubble (1889–1953) had such standing that no-one seriously questioned his value for the constant. As recently as 1936, Hubble had concluded that any further revision of the constant would only be of minor importance.

In the late 1940s the age of the earth became twice as old as the age of the universe, Figure 19.3!

So, the blame was levelled at radioactive dating. Even in 1949 it was considered highly improbable that observational changes in the value of the Hubble constant would resolve the timescale problem. Some astronomers were again suggesting that the radioactive decay rate had changed with time (yet modern creationists are castigated for the same suggestion!).

Fig.19.3: In the late 1940s the age of the earth became twice as old as the age of the universe!

But the astronomers eventually gave in. In the 1950s new measurements of the Hubble constant extended the age of the universe and at last it was 'safely older than the age of the Earth.' This dramatic episode again illustrates that the age issue is a battle of wills and beliefs, and not a scientifically measurable parameter.

PATTERSON TAKES THE PRIZE

As readers would expect, Lewis reveals the answer to the age question just before the end of her book. She relates that Clair Patterson (1922– 1995) 'goes down in the history books as the man who finally dated the true age of the Earth. Wild miracle, finally achieved.'

But here we see an ironic twist. Patterson did not date the earth primarily using earth rocks. His key evidence came from meteorites! What have meteorites to do with the age of the earth?

Remember that before anyone can calculate an age for anything, they have to assume its history, namely how it formed and what has happened to it from that time to the present. Early in the 20th century, T.C. Chamberlin had developed the idea that the earth had formed by the accumulation of cold, solid particles and rocks he called 'planetesimals.' By the 1950s this explanation was widely accepted, and meteorites were considered to be junk left over from when the earth formed.

The number calculated from the meteorite data based on these assumptions gave an age of 4.55 ± 0.07 Ga, the age Patterson announced in 1956, Figure 10.4, and which is still accepted today, Figure 19.4.

Fig.19.4: The presently accepted age of the earth of 4.55 billion years was calculated by Clair Patterson from ... meteorites. Holmes initially said the method was 'unsound in principle'

At first Holmes was not Enthusiastic with the Method

'to use the isotopic composition of lead from iron meteorites as part of the basic data for calculating the age of the earth or its crust, is unsound in principle … the correct procedure is to use terrestrial materials.'

That of course raises one very obvious problem. As Lewis explains,

'If there was no genetic relationship and the Earth and meteorites had not formed at the same time from the same material, then the primeval lead of meteorites would not be that of the Earth; thus, there would be no point of trying to determine the age of Earth from meteorites, and everyone would be back to square one.' To answer this challenge, Patterson produced a graph in 1956 showing the isotopic composition of lead from four meteorites and lead from modern ocean floor sediment. Because the ocean floor sample plotted on the same line as the meteorites, Patterson argued that they all formed from the same cosmic material.

That settled the matter, and the age of 4.55 Ga is universally quoted. Yet, as more ocean floor sediment has been analyzed, it has been found that they do not all fall on the straight line but plot all over the place.

Another problem concerns a view developing 'that the lead isotope clock of the Earth may have been reset by the formation of the Earth's core.' In other words, the consensus history of the earth is different now from what Patterson assumed, yet his result is still accepted as the true age of the earth. As Lewis muses, 'Patterson's results were more fortuitous than was realized fifty years ago.' This raises a question: if we know Patterson's assumptions are wrong, why should we believe that his answer is right?

Clearly the age of the earth is not a scientific issue but a religious and philosophical one.

Why no More Changes?

Lewis's book highlights another fascinating insight into how science works. In the first fifty years of the 20th century, the age of the earth increased from 20 million years to 4,550 million. But in the second fifty years the age has not changed at all. Why?

Some would argue it's because scientists have discovered the correct answer. But how would anyone check? It could also be argued that the changing age was driven by changing cultural and philosophical values in the West. Holmes lived through a period that saw the progressive development of an all-encompassing naturalistic philosophy in Western thought. All supernatural actions by a Creator God were ruled out; only naturalistic explanations were allowed.

The key parameter for every naturalistic explanation is time:

'Time is in fact the hero of the plot ... given so much time the "impossible" becomes possible, the possible probable and the probably virtually certainly certain. One has only to wait: time itself performs miracles.'

In the first fifty years, every academic discipline was developing its naturalistic models. The age of the earth was the crucial parameter in every case: in geology, biology, astronomy, cosmology, geography, archaeology, anthropology, history and so on. Enough time for one discipline was often too little for another—hence the Kelvin and Hubble conflicts. With the present state of play, anywhere between 3 and 7 billion years would probably be suitable. So, 4.55 billion is a happy choice—and it looks precise and authoritative.

4.55 billion years is comfortably less than the age of the universe, allowing enough time for the big bang, stellar evolution and the origin of our solar system. It is also allegedly old enough for geological evolution, for the chemical evolution of the first living cell, for the evolution of life, and for landscape evolution, etc. So, by the mid-20th century the jostling between disciplines had settled down—the different naturalistic models appeared to be meshing together. Everyone had enough time to work with, and there was nothing to gain by changing the number.

IT AFFECTS ME PERSONALLY

The age of the earth is not just an academic issue. As Lewis states,

'By knowing the age of Earth rocks, Moon rocks and rocks from other planets we ... are more able to understand our place in the order of things, our relationship with other celestial bodies. It helps us to navigate our way around the Universe and build up a picture of why we are here at all.'

That, of course, is a religious question involving the meaning of life. In Mozambique in his early twenties, Holmes wrote home about the stars: 'I felt somehow what a fearful meaningless tragedy the whole Universe appeared to be.' If naturalism is true, then Holmes was right—there is no meaning to this Universe.

However, the Bible reveals the true history of the world and why we are here. There is a purpose for this universe, and for every human life. That's why the age of the earth is a critical issue for the Christian worldview. Long ages destroy the credibility and message of the Bible.

I enjoyed Lewis's book because she so vividly demonstrated that the billion-year age of the earth is subjective and arbitrary. Thus, ***it is perfectly valid scientifically to start with the biblical data on the age of the earth and interpret the scientific evidence accordingly***. In reality, the only sure way of knowing the age of anything is by reliable eyewitnesses.

ARCHAEOLOGY CONFLICTS WITH THE BIBLE

"Archaeology team finds 9,000-year-old artefacts in NewBo, Cedar Rapids, Iowa, USA, neighborhood" declares an article from a recent edition of *The Gazette,* Figure 19.5. As biblical creationists, we are used to seeing claims of 'millions and billions of years' for fossils and rocks. These supposedly tell of long epochs of time and evolution and we have become adept at putting such claims into biblical perspective.

Fig.19.5: A flint arrowhead, (similar to the one pictured above), found in NewBo, Cedar Rapids, Iowa, USA, is claimed to be 9000 years old. How do we understand such claims from a biblical perspective?

According to the Bible, there was no death or bloodshed of *nephesh chayyāh* - life before Adam and Eve sinned. Hence, the fossils contained in rocks should be understood as post-Fall—mostly arising from the Genesis Flood—because fossils are preserved dead things.

But what of claims of civilizations that, according to the biblical timeline, would pre-date the Flood?

Is it reasonable to accept that a wheel discovered in Slovenia is between 5,100 and 5,350 years old, or that agriculture flourished and building projects were undertaken 12,000 years ago?

Are these dates still in the biblical 'ball-park'?

Where do we draw the line when archaeologists claim that the oldest pottery is 18,300 years old, or the remains of "Mungo Man", (the first reported Australian human), are 62,000 years old? Are these more recent 'dates' derived by more reliable methods than the highly questionable radiometric dating techniques used to argue that rocks are millions of years old?

How Young? How Old is Young?

Gets a bit confusing, doesn't it?

However, we're *young earth creationists,* right?

Wrong; scripture never uses the phrase 'young' to describe the earth, or humanity. Let me give a few examples to put things in biblical perspective. Peter the Apostle, in 2 Peter 2:5, uses the Greek word ἀρχαῖος (*archaios*) to describe the world that was destroyed by the Flood of Noah. In our English Bibles that word is translated 'ancient' (in 20 versions), or 'old' (in 6 versions).

When Moses, in Deuteronomy 33:15, blessed the twelve tribes of Israel with their possessions of land, he described the hills using the Hebrew word קֶדֶם(*qedem*) which is translated 'ancient' (in 25 versions) and 'old' (in 1 version). The Bible does not indicate 'youth' when referring to the earth, for example, the hills and mountains are described as being very old (e.g., Habakkuk 3:6; Ezekiel 36:2).

So, according to the Bible, the earth close to the time of Noah's Flood is described as being 'ancient' or 'old'. The Hebrew Bible describes the nations which were considered to be old at the time of Israel's exile—for instance Jeremiah 5:15 uses the Hebrew word עוֹלָם(*olam*), when referring to the antiquity of Babylon. 1 Chronicles 4:22 contains genealogical information relating to the Moabites which is described as being from 'ancient records' by the Hebrew word עַתִּיקִים('at-tî-qîm) which means 'old'. Therefore, we can say that, according to the Bible, at 6,000 years old the earth and humanity are ancient, not young.

EARTH HISTORY FROM THE BIBLE

Genesis 5 gives chronogeneological information for human history from Adam to Noah, and the Table of Nations in Genesis 10 gives Noah's three sons' family history. Genesis 11 gives the account of the Tower of Babel along with Shem's family line to Abraham. This has been extensively studied elsewhere, but it is useful to see the chronology of Adam to Abraham.

So, we can see that the period from Creation to the Flood is roughly 1,650 years. Then from the Flood to the birth of Abraham is a further 350 years, Figure 19.6.

Accordingly, the Bible gives us the chronological framework through which we can understand Earth history and the timing of major events.

Believing Scripture to be inerrant, we judge the claims of secular historians and archaeologists against this record. Needless to say, this view is not popular in academia. Indeed, such a presuppositional approach would be considered heretical!

The Bible as history was thrown out of academia post Darwin, first by liberal scholars in the secular universities, starting with a denial of Mosaic authorship of Genesis, and now, sadly, by most evangelical Bible colleges. Such a state of affairs has destroyed the faith of many.

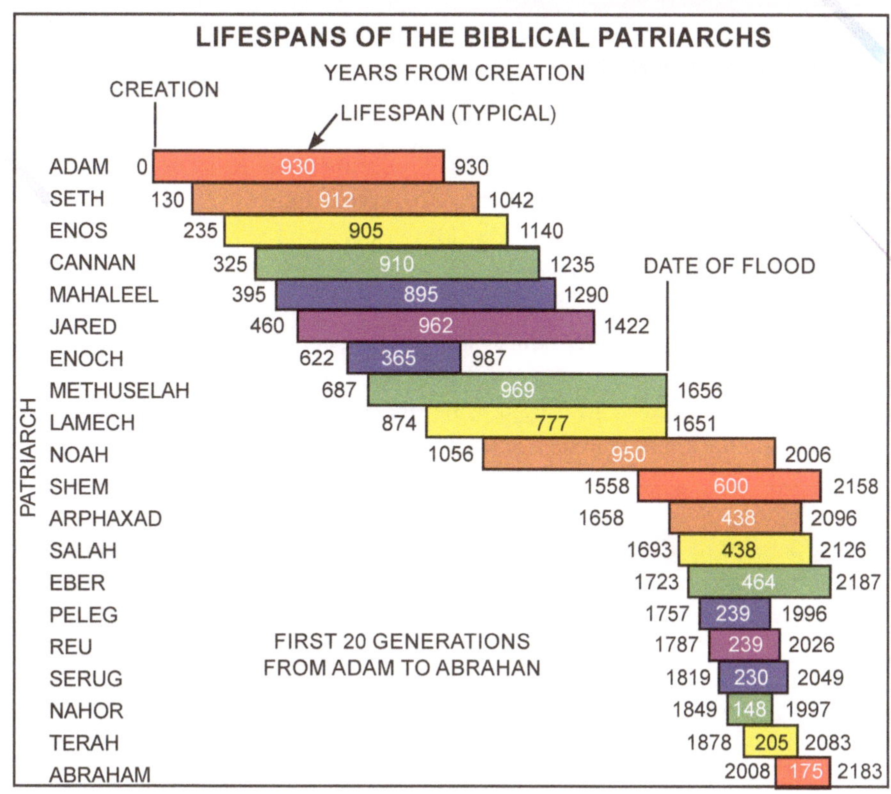

Fig.19.6: Genesis chronogenealogies derived from Genesis chapters 5, and 11 when graphed, show earth's history from Adam to Abraham is approximately 2009 years

So, what of the claims of "9,000-year-old" artefacts found at NewBo, Iowa?

This is clearly of an age far-older than the biblical age for Earth itself, so must be rejected. The artifacts mentioned are hundreds of pieces of knapped flint, as well as a handful of intact spear points. According to the article, "Nine thousand years ago, a group of men sat around a fire in what now is the NewBo neighborhood and repaired their hunting spears."

The researchers, David Benn and his team from Bear Creek Archaeology, analyzed the available evidence—ancient soil, charcoal fragments, flint shards and spearheads—and inferred their relationship to post-glacial (post-Ice Age) soil reported to 'date' from 10,000 BC. This was extracted from a test hole which turned up "an ancient projectile point, from a time when humans were first settling this part of North America."

From a biblical perspective, we would place these finds as post-Flood, as the Flood was the cause of the Ice Age. We can narrow the date down further because humans first settled in North America post-Babel, an event which took place between 101–340 years after the Flood. As for biblical dates for the migration of people groups to their present locations, the details are still being worked out and research is ongoing.

ARRIVING AT A DATE FOR THE NEWBO ARTEFACTS

The studies mention charcoal left by hunters which could offer the opportunity for carbon-14 dating. However, the article was not clear on the source of the 9000-year claim, and the final archaeological report is not available as of the date of this writing. The local geology is described as loess, sands and gravels, and in the area of the dig site the unit is said to encompass "deposits that accumulated primarily during the late Wisconsinan." This is a period believed to represent the end of the last ice age in north America, where multiple ice ages were supposed to have occurred.

However, the evidence better shows that there was only one major Ice Age. Dates for the late Wisconsinan (LW) are derived from radiocarbon dating of wood found in the sediments.

The authors of a paper that offer a chronology for the LW in middle north America admit that dates derived from sediments are often contaminated by "older radiocarbon resulting in chronologic confusion.

By using only dates from wood, much of the confusion disappears." But how much of the confusion disappears is dependent upon the presuppositions of the investigators. The area in question for Iowa is dated at a supposed "12,300 years BP."

It can be noted from photos from the *Gazette* report that layers were identified "between periods" in the exploratory pit dug where the artefacts were recovered. No doubt these layers were used to extrapolate a date up to the -9000year figure using the datum figure of carbon. 12k years given by geological reports from the area.

This was all achieved through assuming gradual deposition of sediments (uniformitarianism). However, it must be stressed that the archaeologists were not there to observe how the sediments were built up over time and how long it took to emplace them; this is all subjective guess work. But what of the reliability of the carbon-14 dating method itself?

Carbon-14 Assumptions

Carbon-14 dating suffers from the same category of unreliability as all dating methods—i.e., the experimenter was not there to observe its initial formation, nor its history to the present. In the case of ^{14}C this is essential. Several factors indicate the ^{14}C clock is unreliable.

Firstly, different species of plants take up ^{14}C at different rates, and this (if at all possible) has to be corrected for.

Secondly, the atmospheric ratio of carbon-12 to carbon-14 has not remained constant; for instance, in recent history the industrial revolution and also atomic testing has changed the ratio.

The earth's magnetic field has not remained constant, and this affects the carbon-12 to carbon-14 ratio by changing the number of cosmic rays entering the earth's atmosphere. The cosmic rays displace neutrons and it is these energetic neutrons which convert nitrogen into carbon-14.

^{12}C 'contamination' can arise from volcanic carbon dioxide which alters the carbon ratio in the wood of trees growing in the area of a volcano, making the wood appear older. The end of the Ice Age also affected the atmospheric ratio of carbon due to the release of carbon from cycling of fresh water with saline.

And lastly, the Flood drastically changed the carbon ratio by burying unquantifiable amounts of ^{12}C in vegetation, thus giving an inflated age to any sample tested.

In short, to base one's faith upon such fallible methodological assumptions is foolhardy indeed; but sadly, many have had their faith shipwrecked on the assumptions of such faulty methodology.

Carbon-14 and Biblical Archaeology

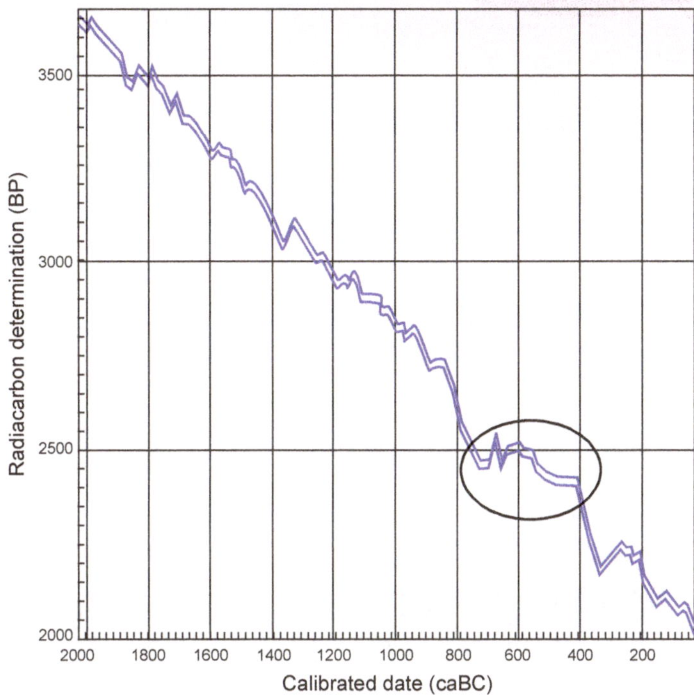

Fig.19.7: The Hallstatt disaster/plateau showing 'flat' curve giving false dates. Curtsy ddd.uab.cat

With respect to using ^{14}C to build chronologies for history involving biblical events the stakes become much higher. For instance, dating of Jericho's fall using ^{14}C in 1995 has simply brought confusion.

Firstly, the calibration errors (standard deviation) of carbon-14 'dates' are often beyond the window required to settle a site's age—to within 50 years—and that is often where the argument lies.

Secondly, there is an acknowledged problem with ^{14}C dating for the period of 400–800 BC. This is known as the "1st millennium radiocarbon disaster area", as ^{14}C produces obviously false dates for this time period, Figure 19.7. The flat area on the graph known as 'The Hallstatt disaster/plateau'.

From the graph it can be seen that no amount of calibration will cause the graph to yield true ages between this archaeologically significant time period. Michael Baillie, a dendrochronology expert, has said of this phenomenon,

"The immediate conclusion is that it is impossible to sensibly resolve the radiocarbon dates of any samples whose true ages lie between 400 and 800 BC. This is a catastrophe for Late Bronze Age/Iron Age archaeology although one which has been predicted for some time."

It is reasoned that this calibration difficulty was produced by an increase in the atmospheric radiocarbon produced by a changing dry-warm to humid-cool climate driven by solar, cosmic ray and Earth magnetic field changes around 700 BC.

Dilemmas in dendrochronology: an attempt to extend the calibration range of ^{14}C

Analyzing tree rings for yearly growth would seem to be a sure way of fixing chronologies. It is widely held that trees develop rings that mark a complete cycle of seasons, where, each year, there would normally be expected a pair of light and dark rings (representing Spring and Autumn wood respectively), in the tree's life, thereby providing a method to calculate a tree's age. Once a tree ring pattern (which relies upon the thickness of rings, not just their number) has been established, this can be compared to older trees and archaeological timbers, so as to extend chronology back in time. This is the theory of dendrochronology and has been explained thus:

"Once it had been shown that a distinctive ring was commonly produced in a given year … a similar ring found in a similar position in another tree could logically be used to date … natural internal markers [which] have been extensively used in dating individual rings, especially in Europe and America."

Such uniformitarian approaches have been used to construct extensive chronologies of Britain using archaeological wood remains, specifically in Ireland.

Fig.19.8: *Pinus longaeva - A specimen of Bristlecone Pine (Pinus longaeva), located in the White Mountains of California. Curtsy - wikipedia.org*

Significantly for biblical creationists, a timeline for the growth of bristle-cone pines (BCPs) (*Pinus longaeva*) Figure 19.8, growing in the White Mountains of eastern California, has been calculated at 8,700 years. One tree dubbed 'Methuselah' has been tree-ring counted to 4,600 years, which, assuming the chronology of the Masoretic text for the Old Testament, places it well before the Flood, which is clearly incorrect. Creationists Ministry International has pointed out that BCPs can grow multiple rings per year, due to their dry environment. Extended chronology relies upon correlation of prone dead wood, with similar ring structures, which are then dated using ^{14}C. Where overlaps with 'identical' ring structures are found, the chronology is considered to be extended.

This is a highly subjective endeavor, which also implies prone wood lay around on the ground without rotting for thousands of years, which is demonstrably false. The entire method of ^{14}C dendrochronology correlation is therefore an exercise in circular reasoning.

As to calibrating ^{14}C to these extended tree-ring chronologies, further uniformitarian assumptions are made about the unobserved past, which do not take into account either the Ice Age, or the Flood. Also, evidence from fossil tree-rings shows that climates were warmer and wetter after the Flood and subsequently dried and cooled to present-day levels, thereby affecting tree ring growth in a non-uniform way.

Significant doubts were raised over the uniformitarian guiding principle behind dendrochronology at a 2015 Association of American Geographers Annual Meeting, Chicago.

The presentation by Dr Henri D. Grissino-Mayer, was provocatively titled, "The long, steady decline of uniformitarianism in dendrochronology: what if the present is no longer the key to the past?" The presentation discussed evidence that change in tree growth rates (since 1963) is due to non-climatic factors. Computer software developed to analyze the statistical factors between climate-tree growth relationships over time suffered from "divergence", i.e., the models used to calibrate the data were themselves calibrated, resulting in data that no longer fitted the model's predictions, showing climate-tree growth relationships had "shifted".

The following conclusions of Grissino-Mayer's paper effectively bring the entire methodology of dendrochronology into doubt. He states:

1. Uniformitarianism/uniformity as a principle may actually be an archaic assumption for tree-ring research in which we analyze trees which are (non-linear) dynamical systems.
2. Tree-ring chronologies which express a significant response change with climate should be used with caution (or in some cases not at all) for such large-scale reconstructions of past temperatures since it is not possible to quantify whether such nonlinear response changes have also occurred in the past. – Wilson *et al.* 2007 in *JGR-Atmospheres*
3. Dendroclimatologists should evaluate whether the climate-tree growth relationship is stable over time.
4. If the climate-tree growth relationship is not stable, then any reconstruction that arises from that relationship may be uncertain and suspect.
5. Ironically, though, uniformitarianism supports temporal instability, i.e., temporally unstable relationships in the present therefore also occurred in the past![28]

Dr. Grissino-Mayer's five concluding remarks sum up what creationist researchers have been saying for years regarding the faulty thinking of the uniformitarian assumptions behind the method. He has, in effect, shone a light on the 'dark-art' of dendrochronology and found it wanting.

If researchers were not there to observe the tree's initial environmental conditions, and measure the growth and relationships to changing seasons through history, then any models built on simple linear factors will be wrong. When major events such as the Ice Age and the Flood are ignored, the history derived from the dendrochronology calculations must be false. The uniformitarian assumptions of historical constant rates are too simplistic; therefore, the science of dendrochronology is no threat to biblical history, which can accommodate the data.

As for the claims of great antiquity for the earth and humanity that far exceed that of the biblical time line, we must treat them with extreme skepticism. The dates provided by the archaeological reports, and often trumpeted in the media, are based upon seriously flawed methodologies.

We have no reason to abandon biblical inerrancy and the true account of Earth history as provided in the scriptures. The investigators' own presuppositions have driven their conclusions, and the flaws within the methodologies invalidate any claims of reliable clocks upon which to build a reliable chronology.

Accordingly, as biblical creationists, we need to be aware of these factors and carefully qualify secular claims of great antiquity with an informed and reasoned critique of the data. Particularly it is necessary to separate out facts from interpretation driven by the presuppositions of the investigator.

Scripture will always remain the final authority by which we measure and compare data and by which we build chronological models of Earth history. Models and secular chronologies may come and go, but the Bible will always remain true.

RADIOMETRIC DISCREPANCIES

Bird Footprints

Using well-known radioisotope technology, scientists dated the Santo Domingo rock formation in Argentina at 212 million years old. This happened to agree well with a nearby geologic formation that was also radiometrically dated.

The radiometric date of the Santo Domingo formation also agreed with the dating based on fossil wood found entombed in the rock. This wood came from an extinct species of tree conventionally believed to have existed around 200 million years ago.

Well-preserved and abundant tracks were also found in the rock, similar in appearance to bird tracks, Figure 19.9. The scientists, who assert that the earth is billions of years old, concluded that the footprints must have been made by an unknown species of a small bird-like dinosaur, because according to Darwinian theory birds weren't supposed to be around 212 million years ago. The results were accepted and published by the science journal *Nature* in 2002.

Fig.19.9: Birds Footprints Invalidated Radiometric Dating

DATING DISCREPANCY

But recently, a different group of long-age-believing scientists took a fresh look at the bird-like dinosaur footprints and concluded that they were indeed made by birds after all—actually, by the familiar sandpiper of today, a small bird common to wetlands, grasslands and coastal habitats around the world.

Many people alive today have seen identical tracks in the sand along a river bank, or at the beach. Realizing that something was very much amiss, the new group asked for further radiometric dating. The new radioisotope date they received gave an age of 37 million years—a massive *175 million years* younger than the original date. The scientists were unperturbed, and the results were again accepted for publication.

The first dating attempt used the Argon/Argon method, and the second used the Lead/Uranium method. In both dating attempts, the date received was very close to what was expected. In the first, the geologists already believed the formation to be around 200 million years old on the basis of an extinct fossil tree—and they received a date very close to this from the laboratory. In the second, it was already believed, due to the presence of sandpiper tracks, that the formation had to fit within the neo-Darwinian timeline for bird evolution—and the geologists again received the date they were expecting.

An episode of geologic thrust-faulting in the past was invoked as a 'just-so' story in an attempt to explain the 175-million- year difference in dates. Such faulting can result in older strata being pushed on top of younger strata, and older rocks, it was said, were mistaken for younger rocks when the first dating was done. However, the explanation is based on circular reasoning, and the scientists admit that the evidence for such faulting is "subtle." This shows how excuses can always be found by geologists who try to avoid the conclusion that long-age radioisotope dating techniques are questionable. Although not widely known, many bird fossils have been found in 'dinosaur era' rocks.

NOT AN ISOLATED CASE

The 175-million-year dating discrepancy for the Santo Domingo formation is not an isolated case, but adds to the growing list of evidence that long-age radioisotope dating does not give real dates at all. For instance, there is also the case of the hominid fossil KNM–ER 1470, found in East Africa beneath a layer of volcanic rock that was first radiometrically 'dated' at 212–230 million years old.

The date was rejected by paleoanthropologist Richard Leakey. He already believed that the rocks underneath the volcanic layer had to be between 2 and 5 million years old, because the australopithecine and mammal fossils found in them indicated such an age to him according to his pre-conceived evolutionary 'model' or timeline. So, he requested new radioisotope dating for the rock layers above and below the fossils, and received 'dates' that fitted perfectly within the range of his already-formed opinion on the age of the rock.

The overlying volcanic layer of rock was given a new 'date' of 2.61 million years—despite the first radioisotope date being more than 200 million years older. Because it was found below this volcanic layer, Leakey estimated his newly-discovered and soon-to-be-famous fossil KNM–ER 1470 was 2.9 million years old. This date was popular for a while, but was eventually challenged by other paleontologists using further radioisotope dating. Today KNM–ER 1470 is generally believed by long-agers to be around 1.9 million years old—a million years younger than Leakey's official 'date'.

Like the sandpiper, Figure 19.9, track discovery in the Santo Domingo formation, Leakey's hominid fossil discovery resulted in a massive change in the radiometric date for the rock formation. Which of these 'dates' are real? All of them are based on man's shifting opinions.

THE VERACITY OF LONG-AGE RADIOISOTOPE DATING

A lot of people, including many scientists who are not geologists, believe that long-age radioisotope dating provides a reliable and empirical measurement of real age. However, the reality is that such 'dating' is based on the types of fossils in the rock. Long-age-believing scientists decide on the age of rock formations according to their naturalistic, secular philosophy, using pre-conceived fossil ages that fit evolutionary assumptions and which trump all other considerations. What if a rabbit fossil was found in 'Precambrian' rock?

The evidence would either be rejected, or explained-away, or the radioisotope 'date' for that rock formation would simply be altered.

The argument that fossils date rocks and rocks date fossils is circular. The fossils are already slotted into an orthodox version or 'model' called the geologic column, that describes an alleged history which is a philosophical construction of fallible man. Before any radioisotope dating is done, laboratories routinely ask geologists (or paleontologists, as the case may be) for their opinion on the age of the rocks. Rocks that have been observed forming in recent times have been radioisotope dated, and the resultant 'dates' were greatly inflated when the laboratory wasn't first informed of the known age of the rock.

Sometimes long-age-believing scientists admit that radioisotope dating has significant problems and weaknesses. In their view this is mainly to do with the great difficulty in finding rock or crystal samples that have not been weathered, altered, or derived from older rock. But as "Marvin Lubenow" put it, "the practical matter of selecting rock samples that can be proved pure and uncontaminated requires an ***omniscience*** beyond humans." The bottom line is that long-age radioisotope 'dates' are demonstrably unreliable and ever-changing. They aren't real.

The Biblical View

If radioisotope dating is empirical science, it must be able to stand or fall on its own consistency of results. Long-age radioisotope 'dating' can't change wildly by hundreds of millions of years, according to geologist's changing opinions, and still be considered a true, empirical method of dating.

The Santo Domingo formation is consistent with large volumes of sediments having been rapidly deposited by massive water currents whilst volcanic eruptions were intermittently occurring.

If the secular radiometric 'dates' can be dismissed because they do not fit with evolutionary ideas, they can also be dismissed because they do not fit with what the Bible teaches about the age of the earth. The Bible preserves the historical account of a year-long watery catastrophe that affected the whole world about 4,500 years ago. The physical characteristics of the Santo Domingo formation are completely consistent with this.

DINOSAUR BLOOD DISCOVERED

Actual red blood cells in fossil bones from a *Tyrannosaurus rex*? With traces of the blood protein hemoglobin (which makes blood red and carries oxygen)? It sounds preposterous—to those who believe that these dinosaur remains are at least 65 million years old.

It is of course much less of a surprise to those who believe Genesis, in which case **dinosaur remains are at most only a few thousand years old.**

In a recent article, scientists from Montana State University, seemingly struggling to allow professional caution to restrain their obvious excitement at the findings, report on the evidence which seems to strongly suggest that **traces of real blood from a *T. rex* have actually been found.**

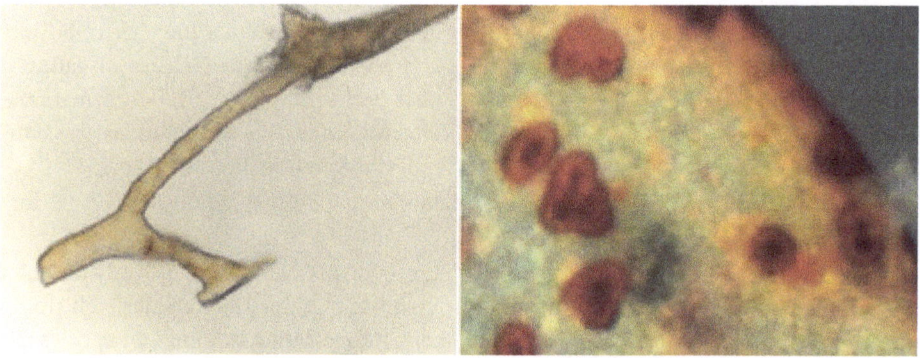

*Fig.19.10: These photos are of a later (2005) find by **Schweitzer** which produced soft tissue, in addition to strengthening the red blood cell identification—Curtsy - **M. H. Schweitzer***

Left: The flexible branching structures in the *T. rex* bone were justifiably identified as "blood vessels." Soft tissues like blood vessels should not be there if the bones were 65 million years old, Figure 19.10.

Right: These microscopic structures were able to be squeezed out of some of the blood vessels, and can be seen to "look like cells" as the researchers said. So once again there is scope for Dr Schweitzer to ask the same question, Figure 18.10.

"How could these cells last for 65 million years?"

The story starts with a stunningly preserved *T. rex* skeleton unearthed in the United States in 1990. When the bones were brought to the Montana State University's lab, it was noticed that 'some parts deep inside the long bone of the leg had not completely fossilized.' To find un-fossilized dinosaur bone is already an indication more consistent with a **young age for the fossils**.

Let **Mary Schweitzer**, the scientist most involved with this find, take up the story of when her co-workers took turns looking through a microscope at a thin section of this *T. rex* bone, complete with blood vessel channels.

'The lab filled with murmurs of amazement, for I had focused on something inside the vessels that none of us had ever noticed before: tiny round objects, translucent red with a dark center. Then a colleague took one look at them and shouted, "***You've got red blood cells. You've got red blood cells!***"'

Schweitzer confronted her boss, famous paleontologist 'Dinosaur' Jack Horner, with her doubts about how these could really be blood cells. Horner suggested she try to prove they were not red blood cells, and she says, 'So far, we haven't been able to.'

Looking for dinosaur DNA in such a specimen was obviously tempting. However, fragments of DNA can be found almost everywhere—from fungi, bacteria, human fingerprints—and so it is hard to be sure that one has DNA from the specimen.

The Montana team did find, along with DNA from fungi, insects and bacteria, unidentifiable DNA sequences, but could not say that these could not have been jumbled sequences from present-day organisms. However, the same problem would not be there for hemoglobin, the protein which makes blood red and carries oxygen, so they looked for this substance in the fossil bone.

MORE ON FRESH DINO BONE

To claim that bone could remain intact for millions of years without being fossilized (mineralized) stretches credulity. The report here of red blood cells in an un-fossilized section of dinosaur bone is not the first time such bone has been found.

Biologist Dr Margaret Helder alerted readers of *Creation* magazine to documented finds of 'fresh', un-fossilized dinosaur bone as far back as 1992.

The evidence that hemoglobin has indeed survived in this dinosaur bone (which casts immense doubt upon the 'millions of years' idea- is, to date, as follows:

- The tissue was colored reddish brown, the color of hemoglobin, as was liquid extracted from the dinosaur tissue.
- Hemoglobin contains heme units. Chemical signatures unique to heme were found in the specimens when certain wavelengths of laser light were applied.
- Because it contains iron, heme reacts to magnetic fields differently from other proteins—extracts from this specimen reacted in the same way as modern heme compounds.
- To ensure that the samples had not been contaminated with certain bacteria which have heme (but never the protein hemoglobin), extracts of the dinosaur fossil were injected over several weeks into rats. If there was even a minute amount of hemoglobin present in the *T. Rex* sample, the rats' immune system should build up detectable antibodies against this compound. This is exactly what happened in carefully controlled experiments.

Evidence of hemoglobin, and the still-recognizable shapes of red blood cells, in un-fossilized dinosaur bone is powerful testimony against the whole idea of dinosaurs living millions of years ago. **It speaks volumes for the Bible's account of a recent creation**.

DINOSAUR SOFT TISSUE

A Stunning Rebuttal of 'Millions of Years'

It has been announced the discovery of what seemed to be microscopic red blood cells, and immunological evidence of hemoglobin in dinosaur bone. Also, response to critic was provided.

Now a further announcement, involving the same scientist (Montana State University's Dr Mary Schweitzer) stretches (pun intentional) the long-age paradigm beyond belief.

Not only have more blood cells been found, but also soft, fibrous tissue, and complete blood vessels. The fact that this really is un-fossilized soft tissue from a dinosaur is in this instance so obvious to the naked eye that any skepticism directed at the previous discovery is completely 'history'.

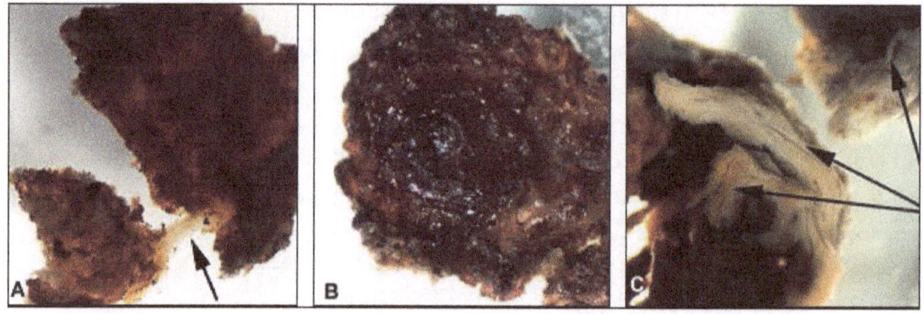

Fig.19.11: Dinosaur Soft Tissue Curtsy - Science via AP-(From www.msnbc.msn.com/id/7285683/)

A: The arrow points to a tissue fragment that is **still elastic**. It beggars's belief that elastic tissue like this could have lasted for 65 million years, Figure 19.11.

B: Another instance of '**fresh appearance**' which similarly makes it hard to believe in the 'millions of years.' Figure 19.11.

C: Regions of bone showing where the **fibrous structure** is still present, compared to most fossil bones which lack this structure. But these bones are claimed to be 65 million years old, yet they manage to retain this structure. Figure 19.11.

One description of a portion of the tissue was that it is 'flexible and resilient and when stretched returns to its original shape.'

The exciting discovery was apparently made when researchers were forced to break open the leg bone of a *Tyrannosaurus rex* fossil to lift it by helicopter. The bone was still largely hollow and not filled up with minerals as is usual. Dr Schweitzer used chemicals to dissolve the bony matrix, revealing the soft tissue still present.

She has been cited as saying that the blood vessels were flexible, and that in some instances, one could squeeze out their contents. Furthermore, she said, 'The microstructures that look like cells are preserved in every way.' Akso, she is reported as commenting that 'preservation of this extent, where you still have this flexibility and transparency, has never been seen in a dinosaur before.'

It appears that this sort of thing has not been found before mainly because it was never looked for. Schweitzer was probably alert to the possibility because of her previous serendipitous discovery of *T. rex* blood cells. (It appears that the fossils were sent to her to look for soft tissues, prior to preservative being applied, because of her known interest.) *In fact, Schweitzer has since found similar soft tissue in several other dinosaur specimens*!

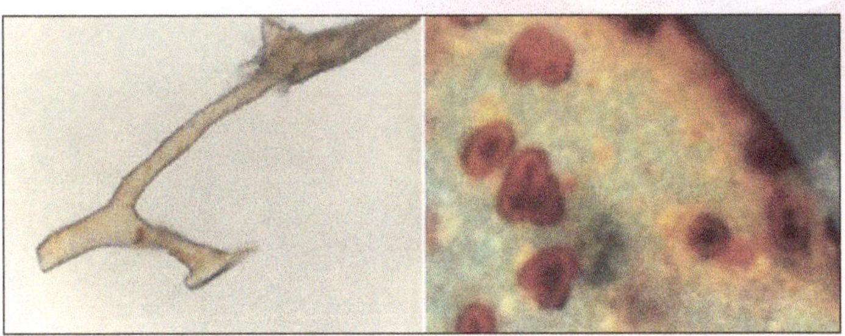

Fig.19.12: T. rex bone 'blood vessels.' Soft tissues like blood vessels Curtsy - M. H. Schweitzer

Left: The flexible branching structures in the *T. rex* bone were justifiably identified as '**blood vessels**.' Soft tissues like blood vessels should not be there if the bones were 65 million years old. Figure 19.12.

Right: These microscopic structures were able to be squeezed out of some of the **blood vessels**, and can be seen to '**look like blood cells**' as the researchers said. Figure 19.12.

Therefore, once again there is scope for Dr Schweitzer to ask the same question, 'How could these cells last for 65 million years?'

The reason that this possibility has long been overlooked seems obvious: the overriding belief in 'millions of years.' The long-age paradigm (dominant belief system) blinded researchers to the possibility, as it were. It is inconceivable that such things should be preserved for (in this case) '70 million years.'

Will they Now be Convinced?

Unfortunately, the long-age paradigm is *so* dominant that facts alone will not readily overturn it. As philosopher of science Thomas Kuhn pointed out, what generally happens when a discovery contradicts a paradigm is that the paradigm is not discarded but modified, usually by making secondary assumptions, to accommodate the new evidence.

That's just what appears to have happened in this case. When Schweitzer first found what appeared to be blood cells in a *T. Rex* specimen, she said, 'It was exactly like looking at a slice of modern bone,' But, of course, I couldn't believe it. I said to the lab technician: 'The bones, after all, are 65 million years old. How could blood cells survive that long?' Notice that her first reaction was to question the evidence, not the paradigm. That is in a way quite understandable and human, and is how science works in reality (though when creationists do that, it's caricatured as non-scientific).

However, will this new evidence cause anyone to stand up and say there's something funny about the emperor's clothes? Not likely.

Instead, it will almost certainly become an 'accepted' phenomenon that even 'stretchy' soft tissues must be somehow capable of surviving for millions of years. (Because, after all, we 'know' that this specimen is '70 million years old'.)?

Schweitzer's mentor, the famous 'Dinosaur Jack' Horner (upon whom Sam Neill's lead character in the *Jurassic Park* movies was modeled) is already urging museums to consider cracking open some of the bones in their existing dinosaur fossils in the hope of finding more such 'Squishosaurus' remains. He is excited about the potential to learn more about dinosaurs, of course. But—nothing about questioning the millions of years—sigh!

I invite the reader to step back and contemplate the obvious. This discovery gives immensely powerful support to the proposition that dinosaur fossils are *not* millions of years old at all, but were mostly fossilized under catastrophic conditions a few thousand years ago at most.

Dino Protein Denial

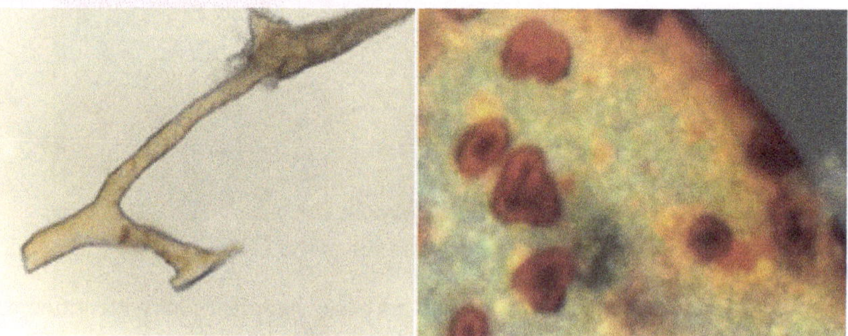

Fig.19.13: These photos are of a 2005 T. rex soft tissue find by Dr Mary Schweitzer—see Still Soft and Stretchy. Curtsy M. H. Schweitzer

Left: The flexible branching structures in the *T. rex* bone were justifiably identified as "blood vessels". Soft tissues like blood vessels should not be there if the bones were 65 million years old. Figure 19.13.

Right: These microscopic structures were able to be squeezed out of some of the blood vessels, and can be seen to "look like cells", paralleling this latest report of the North Dakota hadrosaur. Figure 19.13.

So, once again there is scope for Dr Schweitzer to ask the same question, she asked over a decade ago (see Dinosaur bone blood cells found), "How could these cells last for 65 million years?"

Discoveries of soft tissue and proteins in dinosaur remains are unacceptable to many scientists. That's because if dinosaurs have been extinct for 65 million years, as evolutionary dogma teaches, then cells and even proteins should have decayed long ago.

That view can radically affect how researchers view the evidence.

For example, electron microscopy and X-ray imaging of a well-preserved hadrosaur ("duck-billed" dinosaur) fossil unearthed in North Dakota revealed what was described as "cell-like structures comparable to those of living vertebrates."

Please, observe that University of Manchester researchers Phil Manning and Roy Wogelius were unwilling at that stage to call them "cells". But subsequent analysis of the hadrosaur skin and a claw found amino acids (constituents of proteins) "suggesting that the cell-like structures were indeed cells".

The blatant evolutionary bias of the researchers is highlighted by their reaction to their findings, as reported by *New Scientist*:

"Manning says the presence of amino acids, rather than whole proteins, is a good sign. After 66 million years, proteins in soft tissue should have broken down into amino acids, so finding large proteins would likely be a sign of contamination. The high concentrations of amino acids in the fossil, compared with only traces found in the surrounding sediment, support the idea that they came from the fossil."

In other words, if intact proteins had been found in the hadrosaur fossil, the researchers would have *denied the result*, instead dismissing it as "contamination". Such is the sacrosanct view of the millions-of-years' timeframe, the researchers would rather question their own analysis than the supposed age of the fossil!

This is not the first time that evolutionists have taken such a stance. Evolutionary paleontologist Hans Larsson has actually called for ^{14}C testing ("^{14}C dating should be done") of the "still-soft-and-stretchy" *T. rex* samples reported by fellow paleontologist Mary Schweitzer.

Note that this is not because he wants to "date" the dinosaur remains, but in order to determine whether they have been contaminated by modern microbes! (I.e., the presence of ^{14}C cannot be from a 70-million-year-old *T. rex*, Larsson would reason, because ^{14}C cannot last that long, therefore the ^{14}C is from recent contamination, therefore non-dinosaurian. However, contamination can't explain the widespread ^{14}C in diamonds, which thus can't be billions of years old as evolutionists claim.

The evidence of intact proteins and soft tissue being found in dinosaur remains continues to mount—right in line with the Bible's timeframe of history. The dinosaur fossils date back no more than a few thousand years at most, not millions of years.

INFLATED CARBON-14 DATES

Subfossil Trees

Artificially-inflated ^{14}C dates have been found to occur when trees absorb 'infinitely old' carbon dioxide released into the atmosphere from local, volcanogenic, subterranean sources.

This is not to be confused with wood contamination because the carbon is firmly locked within the wood fibers. A similar effect has long been recognized with the fictitious 'built-in' carbon-14 dates that occur in mollusks when they absorb 'infinitely old' carbon from carbonate rocks.

In addition, creation scientists recognize that the global atmospheric buildup of ^{14}C after the Creation and Flood would have produced artificially-old carbon-14 dates. However, the widespread emanation of ^{14}C-free volcanogenic carbon dioxide after the Flood would have further inflated the carbon-14 dates of tree rings in a systematic manner in many parts of the world.

The carbon-14 dating method is based on the assumption that tiny amounts of radioactive ^{14}C, produced as cosmic rays hit nitrogen in the upper atmosphere, become incorporated within the bodies of living things.

After death, generally no new ^{14}C can enter the body, nor can any ^{14}C leave. Instead, the ^{14}C gradually disappears from the body by undergoing radioactive decay. This occurs at a half-life of approximately 5,700 years. By assuming that ^{14}C in the present atmosphere, and hence in the organism at the time of death, was essentially the same in the past, and that a closed system has existed since the death of the organism, we can compute how many half-lives of ^{14}C have passed, and hence how many years have elapsed, since the organism died.

Based on these uniformitarian assumptions, dates up to about 40,000 years are believed to be attainable. But even under these assumptions, there is ample evidence that carbon-14 dating has serious problems. According to conventional geology, the ^{14}C in once-living objects older than about 100,000 years should have all gone, yet we frequently find objects supposedly *millions* of years old that contain measurable quantities of carbon14-. And even conventionally-believable dating results are often discarded if they conflict with some preferred hypothesis.

ALTERNATIVE GLOBAL-BIOSPHERE CONDITIONS

Creation scientists are willing to leave these uniformitarian mental boxes and thus have studied carbon-14 dating from a decidedly *non*-uniformitarian viewpoint. One creationist model envisions the earth created some six thousand years ago, the Flood about 1700 years thereafter and ^{14}C building up either after Creation or after the Flood. Because most living objects buried during the Flood contained very little ^{14}C when they died, they already possessed inherited carbon-14 dates (usually at infinity, but sometimes at a few tens of thousands of years, as discussed earlier).

Post-Flood organisms successively acquired less extreme 'built-in' carbon-14 dates at the time of death until they eventually converged upon 'real-time' ages a few thousand years ago.

MECHANISM FOR SPURIOUSLY-HIGH ^{14}CARBON DATES

It turns out that there is another mechanism, probably active after the Flood, that creates greatly-exaggerated carbon-14 dates. Unlike the earlier-discussed global processes, it operates at the local level (relative to each living thing), and is particularly successful at altering the ^{14}C content of living trees. Let us consider its revolutionary implications.

a. The ratio of ^{14}C and ^{12}C is measured to calculate the 'age' of matter. **Figure 19.14.**

b. If the tree imbibes its carbon from a lower ^{14}C source than the general atmosphere (e.g., takes up CO_2 from volcanic gases) then the ratio will be higher in ^{12}C than usual and therefore show an incorrect old 'age.' **Figure 19.14.**

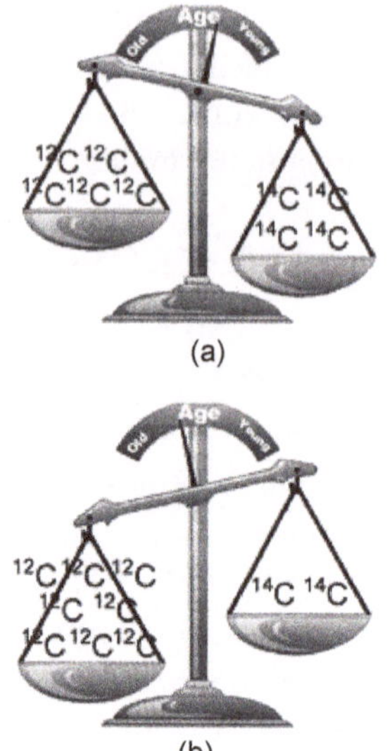

Fig.19.14: Incorrect Old Age of ^{14}Carbon

For the longest time, it had been supposed that, given standard assumptions, carbon-14 dates from properly-decontaminated wood are virtually foolproof. After all, a tree grows a ring each year, and no new structural material is subsequently added to the ring.

When newer, mobile, material is leached away by chemicals, the remaining structural wood fiber undoubtedly contains only the carbon (and hence ^{14}C) that it contained during the year that the ring formed. And, so the reasoning continues, since closed-system conditions are almost guaranteed and since trees get their carbon from the CO_2 in the air, and not from the soil, each ring must reflect the ^{14}C concentration in the atmosphere when it formed.

The key word in the above-described set of assumptions is ***'air'***. Trees absorb whatever carbon dioxide gas is within their vicinity. In the absence of other sources, the only source of CO_2 *is* the atmosphere. However, what other source could there possibly be?

One source is ***volcanogenic*** gases. And, since deep subterranean carbon usually had no prior contact with the atmosphere, it has zero ^{14}C and therefore an infinite carbon-14 age. Now, consider a tree that imbibes half of its CO_2 from the air and the remaining half from local volcanogenic gases. Its concentration of ^{14}C at time of death is only half that of the ambient atmosphere, and hence it dies having a 'built-in' carbon-14 age of 5,700 years (one half-life).

Tuscany, Italy, is probably the first place where 'inherited' carbon-14 dates on wood were described. These dates, much too old to be attributed to any past civilization in Italy, were determined from timbers located several kilometers from a volcano. Since that report, other examples of this phenomenon have surfaced from all over the world.

A recent, detailed study has shed further light on the dynamics of this process. Particularly interesting is the fact that these 'bad' carbon-14 dates do not occur haphazardly, but to the contrary:

'The pattern of ^{14}C depletion in the annual rings is remarkably consistent between all three of the trees cored, suggesting that either changes in CO_2 flux are occurring homogeneously across the entire area of the tree kill, or that trees integrate CO_2 flux very well over relatively large areas.'

Under the right conditions, inherited carbon-14 dates can therefore mimic 'real' ones.

^{14}C DEPLETION AFTER THE FLOOD

All the foregoing examples are infrequent, and localized. However, the situation must have been very different for some time after the Flood. A great deal of 'infinitely-old' carbon dioxide must have been percolating from the depths, all over the world, and over considerable geographic regions, as a result of residual volcanic activity, upper-mantle activity, etc. As the growing plants and trees absorbed much of this ^{14}C-free CO_2 flux, they necessarily acquired quasi-homogenous 'built-in' carbon14- dates—not as an exception, but as a rule.

A large fraction of the 'very old' carbon-14 dates we presently obtain by routine use of the carbon-14 dating method may therefore owe to this mechanism in addition to, or instead of, the earlier-discussed buildup of global atmospheric ^{14}C since the time of Creation or the end of the Flood. Clearly, this volcanogenic CO_2 mechanism deserves further study.

RADIOACTIVE DATING NO PROBLEM FOR THE BIBLE

What about radioactive dating? Doesn't that prove the world is millions of years old? Radioactive dating may be one of the big questions looming in your mind.

However, the idea of an unimaginably old earth did not come from radioactive dating. It was popular long *before* radioactivity was discovered. It came from a geologic philosophy, not a scientific measurement. Figure 19.15.

Note too that radioactive dating is something that most people don't understand. Normal people are not familiar with isotopes, mass spectrographs, rubidium, strontium or half-lives. We find ourselves in the position where we are being asked to trust the specialists, of not being able to check the facts first hand.

Fig.19.15: Isotopes - Photo iStockphoto

However, it's not difficult to understand the basic principles to realize that alleged ages of millions of years have not been measured objectively, but derived from subjective assumptions.

What is Radioactive Decay?

Radioactive dating begins by carefully measuring the concentrations of radioactive isotopes in rocks. Everything is composed of elements and there are about 90 naturally occurring ones, such as hydrogen, carbon, oxygen and iron. Each element comes in different forms, called isotopes, most of which are stable and do not change. Some isotopes, however, are unstable and decay radioactively into other elements. There are many different radioactive isotopes that are used for radiometric dating.

For example, there is a radioactive form of potassium (potassium-40) that decays into argon (argon-40). The unstable **potassium** isotope is called the parent while the **argon** product is called the daughter. There are a couple of different radioactive forms of **uranium** that decay into **lead**. There is a radioactive form of **thorium** that also decays into **lead**. There is an isotope of **samarium** that decays into **neodymium**, and one of **rubidium** that decays into **strontium**.

How does Radioactive Dating Work?

Radioactive dating is often illustrated with an hour glass. The sand grains at the top of the sealed glass are like the atoms of the parent isotope in the rock, and those at the bottom like the atoms of the daughter.

Radioactive decay is where the parent atoms change as a result of radioactive decay into daughter atoms, like the individual grains of sand falling from the top to the bottom of the glass. The hourglass depends on the sand falling at a regular rate. Like an hour glass, it is said, you simply measure the parent and the daughter elements and you can calculate the age.

What was the Starting Amount?

However, an hour glass is only useful if we saw it turned over *and* observed that the bottom glass was empty. In other words, the hourglass only works when we know its ***initial condition***. Unlike the hourglass, we do not know how much of each isotope was in the rock in the beginning. That's because we did not observe what happened in the past when the rock formed. Neither can we travel into the past to make the necessary measurements. All we can do is guess. This is the fatal problem that essentially makes radioactive dating useless as a primary method for determining age.

Geologists don't like to assume the amount of daughter directly (perhaps that sounds like cheating), but they often do, and they call it a 'model' age. Geologists prefer to make *indirect* assumptions. They may assume that different minerals in the rock originally had the same isotopic ratios to start with. Or they may assume that different rock samples from the same geographical area had the same ratio. Each dating method uses different kinds of assumptions to get around this problem for radiometric dating—the deadly problem caused by the fact that we cannot make measurements in the past.

Has the Rock been Disturbed?

Apart from the fatal problem of not knowing the initial conditions, there is another problem that is just as deadly. We don't know what happened to the rock during its 'lifetime'.

An hour glass is only useful if it is not disturbed, Figure 19.16. But after rocks crystallize from molten magma, they can be heated and cooled; they can be affected by metamorphic events and groundwater. These geologic events can cause elements to be gained and lost to the rock. It's like cracking the hourglass and having some of the sand leak out, or other sand leak in. How can we know what disturbances have affected the elements in our rocks? Again, we can only guess.

Fig.19.16: *The Hour Glass Assumption of Parent/Daughter Isotope - Curtsy - Photo iStockphoto*

EVERY DATE HAS TO BE INTERPRETED

Did you hear about the old wood cutter who was bragging about his axe? 'I've had this trusty axe for fifty years,' he said. 'It's only had two new heads and three new handles.' The question is: how old was his axe?

It's much the same with rocks. When a geologist hammers off a sample of rock, he needs to know its history. Different minerals would have crystallized at different times depending on the way the molten magma cooled. Some small pieces of other rock, or even some foreign minerals, may have been carried along by the magma and existed long before the rock crystallized.

Other minerals may have grown inside the rock much later, during a time when the area was heated and metamorphosed. Some minerals may have crystallized even later still when ground waters in the area percolated through the pores of the rock. So, the age of a rock is quite a complicated question, and we first need to know its entire history before we can develop a story to explain the isotopic measurements.

This means that, on its own, a radioactive 'date' is meaningless. Geologists recognize this. You may be surprised to learn that a geologist would never collect a rock at random and send it off for radioactive dating on its own. The result would mean nothing. Every radioactive date has to be interpreted before anyone can say what it means.

What happens is that the geologist will carefully record exactly where he collected the rock. He explores the geology of the area so he can understand the geological history, and where his particular sample fits into the sequence of geological events. He checks out the ages other geologists have assigned to the different rocks in the region. He studies samples of his rock under the microscope looking for clues of how it crystallized, whether it was later heated, deformed, altered or weathered.

Then, when the laboratory sends him the 'date' for his rock, he can decide what the date refers to. Does it represent the time the rock crystallized or when it cooled? Or perhaps the date refers to the time when the rock was heated or deformed or altered, or somewhere between two of these. Or maybe the date refers to an earlier time, a time when the magma melted before the rock even formed. So, the geologist has a lot of options he can choose from as he develops a story to explain the meaning of the date for his rock. He can even combine a number of different explanations to explain his result.

And even after the geologist has interpreted his date and published his interpretation in a journal, another geologist may later decide that there is a problem with that interpretation, and say the date should be disregarded or reinterpreted.

So, radiometric dating never has the final word. It's not objective like the lay-person is led to believe.

Has the Decay Rate Ever Changed?

An hourglass is only useful for telling time if the sand always falls at the same rate. An hourglass can be disturbed if it tips over, is shaken, or gets moisture inside.

Likewise, radioactive dating will only be reliable if the radioactive decay rate of the isotopes has never been disturbed. Each different kind of isotope decays at a regular, repeatable rate called its 'half-life.' It is generally believed that the decay rates for isotopes would never change, even under the sorts of conditions that could be experienced deep inside the earth, or even inside other planets.

However, there is one survey of the scientific literature that refers to more than two dozen experiments where changes in decay rates were reported. Laboratory experiments have quantified, for certain radioactive decay processes, how much the rate is affected by the chemical and physical conditions, but in these cases the changes observed are small. On the other hand, it has been demonstrated in the laboratory that under certain conditions the radioactive decay rate can be accelerated a *billion*-fold.

Some may argue that these sorts of conditions would not apply on the earth, or that the changes are only small in most cases. But in recent years, a group of seven creationist research scientists, called the RATE group, has identified examples in the field that point toward accelerated nuclear decay.

They have also developed a theoretical basis for how accelerated decay could occur.

The fact is that we cannot travel into the past so we cannot know all the different conditions that have existed on Earth and to which rocks may have been subject. So, the idea that decay rates have remained absolutely constant over all time is a belief, not a fact.

And even secular scientists have sometimes proposed that the decay rate changed in the past in order to resolve a disagreement between the age of the earth and the age of the universe.

Not Objective Measurement, but Subjective Assumption

We are all familiar with measuring time so we should easily see that radioactive dating is not everything it's claimed to be. In an Olympic race, for example, the official starts his stopwatch *when* the starting gun sounds. He stops his watch *when* the athlete touches the finish line. He reads the time from his watch.

However, what would happen if he missed the beginning of the race and only saw the finish? It would be impossible for him to measure the time, no matter how accurate his watch. We all know that, so we should all see the inherent problems with radioactive dating.

Every 'scientific' dating method, including radioactive dating, needs to know the initial conditions of the rock. But, unlike the Olympic official, we were not present at the beginning so we can only *assume* how the rock formed and what the conditions were. Not only that, but we must also assume what happened to the rock during its lifetime.

Clearly, radioactive 'dates' are not independently-determined objective measurements of age. Rather, all dates are based on subjective assumptions. And because long-age researchers don't take the Bible's history seriously they make assumptions that are inconsistent with it. That's why their answers contradict the Bible. But the numbers they quote are all based on assumptions and don't disprove the biblical timescale at all.

DATING THE SHROUD OF TURIN

Another look at its Authenticity

Controversy surrounds the Shroud of Turin (hereafter 'the Shroud'), which some say is the authentic burial cloth of Jesus Christ. This cloth shows the front and rear image of a man who appears to have undergone a lot of torture. Due to several lines of evidence, we think that the Shroud of Turin is not the authentic burial cloth of Jesus Christ:

Bible: Our conclusions are primarily based on the biblical evidence, namely that according to John 11:44 and John 20:7 the Jewish custom was to bury their dead using several cloths, not just one. The Jews buried Jesus with a face cloth, which disqualifies the Shroud as being the burial cloth of Christ. Furthermore, Jesus was buried with seventy-five pounds of extremely sticky spices, according to John 19:40, whereas the Shroud shows no signs of them.

Morphology: Several features of the man in the Shroud appear to be distorted, and he is unusually tall, compared to the average height of a first-century Jewish man. Also, he was clearly not *wrapped* in the cloth, as the image does not show the sides of the head or body.

Physical Chemistry: It is also questionable why the blood stains have remained red so long after death.

Nuclear chemistry: Pro-Shroud researchers have always called the reliability of the multiple carbon dates that have been obtained from the Shroud into question. However, despite their attempted re-evaluation of the radiocarbon dates, the only conclusion one can draw from them is that the Shroud is not 2,000 years old. We reject the idea that Jesus' body disappeared from within the Shroud while emitting neutron radiation, which supposedly left traces on the front and rear sides of the Shroud.

Provenance: Many false relics are known from the Middle Ages, including many from the regions of northern Italy and France. This raises the suspicion that the Shroud is also a forgery, since it was first displayed in the 14th century in France. There is no 'paper trail' that gives us a clear chain of custody and it cannot be known that earlier objects with similar claims (e.g., the Image of Edessa) are one and the same.

Manufacturing: It is possible that the image on the Shroud was formed by common biochemical reactions called Maillard reactions. But, even if the Shroud was once wrapped around a human body, this would preclude the body of Jesus because these reactions are associated with decomposition. We should also not overlook the ingenuity of medieval artisans. For example, Leonardo da Vinci was known for his detailed descriptions of anatomy and the mechanical structures that he engineered.

In the end, we do not know how the Shroud was made, but neither do we need to know. We lose nothing if it is not authentic. Even the Apostles did not appeal to physical evidence for the Resurrection. Instead, they appealed to eyewitness testimony. Those testimonies are still with us today, in the pages of the New Testament.

TURIN, NORTHERN ITALY

There is a controversial piece of linen cloth residing in a cathedral in Turin, northern Italy. Called the Shroud of Turin, it is claimed to be the burial shroud of Jesus Christ. Strangely, it bears the full-length frontal and dorsal negative imprint of a man's body, Figure19.17.

The Shroud is a single piece of cloth about 4.3 meters (14.2 ft) long and 1.1 meters (3.6 ft) wide. It was first displayed publicly in the 1350s in Lirey, France. In 1532 the Shroud suffered fire damage in the chapel where it was housed. Since it was folded at the time, this resulted in a series of repetitive burn holes. Patches were then sown on to repair the more damaged sections. In 1578 it was passed into the hands of the Dukes of Savoy, who deposited it at the cathedral of Turin, in northwestern Italy. Further repairs were made to it in 1694 and 1868. In 1983 the Shroud was given to the Vatican, but it still resides in St. John's Cathedral, under the guardianship of the archbishop of Turin.

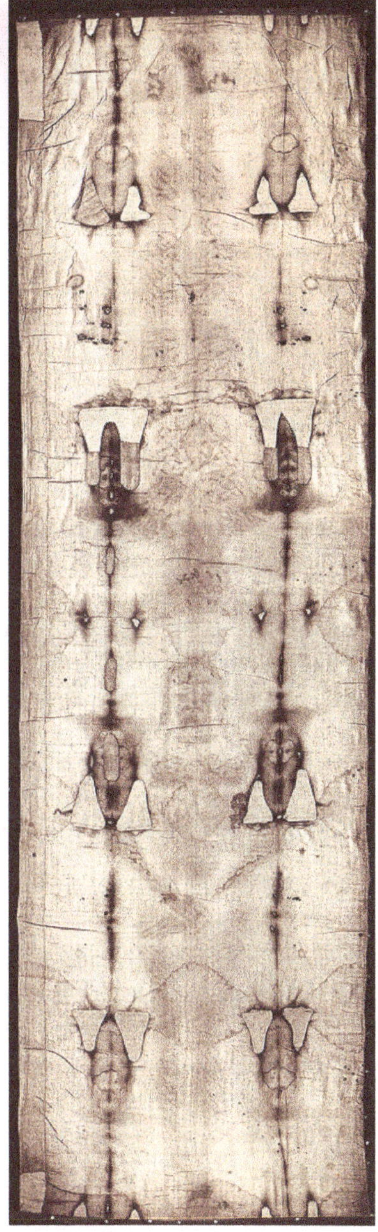

Fig.19.17: *The Shroud of Turin contains a faint dorsal (top half) and frontal (lower half) image of a man, with many features paralleling the Crucifixion. Yet, the historical record of the Shroud is spotty, multiple features on it conflict with the biblical record of events, and carbon dating places it squarely in the medieval era, Figure 19.17.*

Its Authenticity Debated

Its authenticity has always been debated, but this has only gotten worse in recent years. Therefore, creation scientists have taken the time to examine the 'for' and 'against' cases based upon both scientific and biblical evidence. This review depends heavily on two recent books, both of which argue for the authenticity of the Shroud. The first is Mark Antonacci's 2015 book, *Test the Shroud at the Atomic and Molecular Levels.*

Antonacci is a lawyer and founder and president of Test the Shroud Foundation. He is a leading expert on the Shroud of Turin and has spent 30 years studying it. Second is Thomas de Wesselow's 2012 book *The Sign – The Shroud of Turin and the Secret of the Resurrection.*

Wesselow is an art historian who holds a Ph.D. from London's Courtauld Institute of Art. He has researched the Shroud for 12 years.

We also studied *Sacred Blood, Sacred Image, the Sudarium of Oviedo*, by Janice Bennett, who has a Masters of Arts in Spanish literature and a certificate in advanced Biblical studies from the

Catholic Biblical School of Denver. From a somewhat skeptical position we also referenced *Relic, Icon, or Hoax? Carbon Dating the Turin Shroud*, by Harry E. Gove, late emeritus professor of physics at the University of Rochester, New York. Finally, we also read but didn't find much new material in *The Truth about the Shroud of Turin*, by journalist Robert K. Wilcox.

Is it a Genuine Image?

The big question is, does the Shroud really bear the image of Jesus? Is it a genuine burial cloth of some unknown person, or is it just an elaborate forgery? Is it possibly a by-product of naturally occurring chemical processes? Or maybe a combination of these things?

Different groups have different stakes as to whether the Shroud is real. It is one of the most high-profile relics of the medieval Roman Catholic Church and is venerated by many Roman Catholic faithful. Therefore, the Roman Catholic Church has a very real interest in the authenticity of the Shroud. Despite this, the church itself does not make any direct claim that the Shroud is authentic.

On the other hand, atheists and skeptics do not believe the Shroud is real. Their atheistic worldview excludes *a priori* any kind of miracle. Thus, they inadvertently bias themselves when trying to refute the evidence allegedly supporting the authenticity of the Shroud. They fall victim to automatically explaining away evidence which in fact might be real.

Protestants may be least inclined to be biased about the Shroud. Based on the principle of *sola Scriptura*, Protestants hold the Bible as the sole and highest authority pertaining to faith. Therefore, if the Shroud is real, it is just one more proof that Jesus rose from the dead. But if the Shroud is fake, Protestants lose nothing. Catholic doctrine does not hold to *sola Scriptura* but also allows for tradition to help form their doctrine. Catholics *should* lose nothing if the Shroud is not real, as long as they have not gone too far out on a limb in accepting it. The fact of the Resurrection of Jesus is what is at stake, not the Shroud.

THE BIBLICAL EVIDENCE

It would be fitting to start out with the description of Jesus' burial in the New Testament. There are four mentions of Jesus' burial cloth, two in the Gospel of Luke and two in the Gospel of John.

Luke 23:52–53 describes the way Jesus' body was wrapped before he was entombed:

"This man [Joseph of Arimathea] went to Pilate and asked for the body of Jesus. Then he took it down and wrapped it in a linen shroud and laid him in a tomb cut in stone, where no one had ever yet been laid."

Luke 24:12 describes how Peter found the Shroud after Jesus had risen from the dead:

"But Peter rose and ran to the tomb; stooping and looking in, he saw the linen **cloths** by themselves; and he went home marveling at what had happened." (emphasis added)

John 19:40 says:

"So, they took the body of Jesus and bound it in linen **cloths** with the **spices**, as is the burial custom of the Jews." (emphasis added)

John 20:5–7 describes the burial cloth of Jesus in a little bit more detail:

"And stooping to look in, he [the 'disciple Jesus loved'] saw the linen **cloths** lying there, but he did not go in. Then Simon Peter came, following him, and went into the tomb. He saw the linen cloths lying there, **and the face cloth, which had been on Jesus› head, not lying with the linen cloths but folded up in a place by itself**." (Emphasis added)

According to Luke and John, there were multiple pieces of cloth, contrary to the *single*-piece Shroud of Turin. The only other possibility is that Jesus was wrapped in strips of cloth that were smeared with sticky myrrh and aloes, a cloth was placed over his face, then the wrapped body was laid on a separate linen sheet *that the Bible does not mention*. This sheet would then have to be folded over the top of the body, starting at the head. But not only would this be an argument from silence, the sheet should have become stuck to the spice-wrapped strips of linen and the inner cloth layers would have absorbed and otherwise obscured the blood and blood patterns on the body.

The Shroud depicts the face of a man on it, but John 20:5–7 states that Jesus' face cloth was a separate piece of linen, set aside in a place beside the burial "cloths". John 20:5 calls these cloths "τὰ ὀθόνια" (*ta othonia*), which is in the *plural* in Greek. A better understanding, from the Greek text itself, is that the body was wrapped in *multiple* strips of cloth and the face was covered by a separate cloth.

We see this in another New Testament passage that deals with then-current burial customs, and this occurred only a few miles from Jesus' burial site. John 11:11–45 describes the resurrection of Lazarus by Jesus. Verse 44 describes in detail what Lazarus looked like when he came forth from his grave:

"The man who had died came out, his hands and feet bound with linen **strips**, and his **face wrapped with a cloth**. Jesus said to them, 'Unbind him, and let him go.'"

Even if it only specifically mentions his hands and feet, here Lazarus is bound with multiple clothes, just like Jesus was in John 20:7, with a separate napkin around his head.

What did Shroud Researchers Actually Find?

A team of scientists from the "Shroud of Turin Research Project (STURP)" examined almost every single thread on the Shroud. They reported several physical characteristics which they claim have been "coded" into the fabric of the Shroud by some unusual process.

THE IMAGE OF THE MAN IN THE SHROUD

The most notable feature is the face and body of the image on the Shroud, which shows quite a bit of detail. The man's eyes appear to be closed, and he has hair down to his shoulders. STURP scientists counted 130 'blood stains' coming from the man's body. His arms are crossed over his groin area. STURP scientists also claim that the man was scourged by a whip with a dumbbell-like tip, which they claim was commonly used by Roman executioners during the first century. Furthermore, based on the nature of the man's wounds, it appears that he had been carrying a heavy object (possibly the crossbar) to his execution. The man also showed evidence that his feet and wrists (not his hands) were pierced, that he had been stabbed in the side, and that he had worn a crown of spiky objects (possibly the crown of thorns). Streams of blood are visible going down the back of the man's hands.

However, several lines of evidence contradict the idea that the image in the Shroud was that of Jesus Christ. First, the shoulder-length hair should have fallen backwards, since Jesus was lying down in the tomb. Alternatively, his hair would not be free-flowing if his head was wrapped in a separate cloth. Instead, the man's hair seems to be falling to his shoulders due to gravity. STURP scientists claim that they can detect signs of trauma that the man in the Shroud underwent. The man in the Shroud has a full beard, without any hair torn out. Thus, there is no sign of the trauma that would have happened when Roman soldiers tore out wads of Jesus' beard when He was being tortured (Isaiah 50:6). This Old Testament passage is referenced by Matthew 26:67 and 27:30, during which the Roman soldiers strike Jesus in the face and also spit on Him. Even though the two passages from Matthew do not specifically mention that Jesus' beard was torn out, we can still identify Jesus speaking in Isaiah 50:6 where this is specifically mentioned. Also, the famous 'He was pierced for our transgressions' passage in Isaiah 52 and 53 says:

As many were astonished at you—his appearance was so marred, beyond human semblance, and his form beyond that of the children of mankind—so shall he sprinkle many nations. (Isaiah 52:14–15a)

Based on this passage, we wonder how the face in the Shroud can be so free of the trauma we would expect both from the relevant biblical passages and the way Roman soldiers were famous for treating condemned criminals.

The figure of the man in the Shroud also has unusually long fingers and long arms. Normally, a man's arms below the wrist would not cover his groin when lying flat on his back – the wrists would only cover the groin when a person's head and legs are raised when lying down. This would be an unusual posture for someone lying inside a tomb.

De Wesselow thus makes the surprising claim that the man's head dropped about forty degrees on the cross and stayed that way due to rigor mortis. The rigor mortis in the arms, which were outstretched on the cross beam, must have been broken so that they would have fit under the Shroud, so why not the neck? But rigor mortis only begins to set in several hours after death, before which Jesus would have been laid in the tomb, so his argument is invalid.

Strangely, the width of the right leg is twice that of the left leg above the knee on the frontal image, but not on the dorsal image, however this might be due to distortion of the Shroud image based on the way it was (presumably) draped across the body. The man in the Shroud seems to lack a navel. It might be that the image might be too blurred for it to be noticeable. Although, if STURP scientists were capable of discerning small coins as stated below, they should also be able to make out images of a navel as well.

The height of the man in the Shroud is 5 ft 10 in, according to Antonacci, although de Wesselow says it is a full 6 ft, but this doesn't matter much. We must remember that people were shorter in times past, and even today the mean height of Jewish males in different parts of the world is at most 1.71 m (5 ft 7 in). While it is not inconceivable that Jesus was tall for his time, the height of the man in the Shroud makes it less likely that this really was Jesus Christ.

Antonacci writes on pp. 81–82 of *Test the Shroud* that the man in the Shroud was of a physical type found in modern times among Sephardic Jews. This is puzzling, because Sephardic Jews first appeared in the area around Spain only in the second millennium, possibly around as late as the 11th century. If the man in the Shroud resembled a Sephardic Jew, this would exclude the possibility that the Shroud is truly Jesus' burial cloth. The fact that the man in the Shroud resembles a Sephardic Jew means that such a person could possibly have served as the model for a medieval forger to base the Shroud on. Worse, one wonders what, exactly, he is talking about, for there are essentially no physical characteristics that separate Jewish people from non-Jewish people, nor would any be expected. Even general differences would be incredibly difficult to make out in the Shroud image.

THE SUDARIUM OF OVIEDO

Thus, it is holding a contradicting claim to the Shroud. Yet, both it and the Shroud are medieval, according to carbon dating.

The Sudarium of Oviedo, Figure 19.18, is a 34- by 21-inch (0.86 x 0.53 m) cloth kept in a silver chest in the Cathedral of San Salvador in Oviedo, Spain. It has been studied extensively by the Investigative Team of the Spanish Center of Sindonology (EDICES). The term sudarium in Latin means a sweat-cloth and they are generally about the size of a napkin or a hand-towel. The Sudarium of Oviedo is large enough to completely wrap around the head of a person.

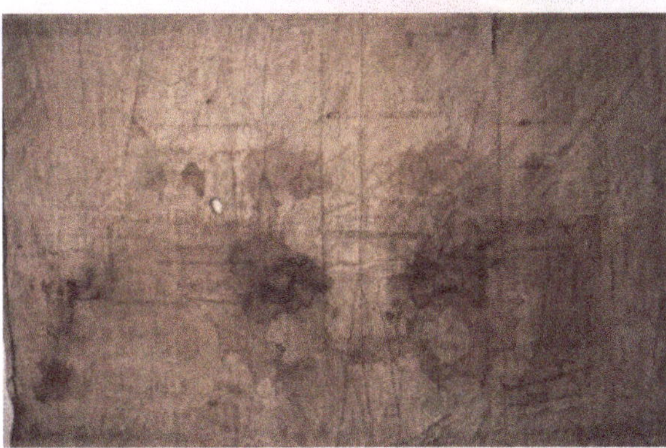

Fig.19.18: *The Sudarium of Oviedo is a blood-stained cloth purported to be the face-cloth from Jesus' burial – Curtsy Shroud.com.*

According to the results of the EDICES research team, the Sudarium originally covered the head of a crucifixion victim and became stained from the blood and pulmonary edema fluid flowing from the victim's nose and mouth. The Sudarium was allegedly fastened to the back of the victim's head by pointy objects. The team suggests this was perhaps the crown of thorns (Sacred Blood, p. 75). Whoever took the body down from the cross would have had to wrap his head in this cloth first. They also claim the Sudarium was re-wrapped around the head after the body of the victim was placed in a horizontal position, remained in place while the victim was transported to a nearby location, then removed and set aside. They claim it did not receive its stains from the Shroud and would not be expected to have the same stain pattern (Sacred Blood, p. 15). In the Bible, it would be the separate napkin that the disciples saw neatly folded beside the other linen garments in the empty tomb (John 20:7).

Strangely, they claim to have found traces of aloe, myrrh, residues of beeswax, vegetal wax, and conifer resin on the bloody parts of the Sudarium (Sacred Blood, p. 69). We have already mentioned the absence of these substances on the Shroud. But this raises a giant question: what explains the presence of aloe and myrrh on the Sudarium, if it was removed from the head when the victim was laid into the tomb, and prior to the body being covered in these sticky substances?

This story is even more questionable when you consider John 11:44, which states that when Lazarus arose from the grave, his face was still wrapped with a cloth. Why did they remove the Sudarium from Jesus' face and not from Lazarus' face? Or, if the Sudarium was just a sweat cloth that was used to cover and/or clean Jesus' head, why was Lazarus' burial different?

Other sources besides EDICES claim that the Sudarium was applied to the face of Jesus after He was covered by the Shroud, others say that it was used to support Jesus' jaw as a sort of chin-band. Furthermore, four churches in France and three in Italy also claim additional or competing portions of the grave cloths of Jesus (Sacred Blood, p. 14). There is no consistent story.

EDICES also claims that there are seventy points of coincidence between the Shroud of Turin and the Sudarium of Oviedo (Sacred Blood, p. 84). However, de Wesselow claims in the Sign (p. 230) that this is far from straightforward. For example, the Sudarium has a taffeta weave and was coarser in texture. This fits with its suggested use for wiping sweat from the face or as a turban. Its weave is different from the herringbone weave of the Shroud.

Like the Shroud, the Sudarium has been carbon dated, this time by two independent laboratories. A lab in Tucson, Arizona obtained a date range from **642** to **869** AD. A lab in Toronto, Canada obtained a date range between **653** and **786** AD (Sacred Blood, p. 78). Yet again, the science does not support the age claims.

Also Like the Shroud, the early years of the Sudarium are surrounded by mystery. In 614 AD (curiously close to the carbon dates) it was allegedly transported from *Jerusalem* to *Alexandria, Egypt* and then to *Spain*, after Jerusalem was invaded by the Persians. Even Bennett admits that this journey is poorly documented and mixed with fantastic legends.

In conclusion, there is no reason to believe the Sudarium is authentic, that it had anything to do with the burial of Christ, or that it has any relationship to the Shroud.

BLOOD STAINS ON THE SHROUD

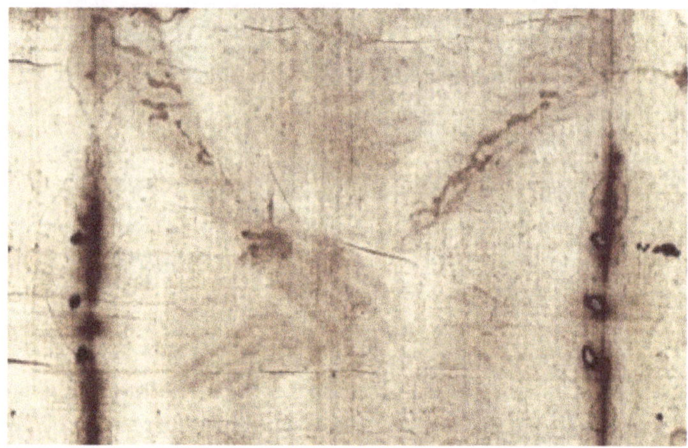

Fig.19.19: *Hand and forearm area of the man in the shroud.*

Antonacci says in *Test the Shroud*, that because the Sabbath was coming, the Jews had little time to wash Jesus' body before wrapping it in cloths and placing it in the tomb (p. 94). If the body was not washed, the Shroud should be smeared with blood, yet the wounds are clear and there is little evidence of 'smearing'. Even more peculiar is the fact that the supposed blood marks on the Shroud are still visibly reddish in color, when it is a well-known fact that blood turns dark brown fairly rapidly after oxidizing in air. Some shroud proponents claim that the still reddish coloration of these blood marks is due to bilirubin, which is a chemically decomposed component of blood that is often pumped out of a person's liver when in shock. Yet bilirubin is yellowish in color and is the chemical responsible for the condition called jaundice, or a yellowing of the skin, Figure 19.19.

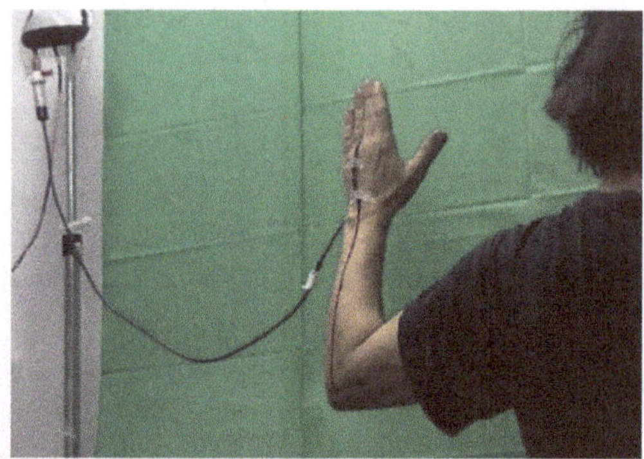

Fig.19.20: *Volunteer simulating blood flowing down his arm in a blood pattern analysis test, taken from figure 4b from Borrini and Garlaschelli, 2019.*

Interestingly, de Wesselow contradicts Antonacci by claiming that the body of Jesus had been washed, in accordance to burial traditions in the Jewish Mishnah (*The Sign*, p. 124, referring to Mishnah Shabbat 23:5). But he also claims that a body would be wrapped *only loosely* in a single sheet called a *sobeb*/*sovev*. However, this provision came to us from sixteenth-century Jewish Law. Let us also not forget that Jewish tradition does not supersede biblical authority, which mentions that Jesus was wrapped in multiple cloths (John 20:7).

Results from blood pattern analysis by Dr. Matteo Borrini and Luigi Garlaschelli suggest that the trickle pattern of the blood on the man in the Shroud do not correspond to the way blood would have flowed from a wounded man, nailed to a cross with his arms at an angle of approximately 45 degrees. Rivulets of blood on the back of the left hand are consistent with a person whose hands are stretched out at 35–45 degrees above horizontal. However, blood stains on the forearm on the Shroud could only come from someone holding their hands nearly vertically, in which case blood would flow all the way down the forearm, instead of at an angle, Figure 19.20.

Coins on the Shroud?

STURP scientists claim that inscriptions from small coins called leptons made imprints on the Shroud of Turin. This was supposedly part of Jewish burial customs in the first century, but this is not mentioned in any of the four Gospels. The scientists even claim that part of the inscription is visible on the Shroud's material. This indeed might sound like convincing evidence, however, looking at the evidence in closer detail raises doubts about whether this is true.

The researchers claim that the letters UCAI from the word "TIBERIOY **KAI**CAROC" (Tiberius Caesar) are visible alongside a curved staff, or *lituus*, on what seems to be a coin imprinted on the Shroud, Figure 19.21. The letter U in UCAI forms the upper two bars of the letter Y in TIBERIOY (*Test the Shroud*). There is an obvious misspelling in the inscription, namely that there is a C instead of a K in the word KAICAROC. This could be due to the minters leaving off the vertical line from the K, creating a letter C instead. STURP scientists claim, however, that they have found four lepton coins with this self-same misspelling, meaning that this is not so uncommon (*Test the Shroud*, pp. 69–72).

While such coins might exist, one study concluded that some religious people are significantly predisposed in detecting what they think are actual words on relics of religious significance, such as the Shroud of Turin. In other words, their 'eyes of faith' may see something that others not only don't see but which might not actually be there. The letters UCAI are short enough to be just a random pattern seen in the cloth.

OHIO SHROUD CONFERENCE

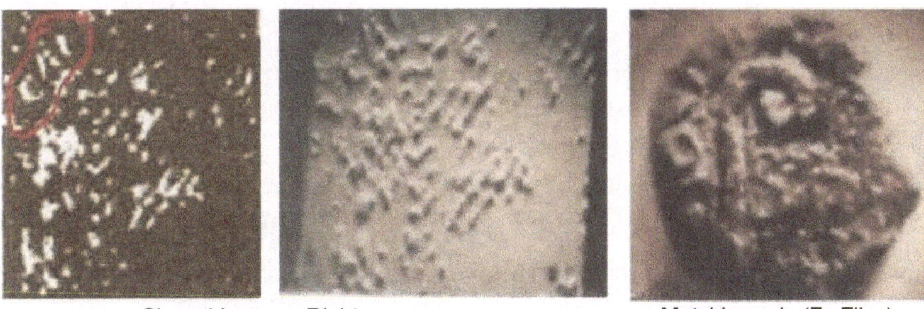

Fig.19.21: The alleged imprint of an inscription from a lepton coin above the right eye of the man in the Shroud. The letters CAI are circled in red (left). Actual lepton coin (right).

Pollen

The scientific analysis of pollen is called *palynology*, and while it cannot put a specific date on an object, it can still provide evidence of where the object had been located, Figure 19.21. What makes this possible is the fact that pollen is very durable and different plant species produce recognizably different pollen grains.

Palynologists (scientists who study pollen) can usually narrow down the pollen to the genus of a given plant species and by comparing plant distribution patterns, especially of rare or geographically restricted species, can thus tell the location from which the pollen originated.

Sticky tape was applied to the Shroud in 1973 by palynologist Max Frei. Interestingly, pollen from twenty plant species was discovered. Initial results claimed these plants are abundant in Turkey, one of the places where the Shroud was supposedly located at on its long way to France, then Italy. But not all pollen researchers agree with Frei's methodology or conclusions. Although there is general agreement about the genera of the pollen found on the Shroud, the species identification is suspect. Silvano Scannerini, a botany professor in Turin says that a more thorough analysis is necessary. An American pro-Shroud group called ASSIST has not been able to do this, despite the fact that they have had Frei's original sticky tapes for over twenty years (*The Sign*, pp. 113–114). In the end, we cannot conclude anything about the palynology of the Shroud.

Spices

Also missing from the Shroud are any traces of spices. Besides wrapping the body in linen *cloths*, Joseph of Arimathea and Nicodemus (the Pharisee of John chapter 3 fame) wrapped the body "with the spices":

"After these things Joseph of Arimathea, who was a disciple of Jesus, but secretly for fear of the Jews, asked Pilate that he might take away the body of Jesus, and Pilate gave him permission. So he came and took away his body. Nicodemus also, who earlier had come to Jesus by night, came bringing **a mixture of myrrh and aloes, about seventy-five pounds** in weight. So, they took the body of Jesus and bound it in linen cloths **with the spices**, as is the burial custom of the Jews." (John 19:38–40) (emphasis added)

Nicodemus had spent a fortune buying 75 *pounds* of sticky resin from the myrrh and aloe trees. We are not exactly certain what the "aloe" of the Bible is (it is not the same thing as the common house plant), but myrrh resin is a sticky, gummy substance that would have definitely left traces behind. Not only would it have bound the linen strips together, but it would not have been possible to keep the spice mixture only on the *inside* of the linen strips. Anyone who has worked with flour, water, and newspaper to make papier-mâché objects knows how messy this process is. Working with the aloe-myrrh concoction would have been much worse. The fact that the Shroud lacks any detectable traces of myrrh, the fact that it is in a nice, flat configuration (not 'stuck together' in any way), and the fact that it is a single sheet all point to it *not* being the burial cloth of Christ.

Antonacci acknowledges the lack of myrrh or aloe in the Shroud (p. 87), claiming that only the inside of the Shroud was examined. But this is specious. Seventy-five pounds were used. There should be abundant evidence for these spices.

THE SHROUD'S WHEREABOUTS BEFORE THE MIDDLE AGES

The whereabouts of the Shroud (its 'provenance') was *unknown* prior to its public display in Lirey, France in 1355. According to some Shroud researchers, it was stolen from Constantinople, probably during the Fourth Crusade (1203–4). The object purported to be the same as the Shroud was known as the *Mandylion* in Constantinople prior to the Fourth Crusade, when many relics were transferred from the East to the West by crusaders.

But it is not at all clear that the *Mandylion* is the same as the Shroud of today. Not only is there a +150 year gap between the Fourth Crusade and the Shroud's appearance in Lirey, but the *Mandylion* was also supposedly displayed in France in the court of King Louis IX in Paris until it disappeared during the French Revolution (1799–1789). During its stay in Constantinople, the *Mandylion* was supposedly displayed for guests of honor in the Eastern emperor's court. It was stored in a container in such a way that only the top fourth was visible (showing Jesus' face), folded over itself and draped on a wooden beam. Thus, it was also known as the *tetradiplon*, Greek for 'folded in four'.

From around the year 550 to 944, the *Mandylion* was purportedly in Edessa, then part of the Byzantine Empire, now in modern Turkey. Assuming the Image of Edessa is the same as both the *Mandylion* and the Shroud, this means that for a whole third of its existence (33 to 550 and 944 to 1355 AD), the whereabouts of the Shroud are unknown. The only historical attestation to the Shroud/*Mandylion*/Image of Edessa, either purported or real, comes from a supposed exchange of letters between King Agbar V of Edessa, who is supposed to have ruled there between AD 50–13, and Jesus Christ Himself. However, even historians and some Roman Catholic authorities reject the authenticity of these letters.

The Shroud of today is almost certainly the same as the one first seen in France in 1355. It is *possibly* the same as the *Mandylion* from Constantinople (although the carbon dates contradict this, see below). And it is *unlikely to be* the Image of Odessa. But that is as far as we can trace it. Considering the biblical descriptions above, we would have to reject the authenticity of the Shroud even if it could be traced back to 33 AD Jerusalem. Yet we cannot get close to that in either time or geography, so all we can say is that the Shroud is ancient, but it is not the burial shroud of Jesus.

The Pray Codex

The four circles that some claim are similar to the holes in the Shroud can be seen at the bottom, just beneath the upraised arm of the person holding the flask. The hole-like marks in the manuscript form a similar pattern to the holes in the Shroud, but they are 90-degrees off. There are additional decorative hole-like patterns on the lower sheet among the red crosses, and on the angel's wing and belt, Figure 19.22.

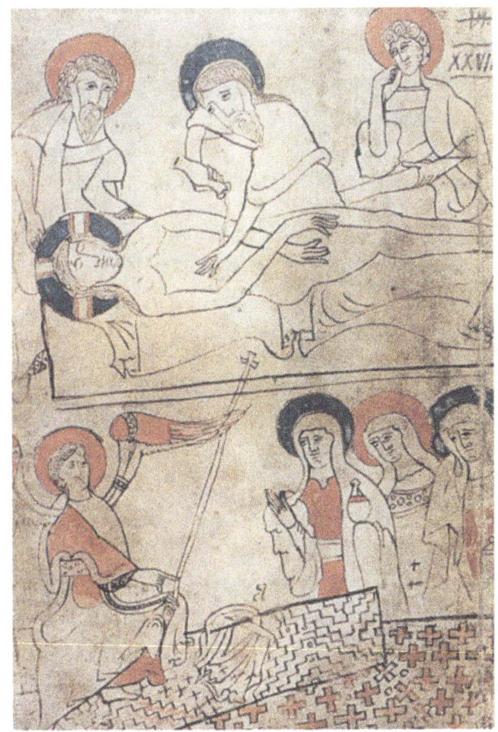

Fig.19.22: *Image of the entombment of Christ from the Pray manuscript. wikipedia.org*

The Pray Codex is a collection of medieval documents written in Hungarian sometime during the late twelfth to early thirteenth century. It was named after György Pray who discovered it in 1770. According to some Shroud supporters, there is an interesting correspondence between the Pray Codex and the Shroud that, they believe, tells us that there may have been some cross-pollination of ideas. During the twelfth to thirteenth centuries the kingdom of Hungary had strong relations with the Byzantine Empire under King Béla III (1148–1196 AD), who as a young man spent eight years in the imperial court at Constantinople. Béla III predates the Fourth Crusade. If the *Mandylion* and the Shroud are one and the same, it is at this time that a drawing of the *Mandylion* could have been made, for later use in the Pray manuscript, which originated from this time period. At least, the presence of the *Mandylion* could have influenced the burial scene in the Pray Codex, or at least that is what some Shroud supporters hope.

One of the documents in the Pray Codex contains a depiction of the entombment of Christ (Figure 19.22). On the dorsal side of the Shroud (the part that shows the back of the man, not the more familiar front image), two sets of four "poker holes" are visible next to the man's upper legs. They form a pattern similar to a knight's step pattern in chess (red circles, Figure 19.23). Some people claim this pattern of poker holes is also depicted on the entombment scene on the Pray manuscript.

There are multiple problems with this, not the least of which is the that the holes are rotated 90 degrees with respect to the pattern on the Shroud. This might indicate that these circles are merely ornaments and no real correlation exists between the two. Similar circles can also be seen on the lower cloth and on the angel's wing and his belt. If the holes are unrepaired burn marks from the 1532 fire, the entombment scene in the Pray Codex could not have been based on them. The fact that the holes follow the scorched fold line is obvious. Alternatively, the Lier Shroud is a hand-drawn copy of the Shroud dating to approximately 1516. It, too, shows the poker holes, but it does not show the many smaller holes associated with the folding pattern that match these holes. Thus, it is possible that the holes were in the Shroud prior to the fire. In this case, they could have been due to wear or to insect damage along the folded edge.

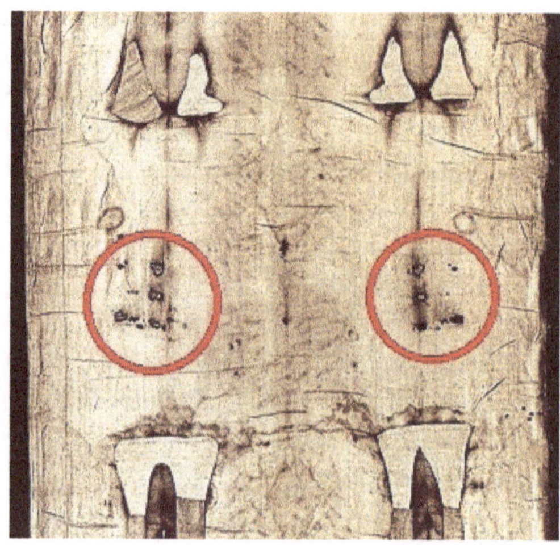

Fig.19.23: *Close-up of the 'poker holes' that make a knight's step pattern on the dorsal side of the Shroud, along a fold line and, as we will see, along the one edge that was scorched during a fire.*

The Shroud and the Pray Codex are approximately contemporaneous in time, although the maximum age of the Shroud (860 years before present, or approximately 1128 AD, Table 19.1), would predate the Pray Codex. But we simply do not know which came first or what ideas were circulating among other artifacts that are now lost to us.

Table 19.1 : C-14 age measurements (in years before present) from the Shroud of Turin. 15 Results for the Arizona samples are averages from two measurements collected on four separate days (see below).

Sample	Individual measurements					Weighted means	St. dev.
Arizona	591±30	690±35	606±41	701±33	-	646±31	17
Oxford	795±65	730±45	745±55	-	-	750±30	32
Zurich	733±61	722±56	635±57	639±45	679±51	676±24	24

DATING THE SHROUD

One of the most common ways of dating a sample of organic material is by carbon dating. Radiometric dating in general may be controversial, but carbon dating can be quite accurate, especially for material in this age range. On April 1988 21 four samples Figure 19.23, were removed from the Shroud to be analyzed. Each sample was about 50 mg in weight and 10x70 mm^2 in size. It is important to note that the samples were taken from the main body of the Shroud, away from patches, but not necessarily far from the charred areas or the obvious water stains.

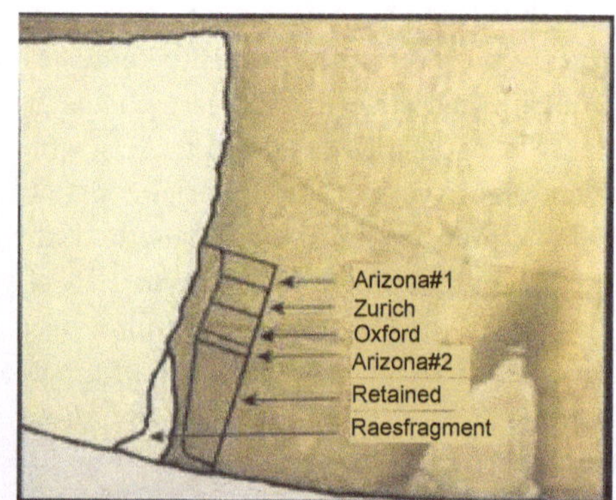

Fig.19.24: *Fragments from the Shroud of Turin which were sent to the three radiocarbon laboratories. Shroud.com*

This area would have been inside the folded-up image when on display, Figures 19.24. It would have been located on the top-most edge, above and to the left of the face. Even though the very edge would have been exposed, it would also have been protected by several layers of cloth. Thus, it may have been handled less frequently than some other sections, perhaps minimizing contamination from modern carbon.

The four samples were sent to three accelerator-mass-spectrometry (AMS) labs in Tucson (Arizona), Oxford (England), and Zurich (Switzerland) for independent testing. Three control samples were also included:

1. a piece of linen from a Nubian tomb dating to the eleventh to twelfth centuries,
2. a linen cloth from a mummy of Cleopatra of Thebes from the early second century, and
3. threads removed from the cope (a type of coat) of St. Louis d'Anjou from the Basilica of Saint-Maximin, France from the turn of the thirteenth century. Table 1 summarizes the age of the results as reported by the journal *Nature*. After applying strict calibration methods, the radiocarbon date of the Shroud ranged from 1260–1390 AD. Importantly, the dates for the other objects were appropriate to their expected timeframe.

OBJECTIONS TO THE RADIOCARBON DATES

The fact that three labs independently dated the Shroud to the Middle Ages and not to the first century (i.e., closer to the date of the linen cloth from the mummy of Cleopatra) is a strong indication that the Shroud of Turin is a medieval relic. Proponents of a first century age of the Shroud know very well that dating the Shroud to the 13th–14th century is a major obstacle for them. Therefore, they have come up with several arguments to refute the radiocarbon dates.

Antonacci reveals to us on p. 310–314 of *Test the Shroud* that the radiocarbon dates from the Arizona lab are actually *averaged* values. In reality, the Arizona lab had performed eight measurements (two per day for four dates from May to June of 1988). These values are shown in Table 19.2.

The lowest age value measured by the three radiocarbon labs was 540 years, which just so happens to come from the Arizona lab. The highest age is 795 years, coming from the Oxford lab. This would mean an age difference of 255 years. This large of a difference between Shroud samples which are only centimeters away from one another indicates that the radiocarbon dates are imperfect. That is why the Arizona lab averaged its age values. But even with a 225-year range, this puts the dates squarely in the late medieval period (1203–1458 AD). The reader should note that measurement ranges like this are entirely appropriate for objects of *known* age.

Table 19.2. Unaveraged C-14 age measurements (in years before present) of the Shroud of Turin by the radiocarbon lab in Arizona (Test the Shroud, p. 311).

	May 6	May 12	May 24	June 2
First measurement	574±41	753±51	676±59	701±47
Second measurement	606±43	632±49	540±57	701±47

Shroud proponents also claim that a large amount of external 14C contaminated the Shroud, making it look much younger than it really is. Outside contamination could come from oils from human skin and soot from candles, among other things. Riani *et al.* analyzed the dates measured by the three radiocarbon labs and concluded that the original sampling was based on poor experimental design. Based on the 12 samples measured by the three radiocarbon labs, they found a decrease of radiocarbon age the farther one gets from the main body of the Shroud. In other words, samples taken closer to the part of the Shroud where the body lay were dated older (less residual 14C, less modern contamination). Samples taken farther away from the center part of the Shroud were dated younger (more residual 14C, more contamination) The sample farthest away from the main body of the Shroud (measured by the Oxford lab) placed the material 750 years before present. This is a little younger than some of the other measurements, but even the oldest measurement is nothing close to 33 AD! According to Dr. Harry Gove, professor of physics at the University of Rochester, who developed AMS technology and witnessed the dating of the Shroud in the Arizona lab, if the Shroud really was from the first century, and contamination skewed the results to produce a younger age, then the shroud samples that they tested would have had to contain as much as up to one-third contamination, which is clearly implausible (Relic, pp. 291–292).

Even though the radiocarbon dates might be somewhat questionable, they are in line with the status quo. When we 'carbon date' objects of *known* historical age, we can often get to within decades, centuries at worst. The Shroud ^{14}C results are what we would expect from such an object, in the condition in which it exists, from the time period in which it first appears.

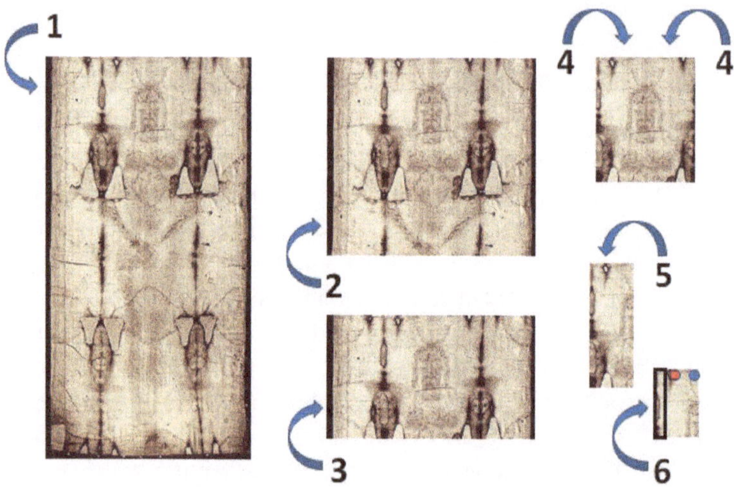

Fig.19.25: Possible folding patterns of the Shroud while on display or while in storage. By paying attention to the repeating patterns on the cloth, it is possible to reconstruct the way it was folded when it was damaged by fire in 1532. We can know that the Shroud was first folded in half lengthwise twice Figure 19.25 *(1 and 2). The lower third* Figure 19.25 *(3) then the sides* Figure 19.25 *(4) were folded back. This created an almost square section that contained the facial image only. Two more folds were then made, one vertical* Figure 19.25 *(5), and one horizontal* Figure 19.25 *(6). By placing a red dot in the only visible burn mark, a blue dot on the only visible water stain, and a rectangle around the only visible scorched edge, all the relevant patterns can then be explained. Importantly, the "knight's step pattern" some claim links the Shroud to the Pray Codex, is found along the burned edge of the folded Shroud. To see how this affects the Shroud, see Figure* 19.26.

NEUTRON RADIATION?

What kind of process or event would be capable of creating surplus ^{14}C? Antonacci claims in *Test the Shroud* that when Jesus rose from the dead, His body would have emitted neutrons into the cloth of the Shroud, thereby converting ^{12}C into radioactive ^{14}C. This is a huge assumption and forms one of the main thrusts of the supposed evidence for an old Shroud. Worse, in the earth's atmosphere, ^{14}C is created when a ^{14}N is struck by a cosmic ray. Upon decay, it turns back into a ^{14}N. ^{14}C is not created from ^{12}C.

Proponents of an older Shroud also advocate testing for the presence of radioactive ^{36}Cl (chlorine, which comes from red blood cells), and radioactive ^{41}Ca (calcium, which commonly occurs in cells). The half-lives of ^{36}Cl and ^{41}Ca are 301,000 years and 102,000 years, respectively. The age of the Shroud is negligible compared to these much larger half-lives. Theoretically, ^{36}Cl or ^{41}Ca on the Shroud would have also been created by this supposed neutron radiation, and they would not have had time to decompose significantly. However, if the Shroud is medieval and was not subject to neutron bombardment, it should have quite a bit of ^{14}C and no ^{36}Cl or ^{41}Ca (since without neutron radiation these atoms would not arise).

If this isn't unusual enough, Antonacci's theory involves Jesus' body *disappearing* during the Resurrection, *along with the 130 blood marks reappearing in the Shroud afterwards.* According to his theory, when Jesus' body disappeared, this created a vacuum, which would have sucked the front and back part of the Shroud into a space filled with neutrons, which were responsible for encoding Jesus' physical characteristics into the fabric of the Shroud. De Wesselow thinks the body was revivified instead of disappearing. Other pro-Shroud researchers are skeptical of Antonacci's radiation hypothesis.

How can we know for sure that neutron radiation caused the imprints of Jesus' physical characteristics into the Shroud when He rose from the dead? The Resurrection is a supernatural event, and therefore it is off-limits to scientific examination. Why neutrons and not some other type of particle, if Jesus' body emitted radiation? Matthew 17:2 says that Jesus' face "shown like the sun", and His clothes became "white as light" at the transfiguration. In Acts 9, Paul describes seeing Jesus after a flash of light. If Jesus emitted any kind of particle, it would have been particles of light rather than neutrons, which are invisible to the human eye. Worse, Peter, James, and John would have been instantly killed by the neutron radiation blast were they standing close enough to Jesus to see anything emanating from his body at the Transfiguration.

Lignin Breakdown

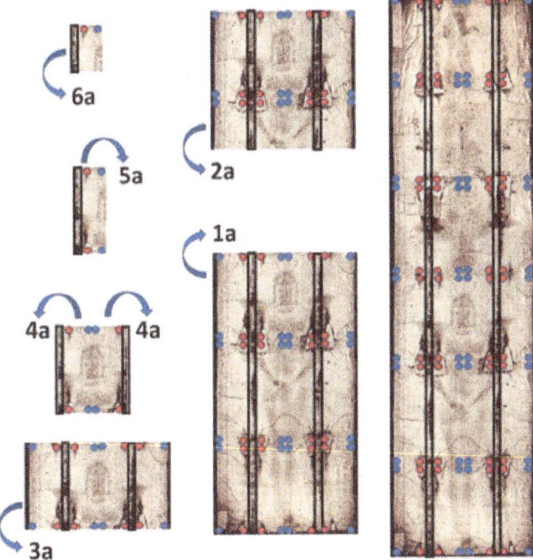

Fig.19.26: Starting with the folded Shroud and paying attention to the location of the burn mark (red dot), water mark (blue dot), and scorched edge (rectangle) allows us to explain multiple patterns across the Shroud when unfolded.

For example, the twelve smaller holes (on the ends and across the center) and the eight larger holes were created from a single burn. The small holes are stacked immediately on top of one another when folded. The larger holes were stacked under the smaller holes in this image, but we do not know which side was 'up' when the molten silver that supposedly made the burn marks dripped onto the Shroud. The section removed for carbon dating is in the lower left corner.

De Wesselow brings up another way to date the Shroud (*The Sign*, pp. 111–112). This method depends on the lignin and vanillin content of the Shroud fibers. One of the breakdown products of lignin is vanillin, and both are molecular components of the cell walls of the plants used to make the Shroud. After formation, vanillin is slowly lost from fibrous material, with a loss of 95% vanillin after 1,319 years at 25°C. De Wesselow concludes that since vanillin was not detected on the Shroud, it cannot be dated to after the year 700 AD. While this evidence speaks against the Shroud originating from the 13th to 14th centuries, it still does not decisively prove that the Shroud was from the first century.

However, even de Wesselow admits that measuring the breakdown of lignin is a relatively imprecise dating method. For example, a lignin sample loses 95% of its vanillin content at 20°C after 3,095 years. A fluctuation of only 5°C results in an age difference of 1,776 years. We know that the Shroud suffered fire damage, which, we assume, could have caused the vanillin to boil off, meaning that the Shroud might be medieval after all. We also wonder why there is no evidence for the continuous breakdown of lignin. It is not like the cloth has been stored in a hermetically sealed vault free of oxygen and water for the past several centuries.

WHERE IS THE VANILLIN?

Other Considerations

There are multiple additional arguments that will not be covered in this review. One deals with the herringbone weave pattern in the fabric. This weave involves a horizontal thread passing under three vertical threads, then passing over another vertical thread. Another one deals with the presence of limestone particles on the Shroud, possibly identifying the burial site. Some other issues include a prior carbon dating result from 1982 and a letter from bishop d'Arcis of Troyes, France, which was written in 1389 to pope Clement VII alleging that the Shroud was a forgery. Suffice it to say that there is much controversy surrounding the Shroud, even among conservative Bible believers.

Medieval Relic?

The Shroud of Turin is not the only potentially false relic from the Middle Ages. We know of *many* others, including:

- The blood of St. Januarius in a vial in Naples, Italy
- A picture of Mary painted by St. Luke in an Augustinian church in Bologna
- A piece of Moses' brazen serpent (Numbers 21:5 – 9; 2 Kings 18:4) in the church of St. Ambrose in Milan
- The table on which Jesus partook of the Last Supper in the church of St. John, Lateran in Rome
- The holy stairs which Jesus walked up to judgment before Pilate
- Water from the prison where the apostles Peter and Paul were kept
- The house in which Mary was born (the so-called Santa Casa), transported from Nazareth to Loretto
- Parts of the veil of Mary
- The 'holy porringer' in which food was made for the baby Jesus

Since fake relics were so common in the Middle Ages, we must take this into account when trying to decide if the Shroud of Turin is authentic. This is not a time in history where we are seeing real historical artifacts with solid provenance and detailed titles of ownership. Instead, this is a time where many, many people were lying about historical events and objects for multiple reasons, including increased prestige and financial gain.

How was it Made?

Since a number of forged relics exist, *it is quite possible that the Shroud of Turin is also a forgery.*

Shroud protagonists may still ask that, if the Shroud of Turin is a fake, how was it then made? This is a good question, and multiple possibilities that we will not discuss here have been proposed. Some of these are interesting, and some much more speculative. Let us remember that the scientific method is used to disprove theories, not prove them. At this point, we can definitely say what the Shroud is not: the biblical burial cloth of Jesus Christ.

Fig.19.27: *Negative and positive Volckringer imprints of a plant on paper. Curtsy Shroud.com*

However, de Wesselow himself suggests a possible solution to this thorny question. In 1942, the French pharmacist Jean Volckringer drew attention to faint imprints made by plants on the paper that they were pressed onto Figure **19.27.** In de Wesselow's book depicts imprints of a partial human hand and lower abdomen and also a plant, both made by natural means. The figure of the human hand and abdomen was made when the abnormal chemistry of patient's urine at the Jospice International hospice in Liverpool, England reacted with the underlying mattress to create a partial outline of his body.

De Wesselow suggests that the imprint is the result of what is known as a Maillard reaction, which occurs between carbohydrates (sugar molecules) and amines (such as amino acids, the building blocks of proteins). Maillard reactions can happen rapidly at low temperatures. These kinds of reactions happen in food chemistry, for example when a bread crust is browned, and within the cloths enveloping Egyptian mummies.

Ammonia and amines such as putrescine and cadaverine from a decomposing corpse may have reacted with the linen cloth of the Shroud to produce the outline of the body. This would contradict the idea that it was Jesus, the Son of God Who was wrapped in the Shroud because of Acts 2:27:

"For you will not abandon my soul to Hades, **or let your Holy One see corruption**". (emphasis added)

Also, let us not try to discredit the ingenuity of people from the Middle Ages/Renaissance time period. For example, Leonardo da Vinci was not only an artist, but also a scientist and engineer. He made detailed studies of the human body, and also devised pre-modern flying machines. It is possible that a little ingenuity combined with a basic understanding of natural processes could have allowed a medieval artist to create something like the Shroud of Turin. There are many possible ways it could have been created, included many not mentioned here. The point is that we don't know how, but neither does the explanation that it formed in Jesus' tomb seem to be valid.

Shroud supporters should not be dismayed if the Shroud is not authentic. Even the archbishop of Turin, Antonio Ballestrero, accepted the radiocarbon dates when they were announced in 1988. In any case, the Christian faith remains 100% intact. After weighing the relevant information, we must draw the conclusion that the Shroud of Turin is probably not the burial cloth of Jesus Christ. We have no good reason to accept it as authentic. Based on the evidence:

- Jesus was wrapped up in multiple burial clothes, not just one.
- A separate head cloth was used, which the Lord took off after He arose from the dead.
- Lazarus was buried in a similar manner to Jesus, with strips of linen for the body and a separate cloth for the head.
- The hair on the man in the Shroud hangs downward and his beard is also intact, both of which contradict Scripture.
- There is no trace of the large quantity of sticky spices with which Jesus is known to have been buried.
- The height of the man in the Shroud does not match that of a first century Jewish man. The arms seem to be distorted and disproportional. And the argument that the head is leaning forward is hard to believe.
- While the ^{14}C dates of the Shroud may be contestable, there is still no positive evidence that the Shroud dates to the first century. Palynology has not helped to clarify anything.
- The historical record is incomplete. At best, there is a 500-year gap between the Crucifixion and the first attestation to the Image of Edessa.

This illustrates the necessity of applying the principle of sola Scriptura (Acts 17:10–11, 1 Corinthians 4:6, 2 Timothy 3:16–17). Even if the man in the Shroud displays several characteristics of a man who was beaten and whipped, similar to how Jesus was treated during His trial and execution, if there is any *other* evidence which contradicts the Bible, the claim must be rejected.

As a final note, we must keep in mind Romans 10:17:

"So faith comes from hearing, and hearing through the word of Christ."

We must also remember 2 Corinthians 5:7:

"for we walk by faith, not by sight."

Antonacci claims, "...if this evidence confirms that the passion, crucifixion, death, burial and resurrection of Jesus Christ were actual events in history, then its implications are mind boggling" (*Test the Shroud*, p. 329).

But if the Shroud is not authentic, we lose nothing. Worse, if the Shroud is false and we cling to it, we risk losing all credibility. Let us not look towards material proof of the Resurrection of Christ. The Apostles certainly did not do so, and there is nothing in the New Testament to indicate they were focused on physical evidence. Instead, every time one of the New Testament writers wanted to substantiate a point, they appealed to eyewitness testimony. They collated that testimony and recorded it in the final 27 books of the Bible. Consider the words of Paul in 1 Corinthians 15:3–8:

"For I delivered to you as of first importance what I also received: that Christ died for our sins in accordance with the Scriptures, that he was buried, that he was raised on the third day in accordance with the Scriptures, and that he appeared to Cephas, then to the twelve. Then he appeared to more than five hundred brothers at one time, most of whom are still alive, though some have fallen asleep. Then he appeared to James, then to all the apostles. Last of all, as to one untimely born, he appeared also to me."

This does not mean we have no evidence for biblical claims. Far from it, as any reader of creation.com should know. What it does mean, however, is that we do not require physical evidence for the resurrection of Jesus. It was the centerpiece of redemptive history and was universally attested to by every New Testament writer. But the eyewitness testimony should be sufficient. Consider what Jesus said to Thomas in John 20:29:

"Jesus said to him, 'Have you believed because you have seen me? **Blessed are those who have not seen and yet have believed.**'" (Emphasis added)

A list of arguments used to support the authenticity of the Shroud, and their refutation

1. If you claim that the Shroud is a fake, you must show us how it was made in the Middle Ages. So far nobody has been able to do that.

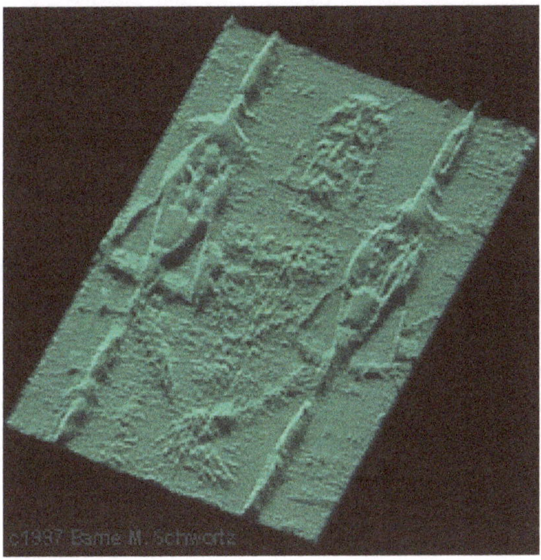

Fig.19.28: VP-8 Analyzer image of the head of the man in the Shroud of Turin. Curtsy shroud. com/78strp10.htm

This is incorrect. This is not how science works. Even if ten possible ways of manufacturing the Shroud have been refuted, we may still find an eleventh way. We could also say that since the pro-Shroud side cannot explain how the Shroud was made in the first century, then it must be medieval, Figure **19.28**.

2. The Shroud is the only known object which encodes 3D information, as demonstrated by NASA's VP-8 Image Analyzer.

The VP-8 Image Analyzer is an analog computer designed in the 1970s that translates shades of black and white on an image into levels of vertical relief that can be visualized on a computer screen. This is perceived to convey 3D information by the human brain.

But almost all photographs encode 3D information. This is because almost all were made of 3D objects in the real world. For example, photos can show objects behind one another, and shading often allows us to easily perceive the 3D shape of an object. Even the reader can do a simple experiment by opening any digitalized photograph in ImageJ and running a 3D Interactive Surface Plot from the Plugins menu. Using this program, one of the authors made a 3D image from a picture of one of his hands.

3. A long strand of cloth from the Shroud was secretly radiocarbon dated in 1982. One end of the strand dated to AD200 and the other end dated to AD1000.

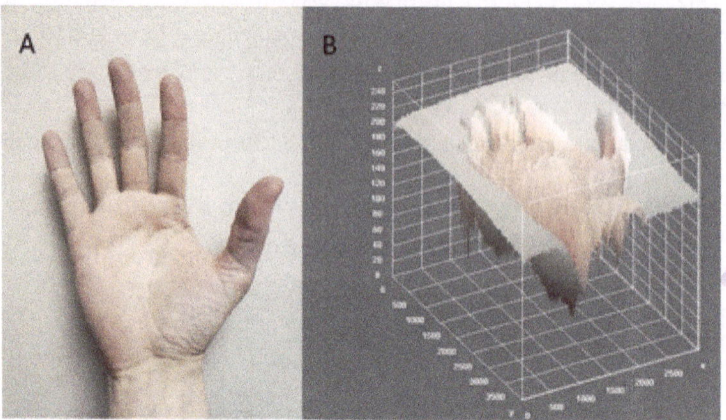

Fig.19.29: **A.** *Image of hand of one of the authors.* **B.** *3D transformation of same image in Image J.*

Reports of the 1982 dating are vague. The secret carbon dating results were never published, and not much else is known about this measurement. We don't know who made the measurement, at what lab, and under what kind of circumstances it was carried out. Hence, this cannot be submitted as scientific evidence, Figure 19.29.

Assuming this was actually a piece of the Shroud, and assuming the dates were reported correctly, if we average the dates from both ends of the strand, we still get a date of AD 600, which is still much too young. Even the AD 200 date is too young. These dates vary much too highly to be acceptable in a scientific publication. As opposed to this, the 1988 radiocarbon dates give a range of only 130 years (1260–1390).

4. The Shroud was contaminated, thereby skewing the radiocarbon dates.

You need to show quantitatively that the amount of contamination (purportedly from the oils on the hands of the priests who handled it, or carbon from the 1532 fire) skewed the radiocarbon dates by 1,300 years. According to Harry Gove, professor of physics at Rochester University, a full one-third of the Shroud samples had to have been contamination for such a large skew in the radiocarbon dates[1]. There would have been so much modern dirt and oils present that the Shroud would have been obscured by the contamination. A picture in his book, Relic, Icon, or Hoax? Shows that the samples were clean (p. 265).

5. The radiocarbon samples came from a part of the Shroud which was "invisibly re-woven" with patches from the Middle Ages, yielding a radiocarbon age from that time period.

This idea is disputed among Shroud researchers and rejected outright by both Wilson and Antonacci. Even if it were true, it implies medieval artists may by capable of things that we would have great difficulty doing today. The Shroud has been patched several times over its lifetime, and these patches are always visible. There is no evidence that the corner used to perform the ^{14}C testing was not original. To turn the argument around, if they could perform this amazing feat of 'invisible reweaving', why could they not perform the amazing feat of making a ghostly negative image on a piece of linen?

6. The image on the Shroud was made by 'coronal discharge'.

In our article, we specifically dealt with the neutron radiation hypothesis put forth by Jackson and Antonacci[2]. This theory is rejected by Ray Rogers. It seems that there are multiple competing hypotheses as to how the image was made in the Shroud, such as coronal discharge, X-rays, biophotonics, and holograms. Shroud supporters cannot give a clear explanation as to how the image was formed. Some authorities even claim that the image on the Shroud was formed very slowly over a long period of time, finally appearing years after the Resurrection.

But does a dead body, which is miraculously resurrected really give off any kind of radiation? According to the Bible, no. Following are several examples of people who were miraculously resurrected by Jesus and the Apostles, yet none are described as emitting any kind of radiation:

- the son of the widow of Nain (Luke 7)
- Jairus' daughter (Luke 8)
- Lazarus (John 11)
- the young girl, Tabitha (Acts 9)
- Eutychus, who fell asleep and fell out of a building (Acts 20)
- The son of the widow of Zarephath (1Kings 17)

Coronal discharge is as speculative as any other theory. Point is, nobody knows how the image formed, but no other biblical resurrections are associated with any such thing.

7. All medieval paintings show Jesus with nails in his palms, yet the Shroud shows Jesus with holes in His wrists. How could a medieval forger know about that? Therefore, the Shroud must be authentic.

The man in the Shroud has his palms pointing inwards, towards himself. Therefore, we cannot say dogmatically that the man had his wrists pierced and not his palms. Yet, either way, this is a non-argument.

8. According to blood expert Alan Adler, the presence of bilirubin in the blood stains on the Shroud was able to keep the blood stains red for 2,000 years.

Bilirubin is orange-yellow in color. Adler claimed that bilirubin mixed with methemoglobin gave the blood stains their pinkish color. But methemoglobin is normally found in only trace amounts in the body, so it is an open question where it came from. Yet, the presence of blood on the Shroud is debatable. For example, there are alleged blood stains where they cannot be. There is a stain on the hair on the dorsal image, but if present, blood would have been matted into the hair and would not be visible as a 'stain'. Chemical analysis of the 'blood' shows that the Soret absorption band at 420 nm is missing. Normally, an absorption at this wavelength indicates the presence of porphyrins, including molecules like hemoglobin and chlorophyll. Iron oxide has been found on the Shroud, but in far larger concentrations (40 $\mu g/cm^3$) than human blood (1 $\mu g/cm^3$). Traces of vermilion have also been found on the Shroud. This is a mercury-containing pigment commonly used in medieval artwork.

9. The stitching pattern of the clothe of some Jewish warriors has been discovered in Masada resembles those found in the first century, whereas it has never been shown in clothes from the Middle Ages.

This claim comes from the textile expert Mechtilde Flury-Lemberg, who claimed that the stitching holding the Shroud together was both unusual and only found in some textiles found at Masada. Gabriel Vial, another textile expert claims that the material from the Shroud is different from textiles found in many first century Roman excavations:

"So far every example studied – and these have come from Pompeii, Antinoe, Palmyra, Cologne, Dura-Europos – has been radically different from the shroud, both from the point of view of the structure (2/2 twill as opposed to 3/1) and the materials used (wool and silk rather than linen). We have to look to the 16th century to find the first example of linen chevron weaving with a 3/1 twill structure, found in the canvas of a painting in Herentals (Belgium). Taking into account the constituent elements of any textile (material, structure, warp and weft density, the textile of which the shroud is composed *is unlike anything presently known to date prior to the 16th century*."

This is more evidence that the Shroud is not from the first century, but rather is medieval.

Others claim the seam that runs the length of the Shroud is unique to Masada. However, this type of seam is quite common, e.g., it is found along the sides of any pair of jeans.

10. Professor Giulio Fanti from the university of Padua has dated the Shroud with several other dating methods,[24] resulting in the following dates:

- Fourier transform infrared spectroscopy (FT-IR) on the fabric of the Shroud: 650 BC–AD 150
- Raman spectroscopy on the fabric of the Shroud: 370 BC–AD 430
- Multi-parametric mechanical test on Shroud fibers: 14–AD 534

While these dates center around the time of Christ, their ranges do seem to be a bit wide, each one covering 500 years or more, with a combined range of 1184 years. Also, these tests are non-standard ways of dating material. In contrast, the range of the radiocarbon dates from 1988 had a much smaller range of only 130 years, which is both more precise and within the expected range for measurements for materials of this age and type.

11. There is so much anatomical detail revealed in the Shroud that it cannot be a forgery.

There are a number of anatomical details visible on the Shroud, but there are some details that are anatomically implausible:

- Blood stains are not matted into the hair.
- The man in the Shroud seemingly has no neck.
- The body should not have produced trickling blood stains after it was supposedly washed.
- The fingers of the man in the Shroud are too long.
- On the dorsal image, the man's right foot is above his left foot, whereas on the ventral image, the two feet seem to point straight.
- The simple consideration that the dorsal image should be stronger than the ventral image, simply because the full weight of the body would have caused a much stronger image, depending, of course, on how the image was formed.

12. Jesus was wearing a chin band to keep His mouth closed. This is the separate facial cloth that is mentioned in John 20:7.

Such a chin band is not visible on the Shroud, and this stands in opposition to the idea that the Sudarium of Oviedo was the cloth that covered Jesus's face. In John 20:7, the facial cloth (σουδάριον) is upon (ἐπὶ) the face of Jesus. The Greek word *epi* (ἐπὶ) can mean any one of ten things: across, against, at, before, by, of position, on, over, to, or upon.[25] The pro-Shroud group has to demonstrate why *epi* can only mean 'around the circumference of the face', especially when the Greek word *peri* (as in perimeter) would normally be used in such a case. Also, the word 'face' in Greek is *opsis* (ὄψις) and not *kefale* (κεφαλή), which means 'head'.

13. The Greek τὰ ὀθόνια (ta othonia) includes the Shroud as well as the facial cloth (the soudarion), hence the 'multiple' pieces of linen.

This is incorrect. The soudarion is mentioned separately from the linen clothes "οὐ μετὰ τῶν ὀθονίων" (John 20:7), which are still in the plural (-ων)[3]. So, the burial clothes covering His body were in multiple pieces, and not a single piece like the Shroud. Also, in John 11:44 we read about how Lazarus's hands and feet were both covered with linen strips. In Greek, "τοὺς πόδας καὶ τὰς χεῖρας κειρίαις", where the word *keiriais* is in the plural. This eliminates the Shroud of Turin as the burial cloth of Jesus Christ. The only other option is that the Shroud was an unmentioned *extra* cloth in the tomb, but then the underlying linen would have absorbed the blood, etc. Or, the hands and feet of Jesus were bound, his head was covered with a sudarium, and then the body was wrapped in the Shroud. But then the facial image would have been obscured and there is no evidence for hand or foot binding in the Shroud image.

14. According to the Antioch Hypothesis of Jack Markwardt, the apostle Peter took the Shroud with him to Antioch in AD44 during the persecutions, and not to Edessa.

This proves that pro-Shroud theorists contradict one another. The so-called Antioch Hypothesis is based on a paragraph found in the now lost apocryphal Gospel of the Hebrews. As cited by Jerome: "When the Lord had given the linen cloth to the servant of the priest, he went to James and appeared to him".[26] First, the Gospel of the Hebrews is a non-canonical book. For example, Origen quotes one of the fragments, where Jesus allegedly says that His mother is the Holy Spirit.[27] Second, it is not certain what this linen cloth is, and who the servant of the priest is. One would think that the Jewish priests would destroy the Shroud, as it would be a sign of Jesus's resurrection, something which the Jewish priests would have taken great care to deny (Matthew 28:11–15).

Markwardt's hypothesis is based mainly on the opinions of scholars and hearsay. Its main pitfall is this: the author claims that the Shroud was hidden from the Romans due to the fear of it being destroyed. However, after Constantine came to power, the Shroud continued to remain in anonymity for hundreds of years.

15. The blood stains on the head of the man in the Shroud fit the pattern of a 'cap' of thorns, as shown by many puncture wounds on the scalp area.

According to Matthew 27:29 Jesus was not wearing a cap of thorns, rather it was shaped like a wreath. In the Greek, the word *stephanon* (στέφανον) is derived from the verb meaning "to twine" or "to wreath". Many Jewish, Syrian and Roman kings and rulers wore circlets or wreaths as opposed to caps. Either way, it would be very hard to say what kind of wounds the thorns would have made. Also, as stated earlier, blood stains would have been matted on the man's head and so should not be visible on the Shroud.

APPENDIX

a. Brown, R.H., Radiometric dating from the perspective of Biblical chronology; in: Walsh, *et al.* (Eds), *Proceedings of the First International Conference on Creationism*, Vol. **2**, pp. 42–57, 1986. For example, suppose that the Flood was 5,700 years ago, during which time a living thing died containing a ^{14}C content 0.125 times that of living things today. At the moment of its death, it already had a 'built-in' carbon-14 'age' of 17,100 years (three half-live periods 'built in': 0.5 x 0.5 x 0.5 = 0.125). Owing to the fact that another half-life of time has actually passed since the Flood, that once-living thing now has a ^{14}C that is 0.0625 that of presently-living things, and a total apparent carbon-14 age of 22,800 years.

b. Giem, P., Carbon-14 content of fossil carbon, *Origins* **51**:6–30, 2001. For a variety of technical reasons discussed in the paper, these occurrences cannot, at least for the most part, be explained away as contamination.

c. Woodmorappe, J., *The Mythology of Modern Dating Methods*, Institute for Creation Research, El Cajon, p. 41, 1999. This, of course, is also true of the dating methods used to obtain much older dates than those presumably obtainable by the carbon-14 method.

d. Note that this contrasts with subfossil mollusks, which often have anomalously-high carbon-14 dates, as reported in earlier creationist literature. The mollusc, unlike the tree, may have additionally absorbed some carbon from dissolved limestone, which has no ^{14}C. Thus the mollusk, at the time of its death, has a shortage of ^{14}C relative to the atmosphere, and hence a fictitiously-high age.

e. Circular, because they would say that the difference in the two radioisotope dates shows the evidence for the faulting ... yet the veracity of *both* the radioisotope 'dates' is unproven, and rests on many assumptions, one of which is that faulting must have occurred for there to be two such discrepant dates. And in any case, the field evidence is inconclusive.

f. Gasser, A., World's oldest wheel found in Slovenia, March 2003; Government Communication Office, Republic of Slovenia; ukom.gov.si.

g. Capuzzo, G., Space-temporal analysis of radiocarbon evidence and associated archaeological record: from Danube to Ebro Rivers and from bronze to iron ages, PhD thesis, p. 107, fig. 21, 2014; ddd.uab.cat/pub/tesis/2014/hdl_10803_283401/gc1de1.pdf.

h. In the first dating, basalt (a volcanic rock) interbedded within the sedimentary rock formation was dated using the 40Ar/39Ar method and yielded an 'age' of 212.5 ± 7.0 million years. In the second, zircon crystals derived from tuff rock (another type of volcanic rock), also interbedded within the sedimentary rock, were dated using the weighted mean 206Pb/238U method, and yielded an 'age' of 37 million years old.

REFERENCES

1. Gowans, A., Archaeology team finds 9,000-year-old artefacts in NewBo neighborhood, The Gazette, 3 August 2018; thegazette.com.
2. The development of agriculture: the farming revolution, National Geographic; https://genographic.nationalgeographic.com/development-of-agriculture/. Return to text.
3. Oard, M.J., Only one glaciation observed in western Alberta, Canada—the ice-age reinforcement syndrome, J. Creation 29(2):12–13, 2015.
4. Clayton, L. and Moran, S.R., Chronology of late Wisconsinan glaciation in middle North America, Quaternary Science Reviews 1(1):55–82, 1982.
5. Batten, D. (Ed.), Catchpoole, D., Sarfati, J., Wieland, C., The Creation Answers Book, Creation Book Publishers, p. 67, 2006.
6. Bruins, H.J. and van der Plicht, J. Tell es-Sultan (Jericho): radiocarbon results of short-lived cereal and multiyear charcoal samples from the end of the Middle Bronze Age, Radiocarbon 37(2):213–220, 1995.
7. James, P., Centuries of Darkness, Pimlico, p. 323, 1992.

8. Studhalter, R.A., Early history of crossdating, Tree-Ring Bulletin, 21:31-35, 1956; (available repository. arizona.edu. Last accessed 21 August 2108.
9. They can be viewed here: cybis.se/forfun/dendro/hollstein/belfast/reports/BelfastADReport/index.htm.
10. Was director of Laboratory of Tree-Ring Science Department of Geography at University of Tennessee, see bio here: https://utk.academia.edu/HenriGrissinoMayer/.
11. From unpublished data by Biermann, C.P., LaForest, L.B., and Grissino-Mayer, H.D. 2015.
12. Grissino-Mayer, H.D., The long, steady decline of uniformitarianism in dendrochronology: what if the present is no longer the key to the past? Presentation given at the Annual Meeting of the Association of American Geographers, Chicago, 2015; http://app.core-apps.com/aagam2015/abstract/4e710bf8a11e39f13afdf3bf9470b0b3, p. 16.
13. The "well-dated" Los Colorados formation, to quote the scientists. They say this rock formation is more than 200 million years old. See Melchor, R.N., de Valais, S., and Genise, J.F., Bird-like fossil footprints from the Late Triassic, Nature 417(6892):936–938, 27 June 2002.
14. Melchor, R.N., Buchwaldt, R., and Bowring, S., A late Eocene date for Late Triassic bird tracks, Nature 495(7441):E1–E2, 21 March 2013.
15. Vizan, H. et al., Geological setting and paleomagnetism of the Eocene red beds of Laguna Brava Formation (Quebrada Santo Domingo, northwestern Argentina), Tectonophysics 583:105–123, 2013. The Santo Domingo formation has been redesignated as two formations in recent years. Part of it is now called the Laguna Brava formation.
16. Batten, D., Modern birds found with dinosaurs, Creation 34(3):48–50, 2012; creation.com/modbirds-dinos.
17. For more on alleged 'hominid' fossil finds, see Wieland, C., Making sense of 'apeman' claims, Creation 36(3):38–41, 2014.
18. This is the museum accession number for the fossil. It stands for 'Kenya National Museum–East Rudolf', field specimen 1470.
19. Lubenow, M.L., The pigs took it all, Creation 17(3):36–38, 1995; creation.com/pigstook.
20. Two of the newer 'dates' (challenging Leakey's) were significantly younger: one was 520,000 years, and another was 1.6 million years. It is interesting to note, also, that Leakey's 'date' of 2.9 million years was supported by five different dating techniques, all in very close agreement, before it eventually fell out of fashion.
21. For long-age believers, localised erosion followed by re-deposition of sediments could explain seemingly out-of-place fossils, as could many other factors; see also Doyle, S., Precambrian rabbits—death knell for evolution? J.Creation 28(1):10–12, 2014.
22. Reed, J.K., Rocks Aren't Clocks—A Critique of the Geologic Timescale, Creation Book Publishers, Powder Springs, GA, 2013.
23. Swenson, K., Radio-dating in rubble, Creation 23(3):23–25, June 2001; creation.com/rubble; and also creation.com/rubble2.
24. Lubenow, M.L., ref. 10.
25. Bird tracks preserved in the rock formation are explainable in the context of the global Flood. See Oard, M.J., Dinosaur Challenges and Mysteries, Creation Book Publishers, USA, p. 127, 2011.
26. Cosner, L., How does the Bible teach 6,000 years? Creation 35(1):54–55, 2013; creation.com/6000-years.
27. Sorensen, HG., 1973. *The ages of Bristlecone pine*. Pensee, 3(2):15-18.Return to text.

28. Bowden, M., 1977. *Apemen - Fact or Fallacy?*, Sovereign Publications, Kent.
29. Osgood, A.J.M., 1981. The Date of Noah' Flood. *Ex Nihilo*, 4 (1):10-13.
30. Anstey, M., 1973. *Chronology of the Old Testament*, Kregel Publications, Grand Rapids, pp.38-45.
31. Moore, A.M.T., 1982. A four-stage sequence for the Levantine Neolithic. *Bulletin of the American Schools of Oriental Research*, 246:13-34.
32. North, R., 1982. The Ghassulin Lacuna at Jericho. *Studies in the Archaeology of Jordan I*, Department of Antiquities of Jordan, Hashemite Kingdom of Jordon, pp.59-66.
33. Beifer-Cohen, A., and Goldberg, G.P., 1982. *Israel Exploration Journal*, 32:185-189.
34. Rollefson, G.O., 1983. *Bulletin of the American Schools of Oriental Research*, 252:28-34.
35. Wainwright, G.A., 1963. *Journal of Egyptian Archaeology*, 49:18.
36. Olami, Ya-Aqov, 1984. Prehistoric Carmel, *Israel Exploration Society*, Jerusalem and M. Stekalis Museum of Prehistory, Haifa pp. 176-177.
37. Bar-Yosef, O., and Tahernov, E., 1970. *Israel Exploration Journal*, 20:148.
38. Kenyon, K.. 1960. *Archaeology in the Holy Land*, Ernest Bean, London, pp.69-70, 273.
39. Kirkbridge, U., 1967. *Palestine Exploration Quarterly*, 1967:12.
40. Clark, 0., 1970. *The Prehistory of Africa*, Praeger, New York and Washington. pp.169 and 193.
41. Gardiner, A., 1961. *Egypt of the Pharaohs*, Oxford, pp.185, 391, 591.Return to text.
42. Emery, W.B., 1961. *Archaic Egypt*, Penguin, p.42.
43. Amiran, I 1978. *Early Arad*, Israel Exploration Society, Jerusalem.
44. Roux, G., 1964. *Ancient Iraq*, Pelican, pp.62, 68.
45. Lloyd. S., and Tuad Sofar, 1945. *Journal of Near Eastern Studies*, 4:264, 276, 279.
46. Bibby, C., 1970. *Looking for Dilmun*, Collins, pp.376-81.
47. We have also reported on Identification of proteinaceous material in the bone of the dinosaur Iguanodon 'dated' to 120 million years old (Connect Tissue Res. 2003; 44 Suppl. 1:41–46).
48. Scientists recover T. rex soft tissue: 70-million-year-old fossil yields preserved blood vessels, www.msnbc.msn.com/id/7285683/, 24 March 2005.
49. Blood vessels recovered from T. rex bone, 24 March 2005, NewScientist.com news service.
50. Kuhn, T.S., The Structure of Scientific Revolutions, 3rd edition, University of Chicago Press, 1996.
51. Some dinosaur fossils could have formed in post-Flood local catastrophes.
52. Hecht, J., Dinosaur "cells" shed light on life 66 million years ago, New Scientist 203(2715):8, 4 July 2009.
53. Soft tissue in dinosaur fossils? The evidence hardens, Science 314(5801):920, 10 November 2006.
54. Schweitzer, M., Wittmeyer, J. and Horner, J., Gender-specific reproductive tissue in ratites and Tyrannosaurus rex, Science 308(5727):1456–1459, 3 June 2005.
55. I do not discuss the significance of several-thousand-year tree-ring chronologies and their calibration of carbon-14 dating. This is a separate issue.
56. Saupe et al., A possible source of error in ^{14}C dates, Radiocarbon 22(2):525–531, 1980.
57. Olsson, I.U., Experiences of ^{14}C dating of samples from volcanic areas, PACT 29:213–223, 1990.
58. Cook et al., Radiocarbon study of plant leaves and tree rings from Mammoth Mountain, CA: a long-term record of magmatic CO_2 release, Chemical Geology 177:117–131, 2001.
59. Hahn, H.-P., Born, H.-J. and Kim, J.I., Survey on the rate perturbation of nuclear decay, Radiochimica Acta 23:23–37, 1976.

60. Huh, C.-A., Dependence of the decay rate of 7Be on chemical forms, Earth and Planetary Science Letters 171:325–328, 1999.
61. Woodmorappe, J., Billion-fold acceleration of radioactivity demonstrated in laboratory, Journal of Creation 15(2):4–6, 2001.
62. RATE stands for Radioactivity and the Age of The Earth.
63. Snelling, A.A., Radioisotope dating of rocks in the Grand Canyon, Creation 27(3):44–49, 2005.
64. Chaffin, E.F., Accelerated decay: theoretical considerations; in: Vardiman, L. et al. (Eds.), Radioisotopes and the Age of the Earth Vol. II, ICR, El Cajon, CA, CRS, Chino Valley, AZ, pp. 525–586, 2005.
65. See also Morris, H.M., Games some people play: the supreme rule of this game is to stifle arguments against evolution—any way you can, Creation 23(4):35, 2001; Wieland, C., The rules of the game: as the 'rules' of science are now defined, creation is forbidden as a conclusion—even if true, Creation 11(1):47–50, 1988.
66. Press, F. and Siever, R., Earth, 4th Ed., W.H. Freeman and Co., New York, p. 40, 1986.
67. Eiseley, L., Darwin's Century: Evolution and the Men who Discovered It, pp. 234, 235, 240, Doubleday, Anchor, New York, 1961.
68. Woodmorappe, J., Lord Kelvin revisited on the young age of the earth, TJ 13(1):14, 1999.
69. Humphreys, D.R., Austin, S.A., Baumgardner, J.R. and Snelling, A.A., Helium diffusion rates support accelerated nuclear decay, in: Ivey, R.L., Proceedings of the 5th International Conference on Creationism, Creation Science Fellowship, Pittsburgh, PA, pp. 175–196, 2003.
70. Patterson, C.C., Age of meteorites and the Earth, Geochimica et Cosmochimica Acta 10:230–237, 1956.
71. Williman, A.R., Long-age isotope dating short on credibility, TJ 6(1):2–5, 1992.
72. Wald, G., The origin of life; in: Physics and Chemistry of Life, p. 12, 1955; quoted in: Morris, J.D., The Young Earth, Creation-Life Publishers, Colorado Springs, CO, p. 41, 1994.
73. See Mortenson, T., The Great Turning Point: The Church's Catastrophic Mistake on Geology—Before Darwin, Master Books, Green Forest, AR, 2004.
74. Antonacci, M., Test the Shroud at the Atomic and Molecular Levels, LE Press, LLC, Chesterfield, 2015.
75. De Wesselow, T., The Sign: The Shroud of Turin and the Secret of the Resurrection, Penguin Group (USA) Inc., New York, 2012.
76. Bennett, J. Sacred Blood, Sacred Image, The Sudarium of Oviedo, Libri de Hispania, Publications about Spain, Littleton, CO, 2001
77. Gove, H. E., Relic, Icon or Hoax? Carbon Dating the Turin Shroud, Institute of Physics Publishing, Techno House, Bristol, United Kingdom, 1996
78. Wilcox, R. K., The Truth about the Shroud of Turin, Regnery Publishing, Washington DC, 2010
79. Catechism of the Catholic Church (para 82), 2nd ed. Vatican: Libreria Editrice Vaticana, 1997
80. Santachiara Benerecetti, A.S. et al. The common, Near-Eastern origin of Ashkenazi and Sephardi Jews supported by Y-chromosome similarity, Annals of Human Genetics 57(1):55–64, 1993.
81. Borrini, M. and Garlaschelli, L., A BPA approach to the Shroud of Turin, Journal of Forensic Sci ence 64(1):137–143, 2019.
82. Oommen, T.V., Shroud coins dating by image extraction. Shroud Science Group International Conference, Aug 14–17, 2008; ohioshroudconference.com/papers/p20.pdf.
83. Jordan, T.R. et al. Seeing inscriptions on the Shroud of Turin: The role of psychological influences in the perception of writing, PLoS One 10(10):e0136860, 2015 | doi: 10.1371/journal.pone.0136860.

84. Runciman, S., Some remarks on the Image of Edessa, The Cambridge Historical Journal 3(3):238–252, 1931.
85. Edessa is a town in SE Turkey very close to the biblical site of Haran (Genesis 11:32 and 12:4) and the ancient archeological site called Göbekli Tepe. Return to text.
86. Leclercq, H. (1907). The Legend of Abgar, In: The Catholic Encyclopedia. New York: Robert Appleton Company. Retrieved May 15, 2019 from New Advent:newadvent.org/cathen/01042c.htm
87. Damon, P.E. et al., Radiocarbon dating of the Shroud of Turin, Nature 337:611–615, 1989.
88. Riani, M., et al. Regression analysis with partially labelled regressors: carbon dating of the Shroud of Turin, Statistics and Computing 23:551–561, 2013.
89. Rogers, R., A Chemist's Perspective on the Shroud of Turin, Florissant, CO, 2008.
90. Rev. John Dowling, A.M., History of Romanism: From the earliest corruptions of Christianity to the present time, Edward Walker, 114 Fulton Street, New York, 1845.
91. Kalb, C., Leonardo's enduring brilliance, National Geographic, May 2019, pp. 56–93.
92. Waldrop, M.M., Shroud of Turin is medieval, Science 242:378, 1988.
93. Wilson, I., The Shroud—The 2000-year-old Mystery Solved, Bantam Press, London, 2010, pp. 21–22.
95. Vial, G., Shrouded in Mystery, HALI (The International Magazine of Fine Carpets and Textiles) 49, 1990; Sourced from medievalshroud.com/the-medieval-weave.
96. Fanti, G., Malfi, P. and Crosilla, F., Mechanical and opto-chemical dating of the Turin Shroud, MATEC Web of Conferences 36:01001, 2015.

Printed at Repro India Ltd.